First
First
58735

BRITAIN'S SCENIC RAILWAYS

THE TIMES BRITAIN'S SCENIC RAILWAYS

Times Books, 77-85 Fulham Palace Road, London W6 8JB

First Edition 2012

Printed in China

British Library Cataloguing in Publication Data
A catalogue record for this book is available from the British Library

ISBN 978 0 00 792671 8
Imp 001

This edition produced for The Book People Ltd, Parc Menai, Bangor LL57 4FB

If you would like to comment on any aspect of this publication, please write to:
Times Atlases, HarperCollins Publishers, Westerhill Road, Bishopbriggs, Glasgow G64 2QT
e-mail: **timesatlas@harpercollins.co.uk**
or visit our website at: **www.timesatlas.com**
www.twitter.com/timesatlas

With special thanks to

nls
National Library of Scotland

THE TIMES

BRITAIN'S SCENIC RAILWAYS

Julian Holland and David Spaven

TIMES BOOKS
LONDON

Thurso
Wick
Dingwall
Kyle of Lochalsh
Inverness
Aberdeen
Mallaig
Fort William
Oban
Crianlarich
Perth
Edinburgh
Berwick-upon-Tweed
Glasgow
Carstairs
Ayr
Stranraer
Carlisle
Newcastle
Middlesbrough
Whitby
Grosmont
Oxenholme
Redmire
Leeming Bar
Pickering
Barrow-in-Furness
Lancaster
Bradford
Leeds
Huddersfield
Barnsley
Llandudno Junction
Manchester
Sheffield
Llanberis
Buxton
Caernarfon
Leekbrook Junction
Matlock
Snowdon Summit
Blaenau Ffestiniog
Kingsley & Froghall
Derby
Sheringham
Pwllheli
Shrewsbury
Norwich
Great Yarmouth
Bridgnorth
Aberystwyth
Kidderminster
Worcester
Sudbury
Hereford
Gloucester
Oxford
Marks Tey
Pembroke Dock
Llanelli
Swansea
Bristol
Swindon
Docklands Light Railway
Minehead
Bishops Lydeard
Tonbridge
Barnstaple
Alresford
Alton
Hythe
Southampton
Dungeness
Exeter
Ryde
Hastings
Liskeard
Gunnislake
Shanklin
Weymouth
Looe
Plymouth
Penzance

CONTENTS

3850

Ex-GWR '2800' Class 2-8-0 No. 3850 heads away from Blue Anchor station with a train on the West Somerset Railway from Bishops Lydeard to Minehead. Privately-owned heritage lines such as this are shown in green throughout the book.

INTRODUCTION

Britain is a relatively small country, yet one of diverse landscapes – and our rail network reflects this remarkable variety. From the sublime coastlines of Devon and Cornwall, through the luxuriant rural landscapes of Kent and Sussex, to the stunning mountains, moors and lochs of the West Highlands, the train provides a safe and civilized way to see the best of Britain's scenery.

Britain's Scenic Railways takes you on an armchair journey along more than fifty of the most scenic routes in England, Scotland and Wales – from electrified inter-city railways, through little-known cross-country links, to charming branch lines. Many of the latter are survivors of the 'Beeching Axe', so named after Dr Richard Beeching, the Chairman of the British Railways Board, who drew up a massive programme of proposed closures for the Government in 1963. Most of the railways in this book are part of the core national network operated and maintained by Network Rail, but our selection also includes volunteer-supported 'heritage' lines (standard gauge, narrow gauge and miniature) – the majority featuring steam haulage – which have prospered in scenic areas.

There is a very long tradition of appreciating views from the train in Britain. As the Victorians increasingly took to exploring hitherto unknown countryside opened up by the rail network, so the rail companies sought to further expand this market with maps and guides – and this tradition has continued to the present day. Britain's diverse geography made its mark on the railway network, with physical obstacles such as mountains, hills,

rivers and sea estuaries presenting a major challenge to the railway engineers – and their endeavours have bequeathed a wealth of scenic riches to the modern railway traveller.

But what do we understand by 'scenic'? Few would disagree that the mountain landscapes of North Wales, Northern England and Scotland – seen to perfection from the likes of the Ffestiniog Railway, the Settle & Carlisle Line and the Highland Main Line – are a real visual treat. Similarly, magnificent seascapes unfold on routes such as those along the Cambrian and Cumbrian coasts, while a more intimate patchwork of rich agricultural land, woodland and hedgerows can be admired in the deep south of England – from the Tonbridge-Hastings railway to the steam-operated Watercress Line. But captivating scenes don't appear just in the countryside, and we include a feature on the swooping gradients and sharp curves of London's Docklands Light Railway, which provides a bird's eye view of some stunning urban scenes – from the ultra-modern to the traditional.

BR Standard Class 4 2-6-0 No. 76079 skirts the beautiful Dyfi Estuary at Penhelig with a northbound 'Cambrian Coast Express' steam special in September 2003.

Britain's Scenic Railways also demonstrates that the attraction of a rail journey comes not just through more distant views, but also from close-up appreciation of the many superb viaducts, bridges and station buildings of past eras, which are still an essential part of day-to-day rail operations. Some routes – such as that from Newcastle to Carlisle, featuring the best of early railway architecture – earn their place in this book on the basis of the rail corridor's infrastructure just as much as the scenery and the surrounding built environment of castles, stately homes and historic towns and villages.

Any selection inevitably involves leaving out some worthy candidates – and we have had a rich collection to choose from. Not all the Cornish branch lines nor all the Welsh narrow-gauge railways could be included, but our samples of both give a taste of what is there to be explored; the East Suffolk Line from Ipswich to Lowestoft has some delightfully scenic stretches; the Newport-Hereford-Shrewsbury railway through the Welsh Marches traverses memorable countryside; and the main line from Edinburgh to Glasgow offers the sight of Linlithgow Palace, the Union Canal, the rolling Campsie Fells and the distant peaks of the southwest Highlands. These are just some of the railways, together with those featured here, which we hope the reader will be inspired to travel.

Each route featured in *Britain's Scenic Railways* comprises four core elements – striking photographs of the railway in the landscape; colour maps of the route corridor; a well-researched commentary on key aspects of the line's history, geography and current operations; and a separate description of the route as it unfolds from start to destination.

We have been privileged to utilize the archive of the former Edinburgh mapmaker, Bartholomew, which for much of the nineteenth and twentieth centuries was renowned for its topographical and railway mapping. Bartholomew's 'layer colouring' perfected the technique of showing height and depth as a succession of subtle colour changes, and this book uses the 'half inch' (1:126,720 scale) series from the early 1960s, so that the reader can trace the many branch lines which linked to our present-day rail routes until the widespread closures following the 1963 'Beeching Report'.

Many of the route features – which vary from two to twelve pages in length, primarily reflecting line mileage – also include other visual aids. For selected mountainous routes and others with 'switchback' alignments, we have incorporated extracts of gradient profiles – cross-sections of the vertical topography of the line – which provide a strong feel for the challenging terrain that gives these railways much of their character. We have also

taken advantage of track diagrams, which illustrate either the complex line infrastructure of scenic railway hubs like Carlisle or the quirkier aspects of modern rail operations on fascinating branches such as Cornwall's Looe Valley Line. The latter has one of the many 'Community Rail Partnerships' – collaborations between the rail industry, local authorities and local communities – helping to shape and promote rural railways across England and Wales.

As *Britain's Scenic Railways* hopes to encourage you to explore more of the country by train – as well as enjoying the best of our railways from the comfort of your own home – some guidance is advisable on the practicalities of viewing the scenery from the train. Traditional 'tail-end' observation cars – typically dating from the halcyon inter-war years – last ran on the pre-privatized national network in the early 1990s, but now can be found only on heritage lines like the Dartmouth Steam Railway in Devon. Almost without exception, heritage lines – and diesel and steam-hauled charter trains operating on the national network – use traditional rolling stock with large picture windows.

On our privatized rail system, with its many different types of train – and these generally designed for urban or inter-urban use – the quality of the viewing experience does however vary from route to route, and even from service to service. Britain has no equivalent to the likes of the purpose-designed panoramic coaches of the wonderful Glacier Express in Switzerland, but many of the modern trains used by Train Operating Companies do afford the traveller a very good view of the passing scene. Particularly worthy of mention are: the Voyager trains which operate on the northern sections of the East Coast and West Coast Main Lines to Edinburgh and Glasgow and on the Exeter-Plymouth-Penzance line though Devon and Cornwall; the Class 156 diesel units which serve a variety of routes the length and breadth of the country; and the Class 170 diesel units operating the Edinburgh to Aberdeen and Inverness routes, amongst others. However, while 125-mph Pendolino trains are a great means of speedy travel from London to Glasgow, their airline-style small windows and high seatbacks do not offer the best way to view the dramatic scenery between Lancaster, Carlisle and Carstairs. But this will not perturb the armchair traveller.

K1 No. 62005 passes Lochan a' Chlaidheimh (Sword Loch) between Rannoch and Corrour at the head of a charter train in October 2009.

Britain's Scenic Railways does not, of course, include those marvellously scenic lines lost in the drastic railway cuts of the 1960s – the Ilfracombe branch, the coastal route from Whitby by Robin Hood's Bay to Scarborough and the Waverley Route through the Scottish Borders, to name but a very few. However, Britain's rail network has been in expansionary mode for decades, with more than 500 miles of new routes opened since 1960. As roads get ever more congested and concerns about the environment and oil prices continue to grow, that expansion is set to continue and is likely to bring some scenic routes back on the discerning traveller's itinerary – from the Oxford to Milton Keynes cross-country line through classic southern English countryside, to the the steep and twisting Borders Railway from Edinburgh to Tweedbank.

But now, starting in England's West Country, it is time to savour our armchair journey along the best of Britain's scenic railways...

HOW THIS BOOK WORKS

Each route feature has a number of components:

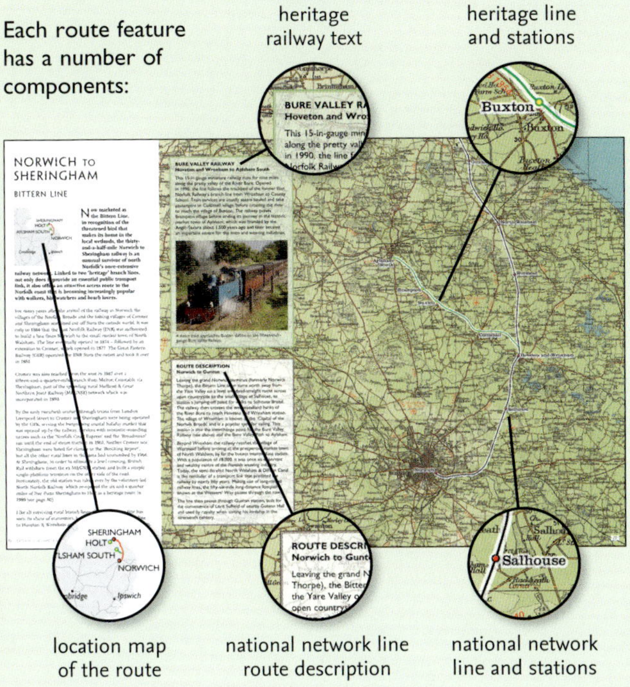

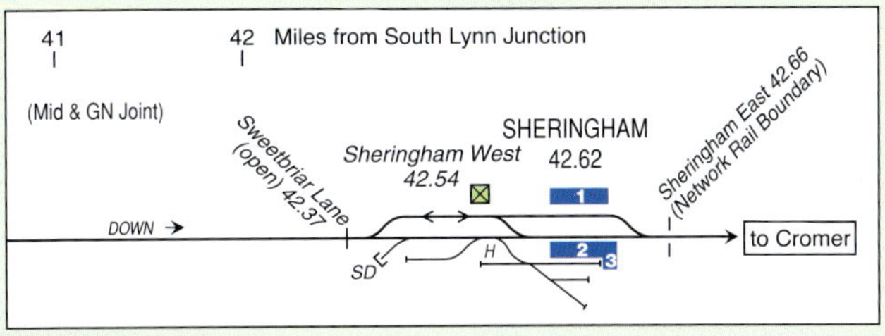

Track diagrams (courtesy of Trackmaps) show the detail of individual running lines and sidings (in black), station platforms (in blue), signal boxes (in green) and route mileages.

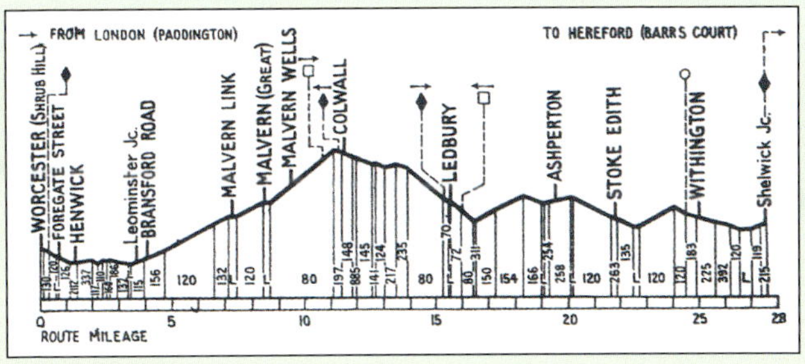

Gradient profiles (courtesy of Ian Allan Publishing) show a vertical cross-section of the topography of a railway, with route mileage marked along the bottom horizontal axis and individual gradients above that, eg '80' = 1 in 80.

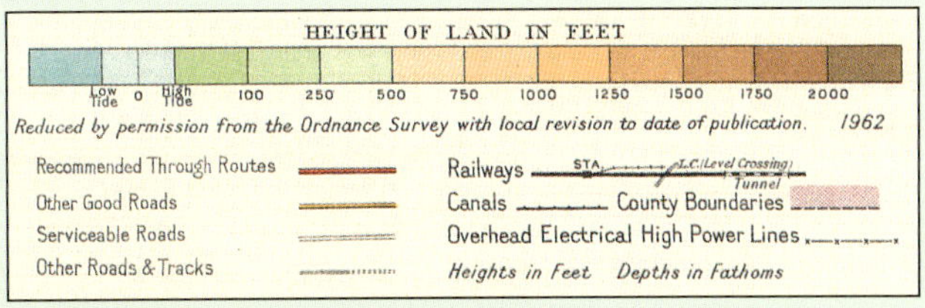

The mapping in this book comes from sheets of Bartholomew's Half-Inch Series, published in the early 1960s. Due to the age of the source material, variations in colour will occur, as well as occasional inconsistencies where sheets have been joined together.

Looking more like Ayers Rock in Australia, these red sandstone cliffs strikingly dominate the Exeter to Plymouth main line near Dawlish Warren.

WEST COUNTRY

ROUTE DESCRIPTION
Plymouth to Doublebois

For the first seven and a half miles the railway weaves its way through the western suburbs of Plymouth, famous for the Royal Naval Dockyard of Devonport and its nuclear submarine base. Separating Devon from Cornwall lies the River Tamar, spanned by Brunel's Royal Albert Bridge. A busy waterway at high tide, as it stretches away to the north it is met by the confluence of the River Tavy flowing in from the northwest, while at low tide, vast mud banks attract wading birds. Along its east bank runs the branch line to Gunnislake (see pages 20–21). Opened in 1961, a modern suspension road bridge now dwarfs Brunel's wrought iron masterpiece, to the south of which the river conjoins with the St Germans River before flowing out to Plymouth Sound and the sea via the Hamoaze, which was for centuries an anchorage for Royal Navy ships.

Entering Cornwall at Saltash, the railway shadows the north bank of the St Germans River before taking its 1908-deviation route to the historic riverside village of St Germans, crossing muddy creeks and the Rivers Lynher and Tiddy over the first of many viaducts that characterize this extended route. Striking off in a northwesterly direction from St Germans, the railway eases its way up through low hills dotted with dairy farms and across narrow valleys via four viaducts before approaching the town of Liskeard over the 150-ft-high, 720-ft-long Liskeard Viaduct.

Far below in the valley, the single-track line to Looe (see pages 22–23) makes its 180-degree loop up from the remote Coombe Junction to Liskeard station. To the west of Liskeard the main line dips down to cross the valley of the East Looe River over Moorswater Viaduct (147 ft high and 954 ft long). To the north lie the former granite quarries of Bodmin Moor, while to the south the wooded East Looe Valley wends its way down to the sea at the fishing village of Looe.

PLYMOUTH TO PENZANCE

CORNISH MAIN LINE

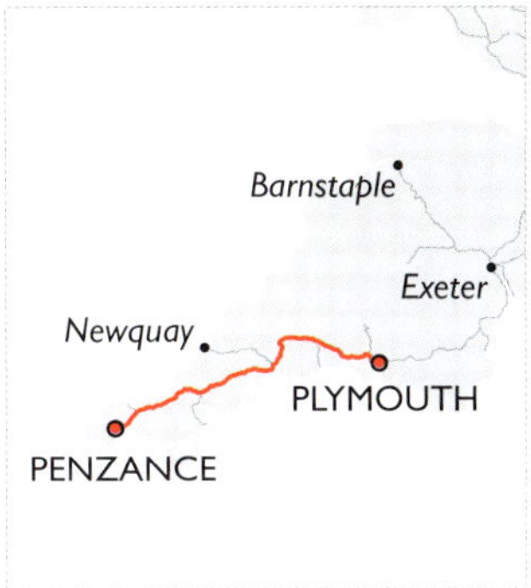

Despite whispers of closure in the mid-1960s, this heavily engineered seventy-nine-and-a-half-mile switchback route has survived mostly intact and, together with its many branch lines, continues to serve the seaside resorts and towns of Cornwall in the twenty-first century.

From the early years of the nineteenth century, a network of horse-drawn tramways and mineral railways sprang up across Cornwall, built to link the tin and copper mining, china clay extraction and granite quarrying industries with ports and harbours. Many of these went on to form the nucleus of the county's modern rail system, including the Bodmin & Wadebridge Railway, one of the earliest steam-hauled lines in Britain.

Later forming part of the main line between Plymouth and Penzance, the Hayle & Portreath Railway (H&PR) opened in 1837 to link copper mines around Redruth with harbours at Portreath and Hayle, although, with inclined planes at Penponds and Angarrack, progress along this route was inevitably slow. This all changed in 1852 when Brunel's West Cornwall Railway (WCR) opened between Truro, Redruth and Penzance. Originally built to the standard gauge, this twenty-five-and-three-quarter-mile railway incorporated part of the route of the H&PR between Redruth and Hayle, bypassing the inclined planes with viaducts.

Opened in 1859, Brunel's single-track Royal Albert Bridge to the west of Plymouth carries the railway over the River Tamar to Cornwall.

ROUTE DESCRIPTION

Doublebois to St Austell

After climbing once more up to Doublebois, the railway embarks on its long descent down the steep-sided wooded valley of the River Fowey, crossing seven more viaducts as it winds its way down to Lostwithiel via Bodmin Parkway station (for Lanhydrock House and the Bodmin & Wenford Railway). With its twelfth-century bridge the small town of Lostwithiel is home to the River Fowey's lowest crossing and the junction for the freight-only china clay line down to Carne Point near to Fowey town.

Climbing away from the river valley at Lostwithiel, the railway dives under the hills through Treverran Tunnel before descending a wooded valley to Par station, junction for the Newquay branch line and gateway to Par harbour for its china clay exports. Leaving Par behind, the railway stretches away from the coast near Carlyon Bay and climbs towards the town of St Austell. With a population of nearly 23,000, St Austell is the largest town in Cornwall and the centre of the region's important china clay industry.

Also engineered by Brunel, the broad-gauge double-track Cornwall Railway (CR) bridged the fifty-three-and-three-quarter-mile gap between Plymouth and Truro and continued a further eleven and three quarter miles to the port of Falmouth. A feat of Victorian engineering, the building of this heavily graded line across Cornwall's deep river valleys included numerous high embankments, tunnels and forty-two viaducts. These viaducts were originally constructed with timber deck spans and bracing supported by masonry piers, though masonry had replaced all the timber by the 1930s. The crossing of the River Tamar near Plymouth was achieved by Brunel's *pièce de résistance*, the wrought-iron single-track Royal Albert Bridge which, when it opened in 1859, finally linked Cornwall with the rest of Britain.

Despite the opening of the CR, both passengers and freight still had to change trains at Truro due to a change of gauge in order to continue their journey on the WCR to Penzance. This operating problem was finally solved in 1866 when the WCR was relaid to broad gauge. The Great Western Railway (GWR) operated both lines from their early years, although the CR was only taken over by the GWR in 1887 and the WCR remained independent until 1948. Brunel's broad-gauge lasted until 1892 when all of the remaining broad gauge lines west of Exeter were converted to standard gauge over just one weekend. Five viaducts between Saltash and St Germans were demolished in 1908 when a new deviation route was opened to the north.

A London (Paddington) to Penzance High Speed Train crosses Forder Viaduct west of Saltash.

Ex-GWR 'King' Class 4-6-0 No. 6024 'King Edward I' and ex-SR 'Battle of Britain' Class 4-6-2 No. 34067 'Tangmere' head across St Germans Viaduct with a steam charter train en route to Penzance.

ROUTE DESCRIPTION
St Austell to Truro

Striking inland away from St Austell, the scenery to the north of the railway undergoes a rapid transformation – for the next five miles the man-made conical waste heaps from vast china clay workings stretch away into the distance, giving the landscape an almost lunar appearance. At Burngullow, a mineral line disappears northwards to serve china clay workings at Treviscoe, while the main line dips down to cross two viaducts over the River Fal and a tributary before climbing again to Grampound Road. Here, the railway heads off southwest towards the county town of Truro, cutting across steep-sided river valleys on three viaducts and dipping under the hills through Polperro and Buckshead Tunnels. This is rich, Cornish farming country, famous for its dairy herds, clotted cream, early potatoes, broccoli and cut flowers. With the cathedral town spread out below, Truro station is reached by traversing two viaducts – the 1,329-ft-long Truro Viaduct crosses the Truro River at a height of 92 ft, while the 969-ft-long Carvedras Viaduct, immediately east of the station, crosses the River Kenwyn at a height of 86 ft. Truro is the junction for the branch line down to the port of Falmouth.

Along with the opening of numerous branch lines serving harbour towns and seaside resorts, the coming of the main line to Cornwall had a significant impact on the county's economy – fishermen, farmers and flower growers could now send their fresh produce on overnight trains to London. But by far the greatest impact was seen in the many seaside villages and towns, which were transformed into thriving resorts. With holidaymakers now able to access Cornwall's scenic coastline and pristine beaches, tourism had become a major industry by the early twentieth century. Greatly assisted by the GWR's publicity department, this reached its peak in the 1930s with trains arriving at Cornish destinations from London, the Midlands and the North. Of these, the best known was the 'Cornish Riviera Express' which, on summer Saturdays, ran in several portions carrying through coaches from Paddington all the way to resorts such as St Ives.

The onset of the Second World War, followed by increasing competition from road transport in the 1950s and early 1960s, saw a steady decline in passenger numbers. Steam traction was replaced by diesel in 1962 and, for a while, the whole future of the railways west of Plymouth was in doubt. Fortunately closure never materialized, although the section from Burngullow Junction to Probus was reduced to single track until its reinstatement in 2004.

Today, First Great Western operates the through trains from Paddington including the 'Night Riviera' sleeper and local services between Plymouth and Penzance, while long-distance services from as far afield as Glasgow and Aberdeen are provided by CrossCountry Trains. Freight traffic is greater than one would expect in a largely rural area, with fuel trains running to Long Rock depot at Penzance, occasional cement trains making the journey to the Lafarge terminal near Liskeard and intensive china clay workings around St Austell and Lostwithiel relying on onward shipment from Carne Point via the freight-only line to Fowey.

Sweeping views over river valleys and tantalizing glimpses of the sea combine with the lunar landscape of the china clay industry and reminders of Cornwall's rich industrial past to make this one of England's most scenic mainline railway journeys.

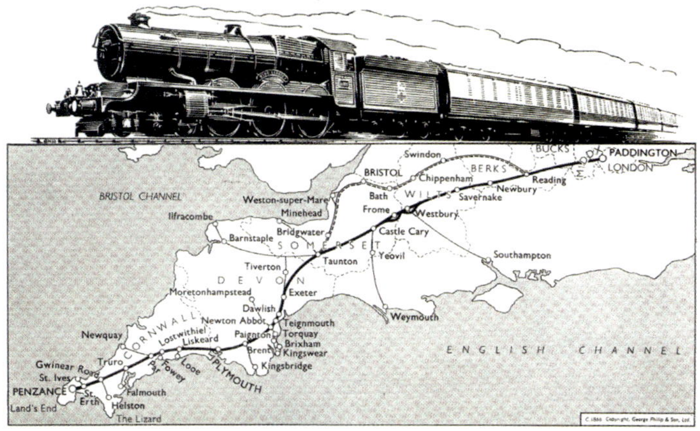

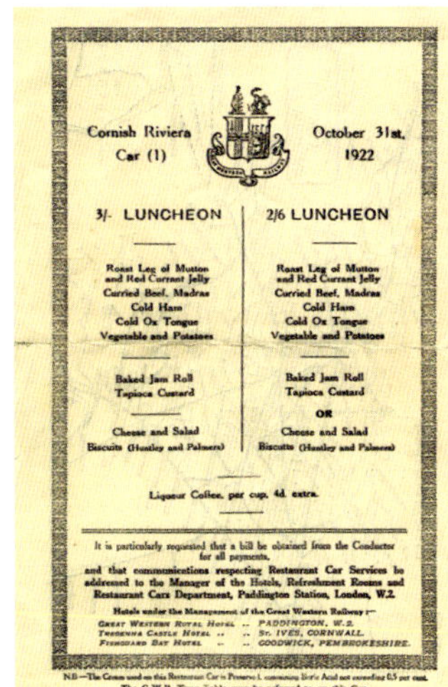
Cornish Riviera Car (1)

October 31st. 1922

3/- LUNCHEON	2/6 LUNCHEON
Roast Leg of Mutton and Red Currant Jelly	Roast Leg of Mutton and Red Currant Jelly
Curried Beef, Madras	Curried Beef, Madras
Cold Ham	Cold Ham
Cold Ox Tongue	Cold Ox Tongue
Vegetable and Potatoes	Vegetable and Potatoes
Baked Jam Roll	Baked Jam Roll
Tapioca Custard	Tapioca Custard
	OR
Cheese and Salad	Cheese and Salad
Biscuits (Huntley and Palmers)	Biscuits (Huntley and Palmers)

Liqueur Coffee, per cup, 4d. extra.

It is particularly requested that a bill be obtained from the Conductor for all payments, and that communications respecting Restaurant Car Services be addressed to the Manager of the Hotels, Refreshment Rooms and Restaurant Cars Department, Paddington Station, London, W.2.

Hotels under the Management of the Great Western Railway :–

Great Western Royal Hotel .. PADDINGTON, W.2.
Tregenna Castle Hotel .. .. St. IVES, CORNWALL.
Fishguard Bay Hotel .. .. GOODWICK, PEMBROKESHIRE.

The G.W.R. Time Table may be referred to on this Car.

The 'Cornish Riviera Express' was the GWR's premier train from London to Penzance.

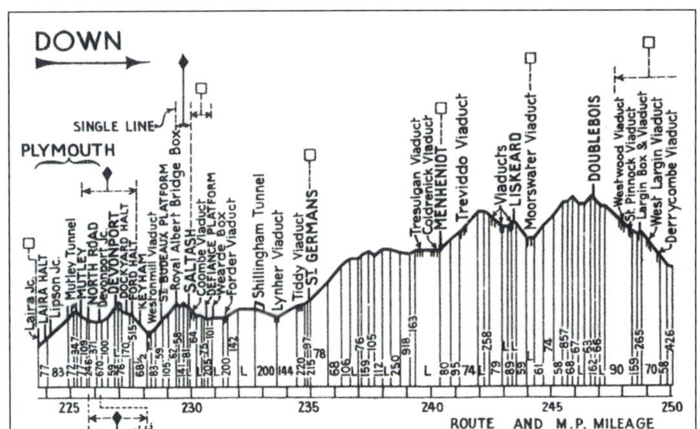

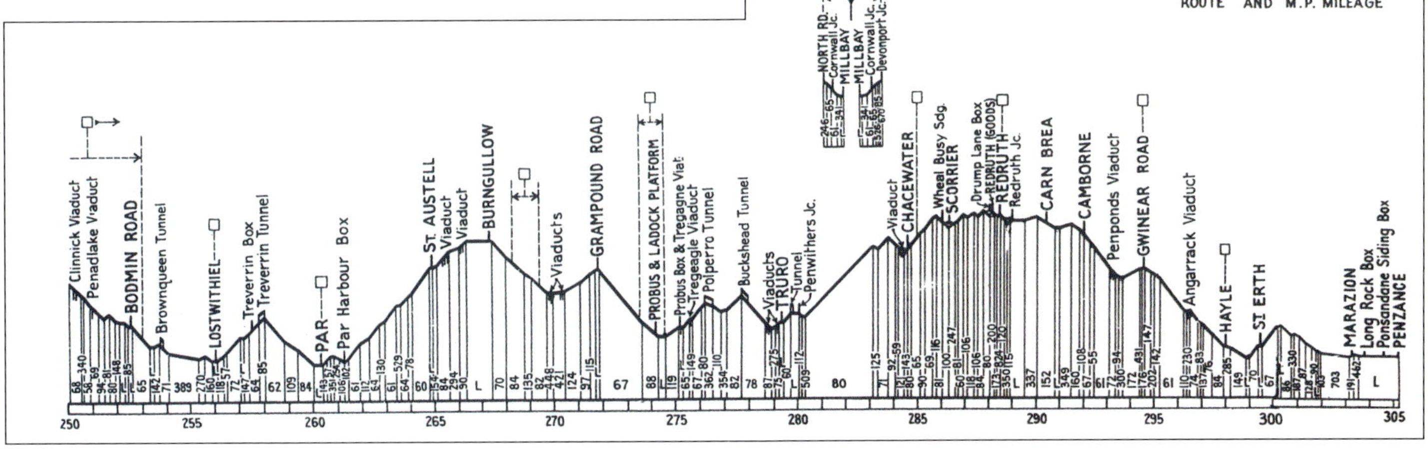

The gradient profile of the switchback Plymouth to Penzance main line.

ROUTE DESCRIPTION
Truro to Penzance

From Truro the railway climbs continuously for nine miles to the former tin-mining towns of Redruth and Camborne, winding its way up through the hills past villages with strange-sounding names such as Green Bottom and Jolly Bottom. Through Redruth and Camborne the landscape is punctuated by the spoil heaps of eighteenth- and nineteenth-century tin mines and their sentinel-like winding house chimneys. To the west of Camborne the railway crosses two valleys via Penponds and Angarrack Viaducts, built by Brunel in 1852 to bypass the inclined planes of the old Hayle & Portreath mineral railway. At Hayle station the harbour and vast expanse of Hayle Sands lie immediately to the north while one mile further west, the station of St Erth is the junction for the branch line to St Ives. From here the railway strikes across this narrow point of the Cornish Peninsula to meet the south coast at Marazion. With the pinnacle of St Michael's Mount to the southeast and the fishing harbour of Newlyn to the southwest, the line makes its final level approach along the sea wall to the overall-roofed station at Penzance. Here, from March to October, the harbour sees the daily departure and arrival of the 'Scillonian III' to and from the Isles of Scilly.

A modern High Speed Train skirts the Cornish coastline as it heads away from Penzance.

PLYMOUTH TO GUNNISLAKE

TAMAR VALLEY LINE

Set in the beautiful Tamar Valley on the Devon-Cornwall border, this lucky survivor of the 'Beeching Axe' runs for half its length along a former main line before following the route of a long-closed narrow-gauge mineral railway into Cornwall.

The Plymouth to Gunnislake branch line started life at its northern end in 1872 as the 3-ft 6-in gauge East Cornwall Mineral Railway (ECR) between Calstock Quay, on the River Tamar, and mines and quarries around Callington. The southern end between Plymouth and Bere Alston was opened by the Plymouth, Devonport & South Western Junction Railway (PD&SWJR) in 1890 – the double-track railway continued to Lydford where it met the London & South Western Railway's line from Okehampton, Exeter and Waterloo. The ECR was taken over by the PD&SWJR in 1891 and converted to a standard gauge light railway – engineered by Colonel H Stephens, the steeply graded line opened in 1908 and was linked to a junction at Bere Alston via a new viaduct over the Tamar at Calstock.

Although the Callington branch and the main line from Plymouth to Okehampton were both listed for closure in the 'Beeching Report', passenger services from Plymouth to Bere Alston and thence to Gunnislake were reprieved due to difficult road access for the proposed replacement buses. The steeply graded section from Gunnislake to Callington closed in 1966, followed by the section from Bere Alston to Okehampton in 1968. The remaining main line north of Plymouth was then singled with trains to and from Gunnislake reversing at Bere Alston.

Today, this single-track community railway is marketed as the Tamar Valley Line. Operated by First Great Western, diesel trains between Plymouth and Gunnislake take around forty-five minutes to make the fifteen-and-a-quarter-mile journey. The lack of any passing loops prevents any increased frequency of trains.

Affording fine views of the River Tamar, this delightful cross-border branch line ends the first part of its journey by reversing direction, before winding around the contours, high above historic quays and wooded valleys, to serve communities still hampered by difficult road access.

The magnificent single-track viaduct over the River Tamar at Calstock was opened in 1908.

ROUTE DESCRIPTION

The Gunnislake branch leaves the Plymouth to Penzance main line at St Budeaux, first diving under Brunel's Royal Albert Bridge and the Tamar road bridge before tracing the east bank of the Tamar and crossing the mouth of the River Tavy on a curving viaduct of eight iron spans and seven stone arches. Still in Devon, the railway passes through the small village of Bere Ferrers before reaching Bere Alston station where trains reverse direction. The railway now makes a short westerly loop high above historic Cotehele Quay before crossing the Tamar into Cornwall on the spectacular Calstock Viaduct – built of concrete blocks at the beginning of the twentieth century its twelve arches carry the railway 120 ft above the meandering river. Until 1934 a steam-powered wagon lift attached to the viaduct was used to lower and raise wagons to and from Calstock Quay.

Until the arrival of the railway, the river was an important transport artery carrying stone and minerals from nearby mines and quarries down to Plymouth. From tiny Calstock station the railway climbs continuously, making a 180-degree turn while following the contours high above the meandering Tamar Valley. Deep down below in the wooded valley lies Morwellham Quay, which was founded by Benedictine monks in the tenth century to export tin, lead, silver and copper ores. By the nineteenth century the wealth of local mines had transformed this into one of the richest mining regions in Europe.

Winding its way northwards, the railway finally rounds a sharp curve to arrive at the small terminus of Gunnislake. To the northeast of the village the sixteenth-century bridge was the Tamar's lowest road crossing until the opening of the Tamar Bridge near Plymouth in 1961. By the early twentieth century mining had virtually ceased around Gunnislake, but thanks to rail transport to markets, a thriving market gardening industry had established itself around the cluster of small villages to the west of the Tamar by the 1930s.

LISKEARD TO LOOE

LOOE VALLEY LINE

This eight-and-three-quarter-mile single-track community railway that runs down the picturesque East Looe Valley to the fishing and resort town of Looe is another lucky survivor of the 'Beeching Axe'.

Until the coming of the railways, tin, copper, lead and granite were carried by packhorse from mines and quarries on Bodmin Moor to the head of the East Looe Valley at Moorswater, west of Liskeard. Here it was loaded onto horse-drawn barges for the journey along the Liskeard & Looe Union Canal for onward shipment from the harbour at Looe. The Liskeard & Caradon Railway (L&CR) mineral line replaced the packhorses in 1844, together with the canal by the Liskeard & Looe Railway (L&LR), which opened alongside its route in 1860.

Passenger services started from Moorswater to Looe in 1879 but the former station was difficult to access and was not linked by rail to the Cornish main line, which passed high above the valley on a viaduct. Eventually, a two-mile circular linking line with a gradient of 1 in 40 was opened in 1901 from Coombe Junction, south of Moorswater, to Liskeard station. Trains between Liskeard and Looe had to reverse directions at the junction, a practice that continues to this day. The Great Western Railway (GWR) operated both the L&CR and the L&LR from 1909 although the former, apart from a short section north of Moorswater, had closed by the end of 1916. Amalgamated with the GWR in 1923, the Looe branch was soon being promoted by the company's publicity department at Paddington for its scenic qualities and quaint seaside destination.

By the early 1960s the line's future was looking bleak – goods traffic, apart from clay trains from Moorswater up to Liskeard, had ceased, and with passenger traffic declining it was listed for closure in the 1963 'Beeching Report'. It was granted a reprieve at the eleventh hour by Labour Transport Minister Barbara Castle in 1966 with difficult road access down the valley for replacement bus services cited as the main reason. Today, with rising passenger numbers, the railway is promoted by the Devon & Cornwall Rail Partnership (the largest Community Rail

A Looe-bound train slows on the approach to Terras Crossing on the east bank of the East Looe River.

Partnership in Britain), assisted by Friends of the Looe Valley Line volunteers. Amongst the line's attractions are the 'Looe Valley Line Rail Ale Trail', linking real ale pubs in Liskeard, Sandplace and Looe. Occasional freight trains from the Lafarge cement terminal at Moorswater continue to use the section from Coombe Junction up to Liskeard.

Diesel train services along the Looe Valley Line between Liskeard and Looe are operated by First Great Western and have a journey time of around thirty minutes.

This charming Cornish branch line, with its unusual reversing operation and fine views across the wooded East Looe Valley, offers visitors a pleasant alternative for travel to the harbour town of Looe during the busy summer months.

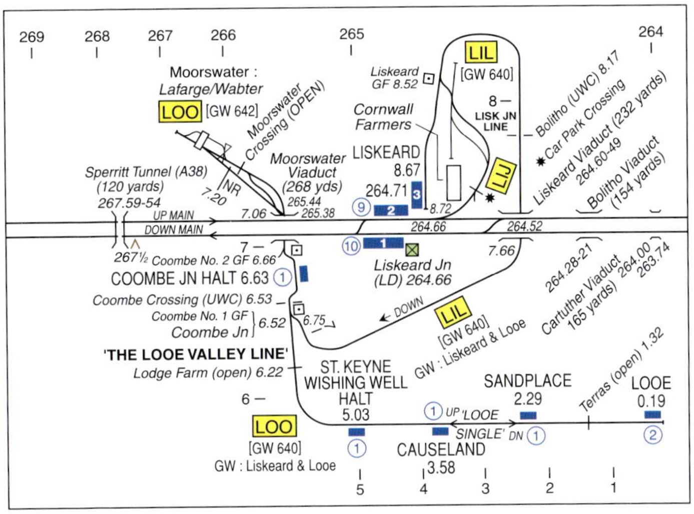

The fascinating track diagram of the Looe branch line.

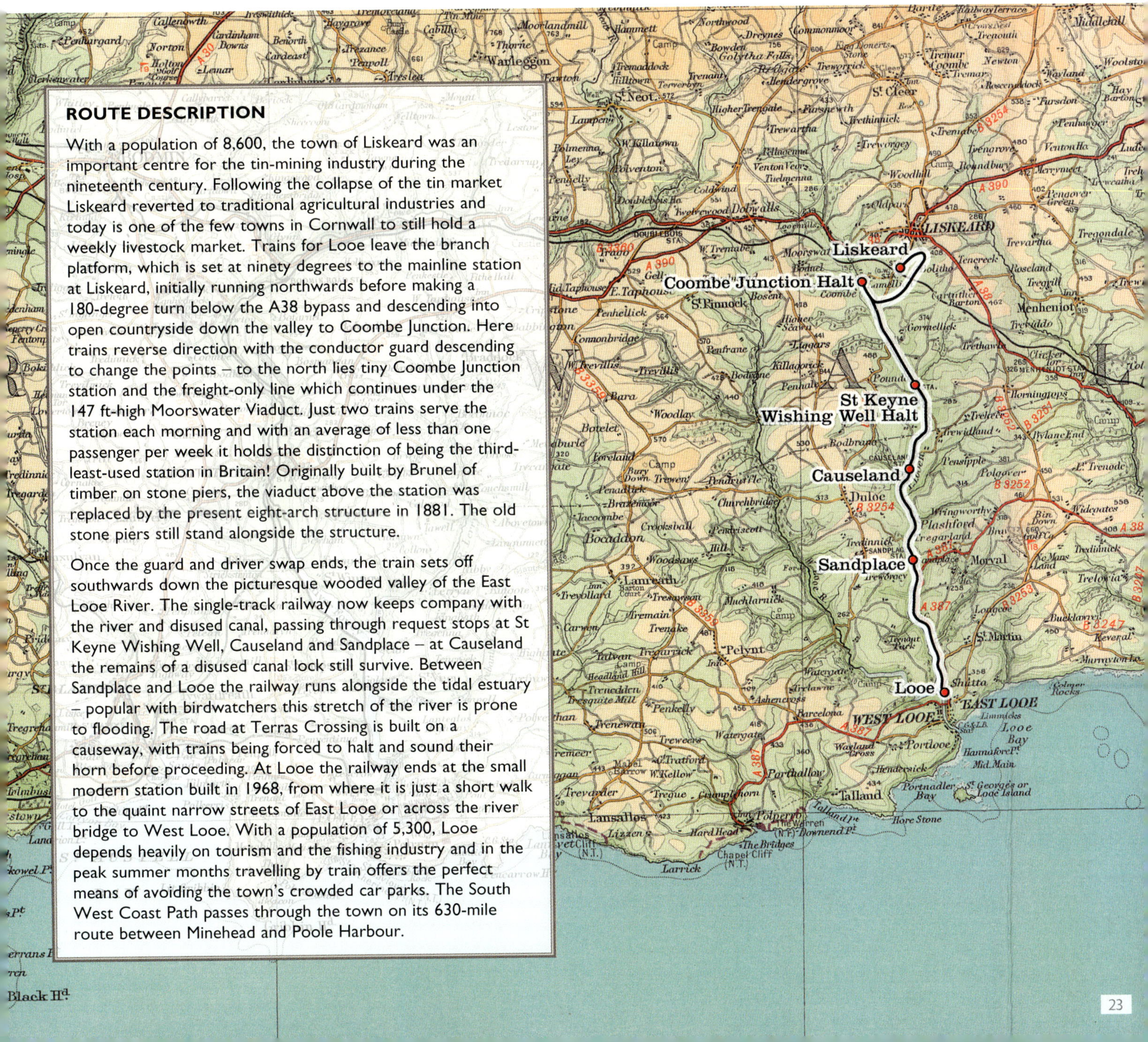

ROUTE DESCRIPTION

With a population of 8,600, the town of Liskeard was an important centre for the tin-mining industry during the nineteenth century. Following the collapse of the tin market Liskeard reverted to traditional agricultural industries and today is one of the few towns in Cornwall to still hold a weekly livestock market. Trains for Looe leave the branch platform, which is set at ninety degrees to the mainline station at Liskeard, initially running northwards before making a 180-degree turn below the A38 bypass and descending into open countryside down the valley to Coombe Junction. Here trains reverse direction with the conductor guard descending to change the points – to the north lies tiny Coombe Junction station and the freight-only line which continues under the 147 ft-high Moorswater Viaduct. Just two trains serve the station each morning and with an average of less than one passenger per week it holds the distinction of being the third-least-used station in Britain! Originally built by Brunel of timber on stone piers, the viaduct above the station was replaced by the present eight-arch structure in 1881. The old stone piers still stand alongside the structure.

Once the guard and driver swap ends, the train sets off southwards down the picturesque wooded valley of the East Looe River. The single-track railway now keeps company with the river and disused canal, passing through request stops at St Keyne Wishing Well, Causeland and Sandplace – at Causeland the remains of a disused canal lock still survive. Between Sandplace and Looe the railway runs alongside the tidal estuary – popular with birdwatchers this stretch of the river is prone to flooding. The road at Terras Crossing is built on a causeway, with trains being forced to halt and sound their horn before proceeding. At Looe the railway ends at the small modern station built in 1968, from where it is just a short walk to the quaint narrow streets of East Looe or across the river bridge to West Looe. With a population of 5,300, Looe depends heavily on tourism and the fishing industry and in the peak summer months travelling by train offers the perfect means of avoiding the town's crowded car parks. The South West Coast Path passes through the town on its 630-mile route between Minehead and Poole Harbour.

EXETER TO PLYMOUTH

After a faltering start as an eccentric 'atmospheric railway', the Exeter to Plymouth line eventually became a vital link in the Great Western Railway's premier broad-gauge route from London (Paddington) to Cornwall. Today this iconic coastal route that tracks the fearsome gradients of the southern edge of Dartmoor, still makes for an unforgettable railway journey.

Following the opening of the broad-gauge Bristol & Exeter Railway (B&ER) in 1844, innovative engineer Isambard Kingdom Brunel turned his attention to his next railway project, the fifty-two-mile broad-gauge South Devon Railway (SDR) from Exeter to Plymouth. While planned to take a fairly level coastal route to Teignmouth, this new railway would then encounter severe gradients as it wound its way across the southern slopes of Dartmoor before reaching Plymouth. After witnessing demonstrations of a unique 'atmospheric railway' in London, Brunel opted to use this new system for the SDR, arguing that it was capable of hauling heavier loads over the difficult terrain of the South Devon Banks than was then possible with adhesion steam locomotives.

Doing away with locomotives altogether, the system centred around a slotted steel pipe laid in the middle of the track – inside the pipe was a piston, connected to the underside of the leading coach by an iron rod, which was sucked along by a permanent vacuum created by steam driven pumping houses set at three-mile intervals along the route. In theory the slot in the top of the pipe was sealed with airtight leather flaps, which first opened and then closed as the train moved forward.

However, in practice the atmospheric system proved to be a resounding failure as the leather flaps were highly unsuccessful in ensuring a permanent air-tight fit – not only did rats find the leather very tasty but during warmer weather the flaps dried and cracked. Before this major problem had been fully encountered the atmospheric railway had already started operating – Teignmouth had been reached in September 1847 and Totnes by July 1848. Then, with mounting maintenance costs taking their toll, Brunel's plans fell into disarray and the atmospheric system was hurriedly abandoned and replaced by adhesion locomotives.

The rest of the route from Totnes to Plymouth had already opened as a conventional railway in May of that year, using locomotives hired from the Great Western Railway (GWR). The SDR was amalgamated with the GWR in 1878 and, having already swallowed up the B&ER two years earlier, this much-enlarged company now controlled the entire route from

ROUTE DESCRIPTION
Exeter to Newton Abbot

Departing from Exeter St David's station in a southerly direction, the railway immediately crosses the River Exe leaving behind the cathedral city of 120,000 souls, closely following the west bank of the ever-widening Exe Estuary, then passing the parkland of fourteenth-century Powderham Castle and Brunel's atmospheric pumping station at Starcross, before reaching the coastal town of Dawlish Warren. Here, the wide estuary with its multitude of small sailing craft, meets the English Channel at a spectacular sand spit, its dunes a nature reserve that is also popular with holidaymakers.

Protected by a sea wall and skirting the coastline close to the high tide mark, the railway passes through the small resort of Dawlish before encountering the crumbling red sandstone cliffs that are a feature of this stretch of coast. South of Dawlish with the sea close at hand, the railway burrows through five tunnels in the space of one mile – this particular stretch as far as Teignmouth is always at the mercy of winter gales and rock falls. Anchored far out to sea in Tor Bay, gigantic oil supertankers await their next call to duty.

Passing through Teignmouth in a series of rock cuttings, the railway then strikes inland, following the north bank of the Teign Estuary, famous for its birdlife and flounder fishing, then passing Newton Abbot Racecourse, before reaching the town's busy junction station. Here, trains for Torquay and Paignton (see page 27) branch off to the south at Aller Junction while the main line to Plymouth strikes inland towards the South Devon hills. North of the station, the once mothballed four-mile branch line to Heathfield has recently come back to life transporting timber.

Ex-GWR 'King' Class 4-6-0 No. 6024 'King Edward I' heads along the sea wall at Dawlish with a Bristol to Kingswear train in August 2011.

Crediton 5 miles
Cullompton 9 miles
Honiton 12 miles
Exeter St David's
Exeter St Thomas
EXETER
TOPSHAM
RIVER EXE
Starcross
Dawlish Warren
Dawlish
EXMOUTH
Teignmouth
RIVER TEIGN
Newton Abbot
NEWTON ABBOT
BABBACOMBE BAY
Torre
CHUDLEIGH
BOVEY TRACEY
Straight Pt
5 Fathom Line

Paddington to Plymouth and thence to Cornwall. Brunel's broad gauge disappeared entirely in 1892 when the route west of Exeter to Penzance was converted to standard gauge in the space of just one weekend.

Until the 1950s the burgeoning growth of holiday traffic during the summer months saw locomotive haulage along the line stretched to the limit. West of Newton Abbot heavy passenger trains such as the 'Cornish Riviera Express' were double headed over the steeply graded South Devon Banks, with freight trains requiring banking assistance up the fearsome 1 in 37 of Dainton Bank. This steam spectacular ended in the early 1960s with the introduction of powerful diesel hydraulics, and the modern HST (High Speed Train) 125s which in turn replaced them in the 1970s still make easy work of this winding and switchback section.

ROUTE DESCRIPTION
Newton Abbot to Totnes

Heading into the South Devon hills the railway soon encounters steep gradients through Dainton Tunnel to the first summit of the line at Dainton. Descending from here down to historic Totnes the railway follows the course of small river valleys, the surrounding hills characterised by picturesque small villages and dairy and cattle farms. At Totnes the railway levels out and crosses the River Dart before entering the station – from here it is just a short walk back across the river to the South Devon Railway's station at Littlehempston (see page 29). Totnes, famous for its unspoilt town centre and nearby Dartington College, is also the lowest crossing point of the tidal River Dart and limit of the popular river cruises at high water up from Kingswear and Dartmouth.

First Great Western and CrossCountry Trains currently operate the diesel trains along the double-track route between Exeter and Plymouth. The route also sees steam charter trains, such as the Bristol to Paignton 'Torbay Express', during certain summer Sundays. Freight traffic includes regular timber trains to a board mill in northeast Wales from the recently reopened Newton Abbot-Heathfield branch, scrap metal from Marsh Barton south of Exeter and nuclear flask trains to Sellafield from Devonport Royal Dockyard.

With its famous coast-hugging route beneath red sandstone cliffs followed by tortuous gradients around the southern edges of bleak Dartmoor, the Exeter to Plymouth line is arguably one of the most scenic railway routes in England.

TORBAY BRANCH
Newton Abbot to Paignton

From Aller Junction, just over a mile south of Newton Abbot, the South Devon Railway (SDR) opened a four-mile broad-gauge branch line to the village of Torre (the station was named Torquay until 1859) at the end of 1848. In 1859 the three-mile broad gauge Dartmouth & Torbay Railway (D&TR) opened from Torre to Paignton with a new station opening at Torquay. The D&TR amalgamated with the SDR in 1872 which itself was amalgamated with the Great Western Railway (GWR) in 1876. The entire branch was converted to standard gauge in 1892. By the beginning of the twentieth century the coming of the railway had totally transformed what were once small fishing villages into thriving seaside resorts – the population of Torbay is today around 134,000.

Once the destination of the 'Torbay Express' on its long journey from Paddington, much of the Paignton branch now runs through the spreading suburbs of Torbay, only glimpsing the sea south of Torquay station and along the short coastal stretch towards Paignton. Paignton station is an interchange point with the Dartmouth Steam Railway (see below).

First Great Western operates trains between Newton Abbot and Paignton. Most services originate from Exmouth via Exeter although the line also sees some long-distance trains from Paddington and Manchester.

BR Standard Class 4 4-6-0 No. 75014 crosses Brunel's Hookhills Viaduct on the Dartmouth Steam Railway with a train for Kingswear.

DARTMOUTH STEAM RAILWAY
Paignton to Kingswear

The six-and-a-half-mile broad-gauge line from Paignton to Kingswear was opened throughout by the Dartmouth & Torbay Railway in 1864, double track to Goodrington Sands Halt, and single track beyond there. A further extension over the River Dart to Dartmouth never materialized although the rail-less town always had its own railway booking office next to the ferry crossing on the west bank. Amalgamated with the South Devon Railway in 1872, the line became part of the GWR in 1876 and was converted to standard gauge in 1892.

Although listed for closure in the 1963 'Beeching Report', the railway survived until the end of 1972 when it was taken over, in working order, by the Dart Valley Light Railway. The railway continued under this ownership until 1991 when the Dartmouth Steam Railway took over operations. While physically linked to the national rail network at Paignton – enabling charter trains from distant starting points to move seamlessly from one line to the other – a change of trains is always necessary at this station when arriving by normal scheduled services.

Although fairly short, this railway offers many scenic attractions. Leaving Paignton station in a southerly direction it soon reaches Goodringon Sands where the line then climbs along the coastline above this popular beach and runs past Saltern Cove to the summit at Churston station. Inland is the ribbon development of south Torbay, its population inflated by holidaymakers and seasonal workers during the summer months.

At Churston, the junction for the short Brixham branch until 1963, the railway takes a southwesterly course down into open countryside and through Greenway Tunnel to emerge on the east bank of the tidal Dart Estuary. Here at low tide the exposed mud flats are frequented by wading birds, while at high tide the river is a colourful picture of small sailing craft and pleasure steamers. With steep wooded hillsides to the east the railway closely follows the estuary for two and a half miles before arriving at the overall-roofed 'train shed' terminus at Kingswear.

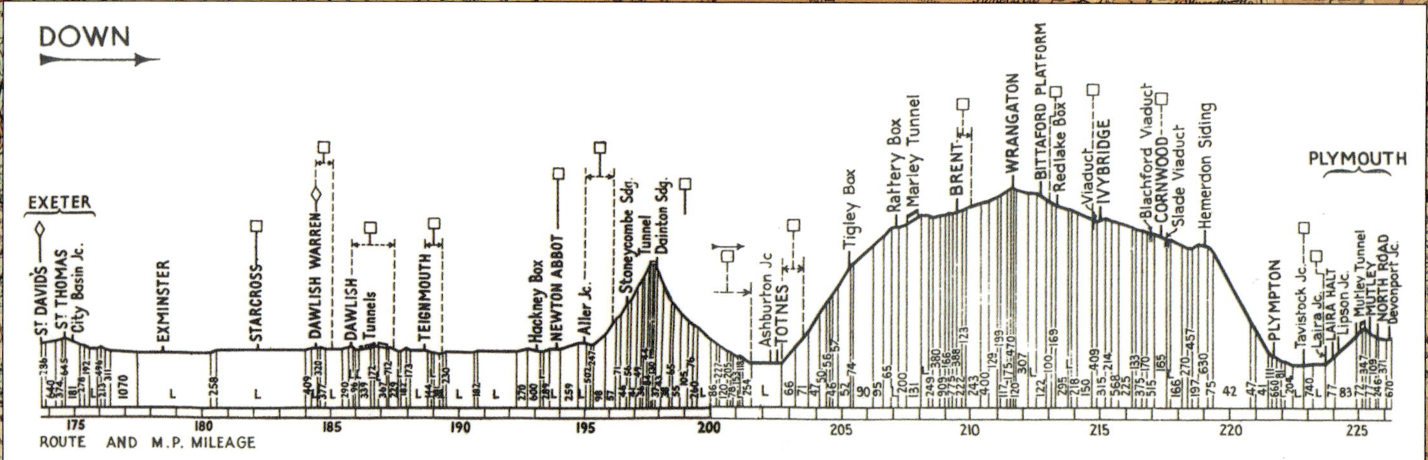

The flat coastal route contrasts markedly with the steeply graded South Devon Banks on this gradient profile of the Exeter to Plymouth route.

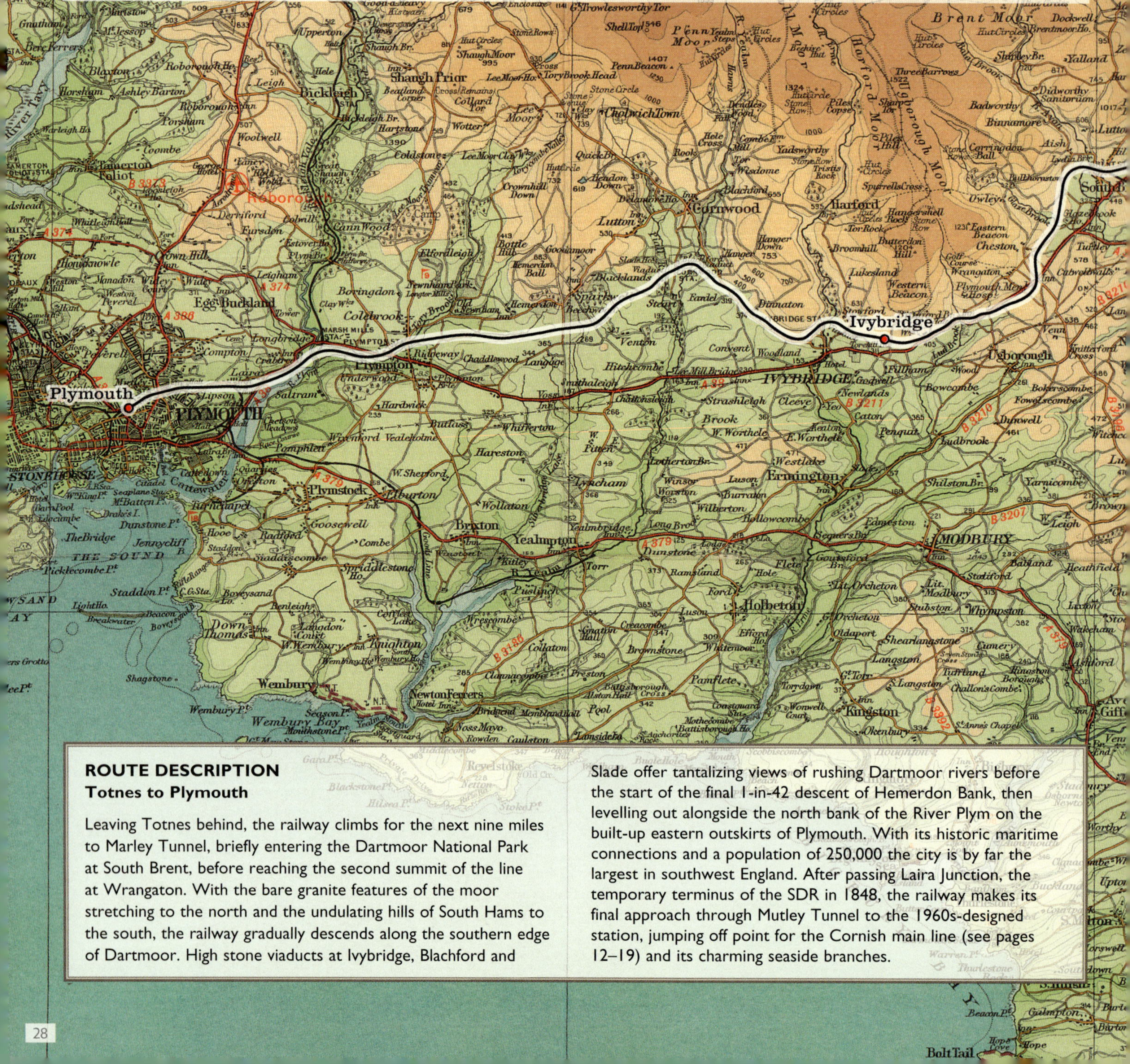

ROUTE DESCRIPTION
Totnes to Plymouth

Leaving Totnes behind, the railway climbs for the next nine miles to Marley Tunnel, briefly entering the Dartmoor National Park at South Brent, before reaching the second summit of the line at Wrangaton. With the bare granite features of the moor stretching to the north and the undulating hills of South Hams to the south, the railway gradually descends along the southern edge of Dartmoor. High stone viaducts at Ivybridge, Blachford and Slade offer tantalizing views of rushing Dartmoor rivers before the start of the final 1-in-42 descent of Hemerdon Bank, then levelling out alongside the north bank of the River Plym on the built-up eastern outskirts of Plymouth. With its historic maritime connections and a population of 250,000 the city is by far the largest in southwest England. After passing Laira Junction, the temporary terminus of the SDR in 1848, the railway makes its final approach through Mutley Tunnel to the 1960s-designed station, jumping off point for the Cornish main line (see pages 12–19) and its charming seaside branches.

SOUTH DEVON RAILWAY

Totnes to Buckfastleigh

The nine-and-a-half-mile broad-gauge branch line from Totnes to Ashburton was opened by the Buckfastleigh, Totnes & South Devon Railway in 1872. Converted to standard gauge in 1892, the company was amalgamated with the GWR in 1897. This sleepy country railway serving small farming communities and villages on the edge of Dartmoor was closed to passengers in 1958 and to goods traffic in 1962. The railway was re-opened as a heritage line by the Dart Valley Railway in 1969. Lord Beeching himself attended the opening ceremony, although his 'Report' of 1963 came after the line had already closed. Improvements to the A38 trunk road led to the Buckfastleigh to Ashburton section being closed in 1971, and since then the remaining seven miles have become a popular journey for both holidaymakers and steam railway enthusiasts. Although physically connected to the national rail network at Totnes, the South Devon Railway as it has been known since 1990, operates from its own station at Littlehempston which is reached by a footbridge over the River Dart from Totnes mainline station.

Steam trains between Totnes and Buckfastleigh are operated by the South Devon Railway and operate from early spring to late autumn.

From its terminus at Totnes, the railway winds its way northwards up the picturesque and meandering Dart Valley, following the river closely on its east bank until just before Buckfastleigh. Passing the sweeping parkland grounds of Dartington Hall, the railway reaches the only intermediate station on the line at Staverton. The picturesque village it serves is famous for its 'Sea Trout' Inn, fourteenth-century church and a fine example of a fifteenth-century river bridge. Continuing up the ever-narrowing and occasionally wooded valley, through typical dairy farming country, the railway finally reaches Buckfastleigh station with its small railway museum and miniature railway. The railway here is now truncated by the concrete fly-over of the A38 dual carriageway, which also visually separates it from the nearby village and its famous Buckfast Abbey.

Ex-GWR 2-6-2T No. 5526 heads a train of vintage coaches along the South Devon Railway near Riverford Bridge.

The Exeter to Plymouth railway's iconic coast-hugging route along the sea wall at Dawlish is often at the mercy of storms.

DARTMOOR RAILWAY

Coleford to Okehampton

The Okehampton Railway opened the stretch from Coleford to Okehampton, on the Exeter to Barnstaple line, in 1871. It was extended over Meldon Viaduct to Lydford in 1874 and eventually became part of the London & South Western Railway's main line to Devonport. Along with branches to Bude, Wadebridge and Padstow, the section west of Okehampton was listed for closure in the 'Beeching Report' and was closed by 1968. The section from Coleford to Okehampton, while not being earmarked by Beeching, lost its passenger service in 1972 although the remaining single line stayed open for railway track ballast from Meldon Quarry and occasional troop trains. Unusually, the line was not taken over by Railtrack on privatization, but was bought privately to support the movement of ballast.

The track is currently owned by Aggregate Industries, but the sixteen-and-a-half-mile line is also used by the Dartmoor Railway, which has operated a predominantly diesel heritage service between Meldon and Sampford Courtenay since 1997. The Dartmoor Railway, together with the Weardale Railway in Co. Durham, is owned by Devon & Cornwall Railways Ltd, a subsidiary of British American Railway Services Ltd. Future plans include extending the heritage service to connect with Exeter-Barnstaple trains at Yeoford.

As part of the Dartmoor Sunday Rover Network, First Great Western operates a diesel train service between Exeter and Okehampton on Sundays from late May to mid-September.

A vintage diesel-electric multiple unit halts at Okehampton station on the Dartmoor Railway.

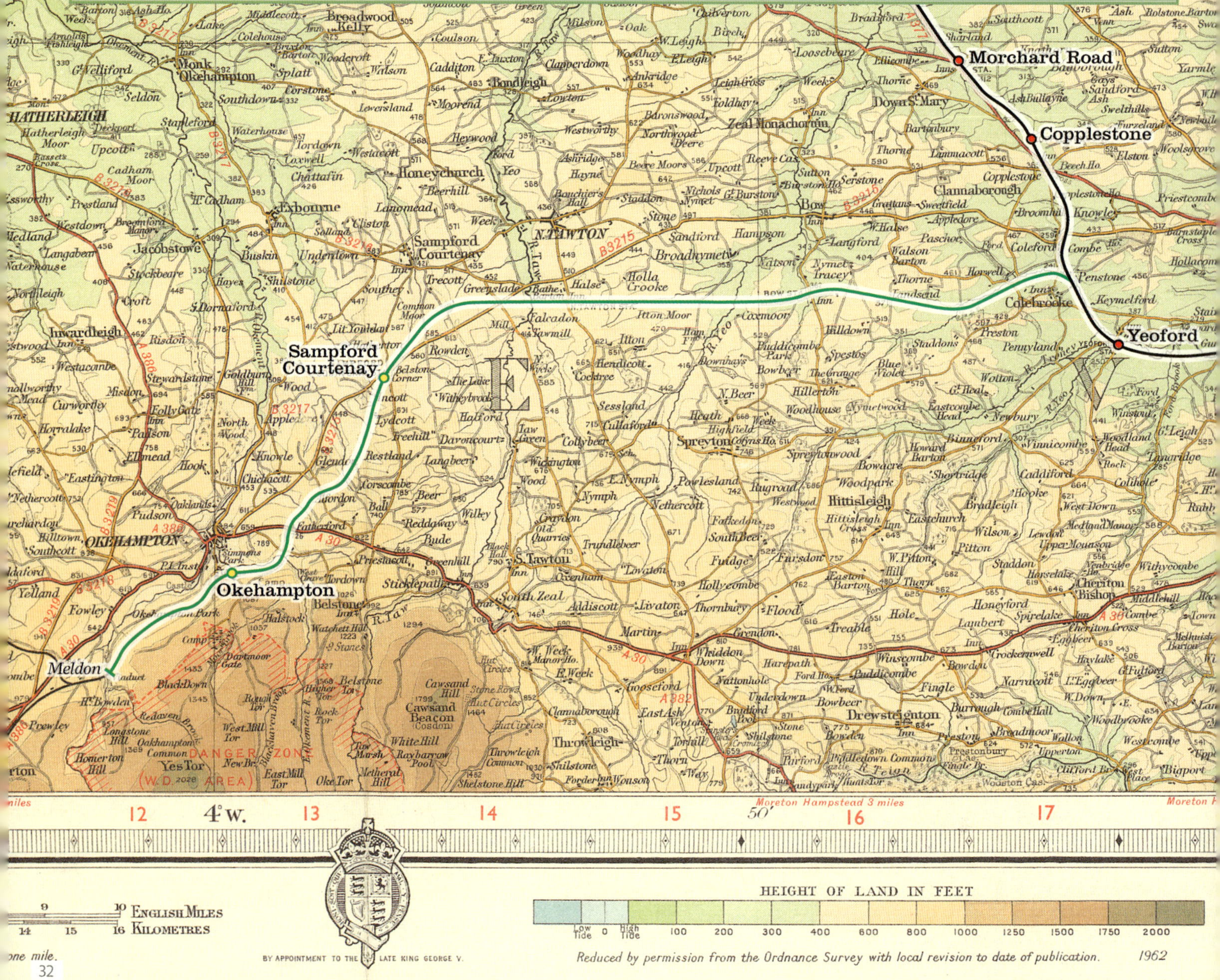

EXETER TO BARNSTAPLE

TARKA LINE

Named after the animal character in Henry Williamson's book 'Tarka the Otter', the thirty-nine-mile Tarka Line is the only survivor of a once-extensive rail network west of Exeter. It winds its way along the pretty valleys of the Yeo and Taw rivers through rich farmland and wooded countryside, linking North Devon and its scattered small villages with the outside world.

What was to become part of the London & South Western Railway's (LSWR) route to North Devon and North Cornwall started life in 1847 when the six-and-three-quarter-mile double-track broad-gauge Exeter & Crediton Railway was completed. However, the line was not opened until 1851 by which time one of the two tracks had been converted to standard gauge. Initially leased by the Bristol & Exeter Railway (B&ER), the LSWR operated it from 1862, converted it to mixed gauge and took it over in 1879. The broad gauge, used by Great Western Railway (GWR) trains, was taken out in 1892.

Meanwhile the North Devon Railway & Dock Company had opened its thirty-two-and-a-quarter-mile single-track broad-gauge line between Barnstaple and Crediton in 1854. It was worked for a year by the B&ER then leased by Thomas Brassey

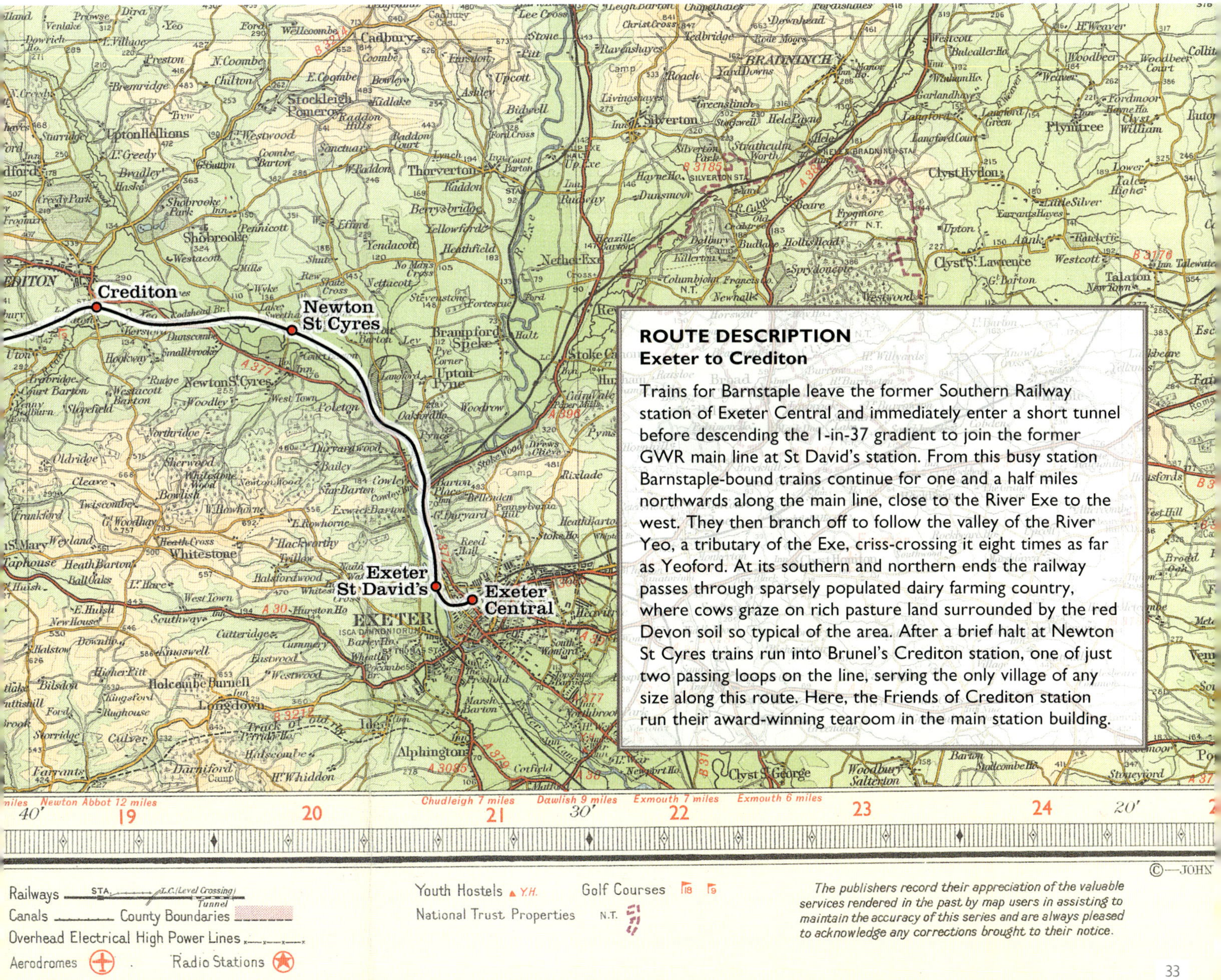

ROUTE DESCRIPTION
Exeter to Crediton

Trains for Barnstaple leave the former Southern Railway station of Exeter Central and immediately enter a short tunnel before descending the 1-in-37 gradient to join the former GWR main line at St David's station. From this busy station Barnstaple-bound trains continue for one and a half miles northwards along the main line, close to the River Exe to the west. They then branch off to follow the valley of the River Yeo, a tributary of the Exe, criss-crossing it eight times as far as Yeoford. At its southern and northern ends the railway passes through sparsely populated dairy farming country, where cows graze on rich pasture land surrounded by the red Devon soil so typical of the area. After a brief halt at Newton St Cyres trains run into Brunel's Crediton station, one of just two passing loops on the line, serving the only village of any size along this route. Here, the Friends of Crediton station run their award-winning tearoom in the main station building.

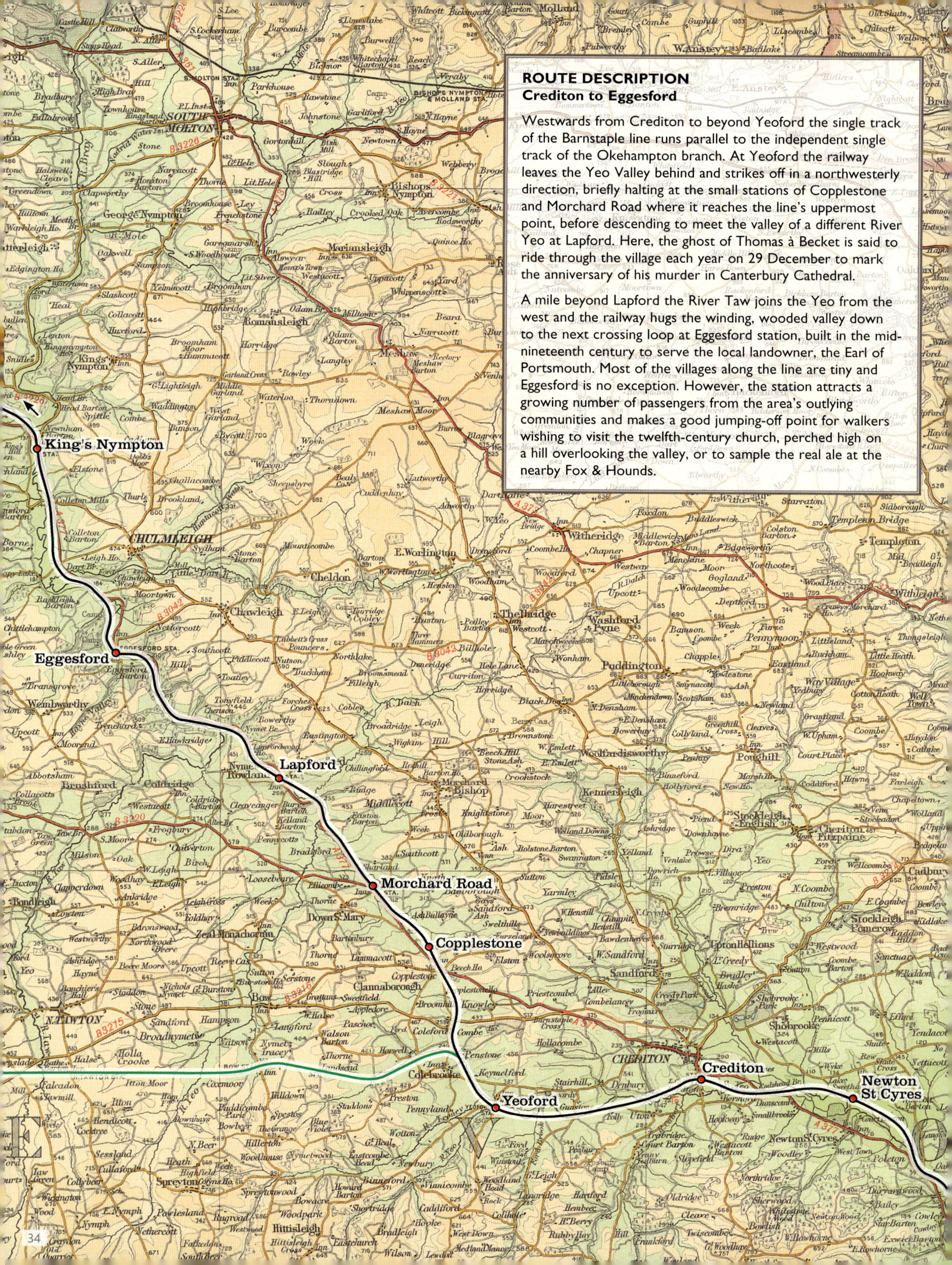

ROUTE DESCRIPTION

Crediton to Eggesford

Westwards from Crediton to beyond Yeoford the single track of the Barnstaple line runs parallel to the independent single track of the Okehampton branch. At Yeoford the railway leaves the Yeo Valley behind and strikes off in a northwesterly direction, briefly halting at the small stations of Copplestone and Morchard Road where it reaches the line's uppermost point, before descending to meet the valley of a different River Yeo at Lapford. Here, the ghost of Thomas à Becket is said to ride through the village each year on 29 December to mark the anniversary of his murder in Canterbury Cathedral.

A mile beyond Lapford the River Taw joins the Yeo from the west and the railway hugs the winding, wooded valley down to the next crossing loop at Eggesford station, built in the mid-nineteenth century to serve the local landowner, the Earl of Portsmouth. Most of the villages along the line are tiny and Eggesford is no exception. However, the station attracts a growing number of passengers from the area's outlying communities and makes a good jumping-off point for walkers wishing to visit the twelfth-century church, perched high on a hill overlooking the valley, or to sample the real ale at the nearby Fox & Hounds.

along with his Bideford Extension Railway before being leased by the LSWR in 1863. It was finally taken over by the LSWR in 1865 and converted to standard gauge in 1876 – the LSWR's link between Exeter and Barnstaple was now in place. The fifteen-and-a-half-mile standard-gauge Okehampton Railway opened between Okehampton and Coleford Junction, one mile north of Yeoford on the Crediton to Barnstaple line, in 1871 (see page 32).

By 1899, extensions from Barnstaple to Torrington and Ilfracombe and from Okehampton to Padstow, Bude and Plymouth had all opened, completing the LSWR's railway empire west of Exeter. Its drooping appearance on maps earned it the nickname the 'Withered Arm'. The coming of the railway along the Taw Valley to Barnstaple soon brought dividends for local farmers who could now send their cattle and milk speedily to markets in London. By the early twentieth century the rail connection via Barnstaple to the growing resort of Ilfracombe saw many thousands of holidaymakers travelling over the line during the summer months.

The LSWR became part of the newly formed Southern Railway in 1923 but the single-track line from Coleford Junction to Barnstaple notably missed out on much-needed improvements. An increase in the length of passing loops to accommodate longer trains failed to materialize and journey times between Exeter Central and Barnstaple Junction were up to 1 hour 30 minutes to cover just thirty-nine and three-quarter miles. While the resorts of South Devon were booming, the failure to make improvements to the railway led to economic stagnation for the seaside resorts and industries of North Devon during the twentieth century. However, there were two trains on weekdays that carried through coaches between North Devon and London Waterloo. Included in these was the famous 'Atlantic Coast Express' ('ACE'), which, apart from the war years, ran continuously from 1927 to 1964. A short-lived train was the all-Pullman 'Devon Belle' from Waterloo which was introduced in 1947 – the train splitting at Exeter Central with through carriages for Ilfracombe or Plymouth. The latter was discontinued in 1949 but the Ilfracombe portion – complete with observation car at the rear – survived the rapidly increasing competition from road transport through the post-war years until 1954.

The 'Beeching Report' of 1963 led to widespread closures in North and West Devon, and by 1972 only a slow stopping-train service between Exeter and Barnstaple survived. Surprisingly,

A Tarka Line train crosses the River Taw near Umberleigh in North Devon.

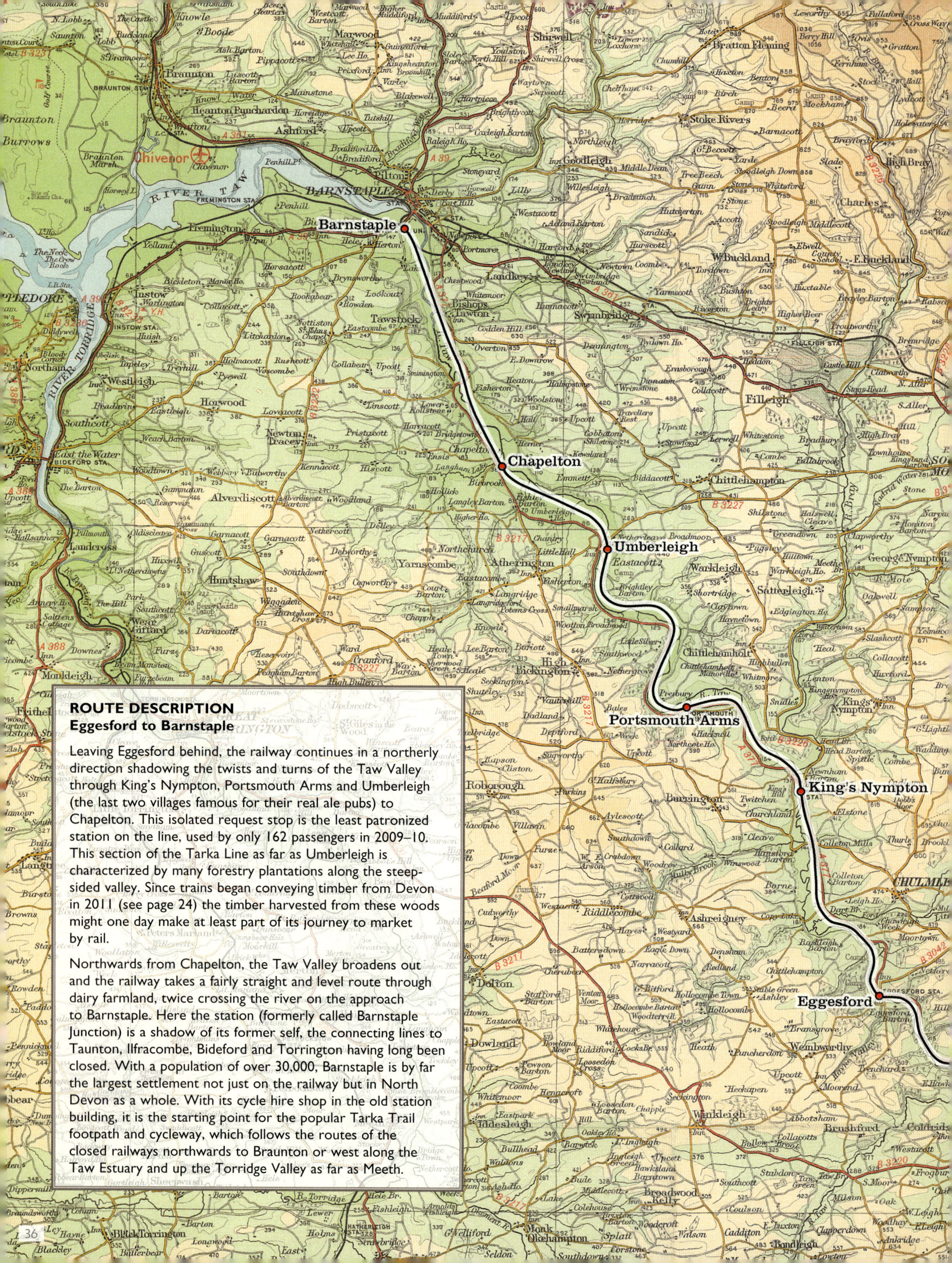

ROUTE DESCRIPTION
Eggesford to Barnstaple

Leaving Eggesford behind, the railway continues in a northerly direction shadowing the twists and turns of the Taw Valley through King's Nympton, Portsmouth Arms and Umberleigh (the last two villages famous for their real ale pubs) to Chapelton. This isolated request stop is the least patronized station on the line, used by only 162 passengers in 2009–10. This section of the Tarka Line as far as Umberleigh is characterized by many forestry plantations along the steep-sided valley. Since trains began conveying timber from Devon in 2011 (see page 24) the timber harvested from these woods might one day make at least part of its journey to market by rail.

Northwards from Chapelton, the Taw Valley broadens out and the railway takes a fairly straight and level route through dairy farmland, twice crossing the river on the approach to Barnstaple. Here the station (formerly called Barnstaple Junction) is a shadow of its former self, the connecting lines to Taunton, Ilfracombe, Bideford and Torrington having long been closed. With a population of over 30,000, Barnstaple is by far the largest settlement not just on the railway but in North Devon as a whole. With its cycle hire shop in the old station building, it is the starting point for the popular Tarka Trail footpath and cycleway, which follows the routes of the closed railways northwards to Braunton or west along the Taw Estuary and up the Torridge Valley as far as Meeth.

none of the intermediate stations along the line were closed and remain open, albeit unstaffed, for business today. Many have seen a significant increase in passenger usage in recent years thanks, in part, to the activities of the Community Rail Partnership.

First Great Western operates trains between Exeter Central and Barnstaple. The railway is single track, with intermediate loops at Crediton and Eggesford where trains cross. There are daily year-round train services on Mondays to Saturdays, with a more limited service (especially in winter) on Sundays, and seasonal train services operate on the branch line to Okehampton (see page 32). Interchange with a wide variety of inter-city, cross-country and local routes occurs at Exeter St David's station.

The Tarka Line is a classic scenic railway journey, closely following the river valleys all the way to the capital of North Devon. Having suffered from a lack of investment for many years it is now benefitting from a line partnership that promotes its numerous attractions and is increasing its popularity with walkers, cyclists and real ale aficionados.

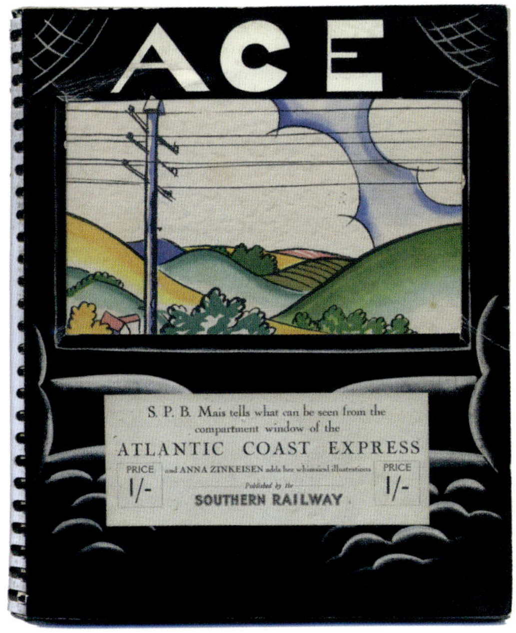

The 'Atlantic Coast Express' ran from London Waterloo to North Devon until 1964.

A Tarka Line train heads along the Creedy Valley near Newton St Cyres.

BISHOPS LYDEARD TO MINEHEAD

WEST SOMERSET RAILWAY

Britain's longest standard-gauge heritage line, the twenty-mile West Somerset Railway, skirts the Quantock and Brendon Hills before taking a coastal route alongside the Bristol Channel to the resort of Minehead.

The fourteen-and-three-quarter-mile broad-gauge West Somerset Railway (WSR) opened between Norton Fitzwarren and the port of Watchet in 1862, and the eight-mile Minehead Railway from Watchet to Minehead in 1874. Operated initially by the Bristol & Exeter Railway, the entire branch was converted to standard gauge in 1882 and subsequently taken over by the Great Western Railway (GWR).

Under GWR ownership both ends of the line were doubled, passing loops were extended and the Minehead terminus was enlarged. Holiday traffic peaked in the 1930s with trains arriving from London and the Midlands on summer Saturdays, but declined again rapidly during the Second World War and with increased competition from road transport in the post-war years. Despite the opening of Butlin's holiday camp at Minehead in 1962 the railway was listed for closure in the 'Beeching Report' and closed in 1971.

In 1976 the newly formed West Somerset Railway Company reopened the line from Minehead to Williton, extending it to Bishops Lydeard in 1979. A connection with the national rail network at Norton Fitzwarren allows charter trains to run through to Minehead.

A small core of permanent staff and hundreds of volunteers operate this single-track railway and with the exception of some off-peak diesel services the trains are steam hauled. Services run on selected days during February, March and December and daily from April to October except on certain Mondays and Fridays. A bus link operates from Taunton station on the national rail network to Bishops Lydeard station.

This is understandably one of the most popular heritage railways in Britain, offering the traveller a slowly unfolding scenic vista of hills and sea and the pleasure of being transported at a leisurely pace by classic steam locomotives.

Relics of a bygone age linger at the trackside while ex-GWR 2-6-2T No. 5542 heads away from Dunster with a train for Bishops Lydeard on the West Somerset Railway.

ROUTE DESCRIPTION

The railway climbs away from Bishops Lydeard through the characteristic rolling West Somerset farmland that fringes the nearby Quantock Hills, passing the country stations of Crowcombe Heathfield and Stogumber, which have been restored to their former GWR glory. In typical country railway fashion they sit some distance from the villages they serve, while Williton station lies on the eastern outskirts of the village.

The Bristol Channel is first glimpsed after passing Doniford Beach Halt, where the line runs along the shore for a short distance before passing Watchet Harbour. From Watchet the railway takes a long loop inland, climbing to Washford to avoid the unstable shale coastline. The ruins of Cleeve Abbey lie to the south with the Brendon Hills beyond, once the destination of a mineral railway that served ironstone mines until 1910. From Washford the railway turns sharply northwards towards the coast where, at Blue Anchor, there are fine views across the channel to South Wales. From Blue Anchor the railway follows a fairly straight and level coastal route towards Dunster, with its dominant castle, before reaching Minehead.

The final approach to the town is marred on the seaward side by the holiday camp but inland the wild expanse of Exmoor National Park beckons on the skyline. Featuring in the 1964 Beatles' film *A Hard Day's Night*, Minehead terminus is the starting point not only for vintage bus tours around Exmoor but also the 630-mile South West Coast Path to Poole Harbour.

BRISTOL TO WEYMOUTH

HEART OF WESSEX LINE

This eighty-seven-mile cross-country route from Bristol to the English Channel takes in the Avon Valley, the Wiltshire Plains and the Ham stone villages and towns of South Somerset before climbing the South Dorset Downs and descending once more to the Jurassic Coast at Weymouth.

The meandering and scenic Bristol-Weymouth line has a long and complicated history. It was opened in stages by the Great Western Railway (GWR) and the short-lived Wilts, Somerset & Weymouth Railway (WS&WR), beginning with Isambard Kingdom Brunel's broad-gauge GWR between Bristol Temple Meads and Bath in 1840. This was extended eastwards through Box Tunnel to Chippenham and on to Swindon and London (Paddington) in 1841. The WS&WR, which was also broad gauge, opened from Thingley Junction (about two miles east of Box Tunnel) to Westbury via Melksham and Trowbridge in 1848.

The rest of the railway to Weymouth was left unfinished in the wake of the financial crash that followed the period known as the Railway Mania. The GWR eventually came to the rescue and took over the unfinished project, opening a single-track line from Westbury to Frome in 1850 and eventually reaching Weymouth via Castle Cary, Yeovil and Dorchester in 1857.

With the Kennet & Avon Canal never far away, the railway threads up through the lush Avon Valley for over ten miles between Bath and beyond Bradford-upon-Avon.

ROUTE DESCRIPTION
Bristol to Bradford-on-Avon

The first part of the Bristol to Weymouth route is along Brunel's high-speed (originally broad-gauge) main line, which opened between Bristol and Paddington in 1841. Weymouth-bound trains leave the imposing overall-roofed station of Bristol Temple Meads in a northerly direction before swinging east to cross the River Avon and passing beneath the city's eastern suburbs through the two St Anne's Park Tunnels. The railway emerges on the south bank of this navigable and meandering stretch of the river, following it closely past the villages of Keynsham with its imposing chocolate factory, and Saltford before approaching the city of Bath through the short Twerton Tunnel. The southern slopes of the Cotswold Hills reach out to the north while to the south lie the northern limits of the Mendips. The railway makes its triumphant arrival to Bath high up on Brunel's grand castellated viaducts, offering fine views across the city with its elegant Georgian terraced houses and imposing Abbey. For a mile to the east of Bath station the railway is sandwiched between the River Avon to the north and the Kennet & Avon Canal to the south before swinging sharply south at Bathampton Junction to enter the Avon Valley.

Maintaining its course between the canal and the river, this highly scenic section winds its way southwards through the steep-sided, wooded Avon Valley. A mile to the south of Claverton the Kennet & Avon Canal crosses the railway and river via the Doric-style Dundas Aqueduct, built by John Rennie between 1797 and 1801. The station at Limpley Stoke is now closed but has been carefully restored by its owner. Here, the river and canal head east along the valley, passing Freshford and tiny Avoncliff (a request stop) on the approaches to the historic town of Bradford-on-Avon. At Avoncliff the canal crosses the railway and river once more on another of John Rennie's aqueducts. With its weir and seventeenth-century Cross Guns Inn, this is a popular destination for daytrippers and walkers.

A Bristol-bound High Speed Train from Paddington crosses the River Avon as it approaches Bath station.

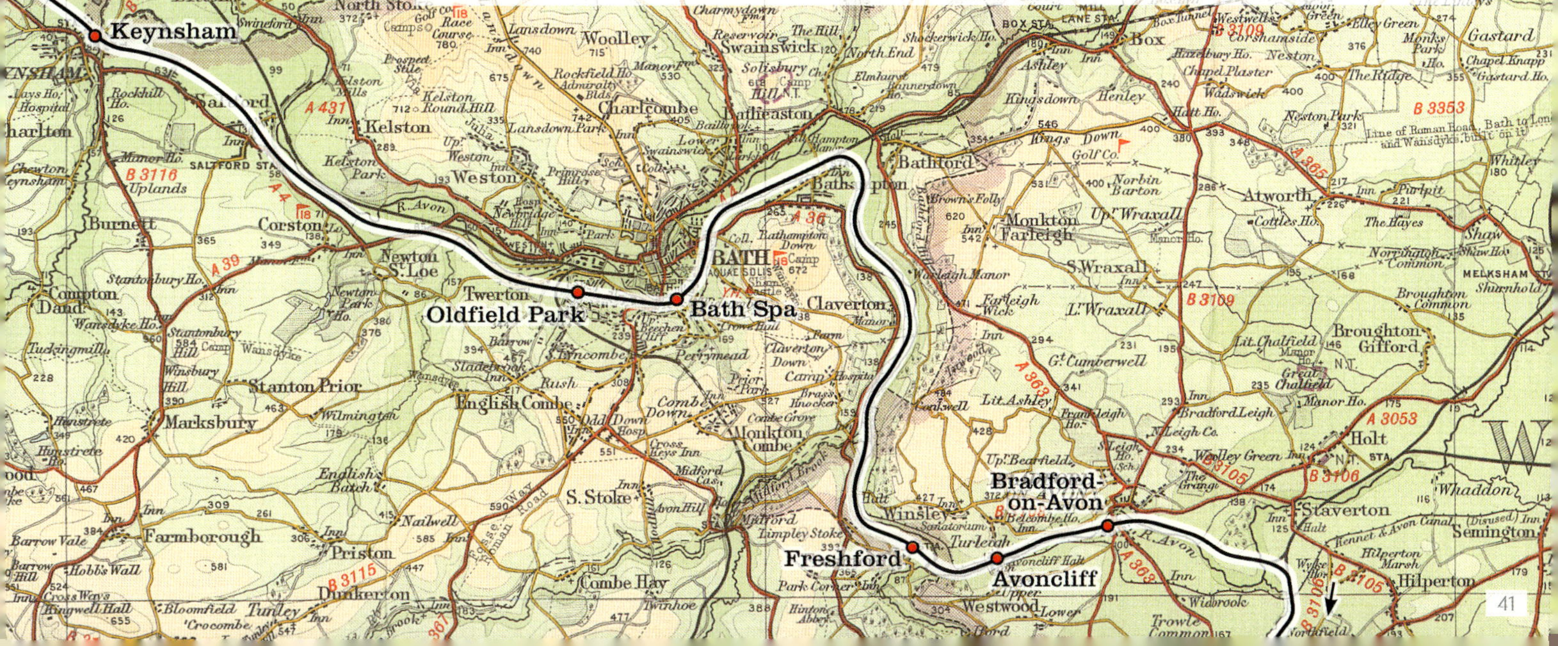

The remaining section of the modern-day route along the Avon Valley from Bathampton Junction (one mile east of Bath) to Bradford Junction (north of Trowbridge) was opened in the same year. Conversion to standard gauge came in 1874 and the whole route had been relaid as double track by 1885.

The building of the Bristol to Weymouth line involved considerable engineering work along two sections. The Avon Valley route from Bathampton Junction to Bradford Junction required seven viaducts and two canal aqueducts, while from Yeovil to Weymouth climbs of gradients as steep as 1 in 50 and four tunnels were needed to cross the South Dorset Downs.

Known as the 'Great Way Round' thanks to the circuitous route it takes from London to the West via Bristol, the GWR took its first steps to developing a direct route to Taunton with the opening of a new double-track line from Patney & Chirton (already served by Devizes-bound trains from Paddington, Reading, Newbury and Hungerford) to Westbury in 1900. Here it joined the existing route to Castle Cary and was complemented by a new line to Cogload Junction, north of Taunton, in 1906. Expresses along this new route to the West were speeded up considerably in 1933 with the opening of lines bypassing Westbury and Frome.

In addition to the regular Bristol to Weymouth stopping trains, the section that ran between Westbury, Castle Cary and Weymouth was used daily until the 1960s by several expresses, including the 'Channel Islands Boat Train' from Paddington. These heavy trains necessitated double heading over the steeply graded sections south of Yeovil and north of Weymouth.

None of the Bristol to Weymouth route was listed for closure in the 1963 'Beeching Report', however the section from Castle Cary to Dorchester West was singled in 1968 with crossing loops retained only at Yeovil Pen Mill and Maiden Newton. A number of intermediate stations and halts were closed in 1966, while the branch line from Limpley Stoke to Hallatrow – immortalized in the Ealing Comedy *The Titfield Thunderbolt* – had closed completely by 1951.

Diesel passenger trains are operated daily by First Great Western along the Heart of Wessex Line between Bristol Temple Meads and Weymouth with more limited services (especially in winter) on Sundays. Occasional charter trains – including steam-hauled services – also operate over the line. There is interchange with a wide variety of inter-city and cross-country routes at Bristol, Bath, Westbury, Castle Cary, Yeovil and Dorchester.

Weaving through numerous river valleys and traversing some of the finest countryside in Somerset, the Bristol to Weymouth line is one of the most attractive, but least known, scenic rail routes in southern England.

ROUTE DESCRIPTION
Bradford-on-Avon to Castle Cary

Heading east from Bradford-on-Avon the railway soon turns to the south, through the scenic farmland of the rolling Wiltshire Plains. The River Avon disappears off to the east just before Bradford Junction, where the single-track line from Thingley Junction and Melksham joins from the northeast. The canal crosses the railway north of Trowbridge and disappears from sight on its easterly journey towards Devizes. The approach to Westbury is overlooked by the famous White Horse, carved out of chalk on the northwest slopes of Salisbury Plain, one and a half miles east of the town. With its constant procession of heavy stone trains from the Mendip quarries and frequent passenger services to Bristol, Salisbury, Portsmouth, Reading, London, Taunton and Plymouth, the station at Westbury is still a hive of railway activity.

Joining the former GWR mainline at Fairwood Junction the railway bears away from Westbury in a southwesterly direction into rural Somerset. It diverts from it at Clink Road Junction to follow the original one-and-a-half-mile loop via Frome station, whose 1850 structure is one of the oldest-surviving through-timber train sheds in Britain. On leaving Frome the railway rejoins the mainline at Blatchbridge Junction before heading up the valley of the River Frome to Witham Friary. Here, the single-track line from Merehead Quarry, the remaining stub of the 'Strawberry Line' from Yatton and the East Somerset Railway at Cranmore, joins from the northwest.

Passing over the watershed the line dives into the valley of the River Brue to Castle Cary, passing through a narrow defile at Bruton close to the town's famous public school.

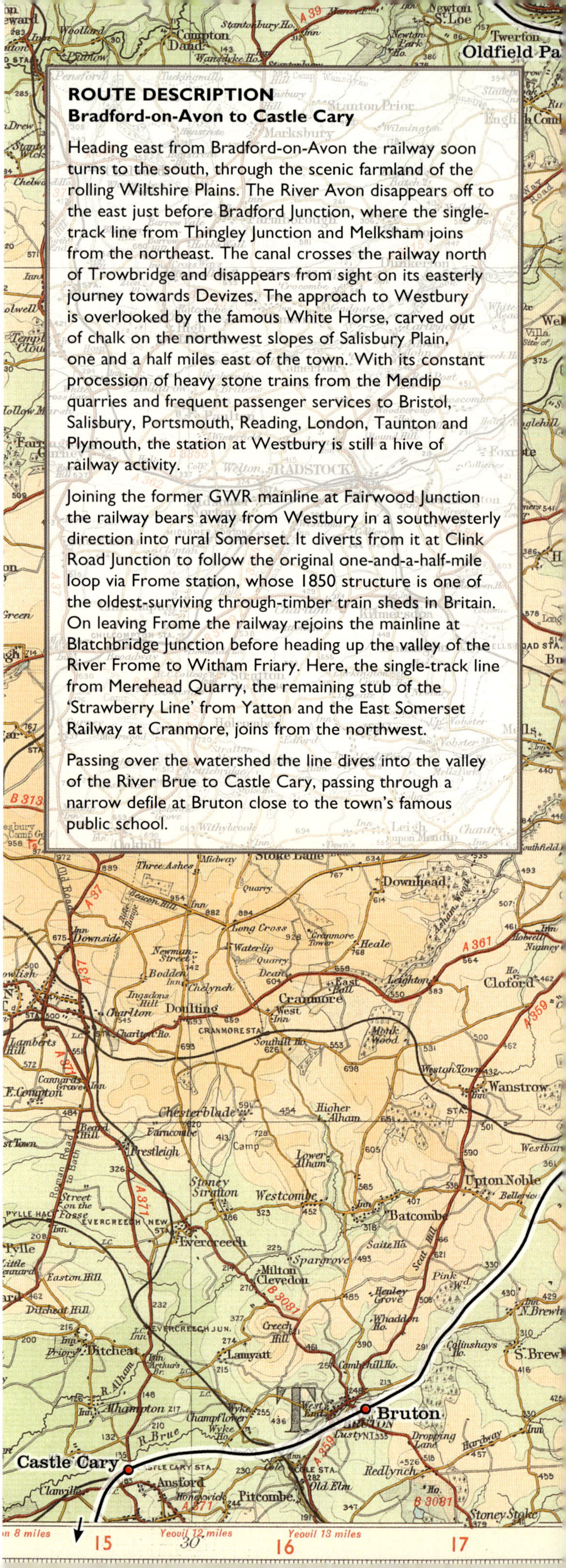

Converted from a High Speed Train, Network Rail's New Measurement Train approaches Bruton on the main line between Westbury and Castle Cary.

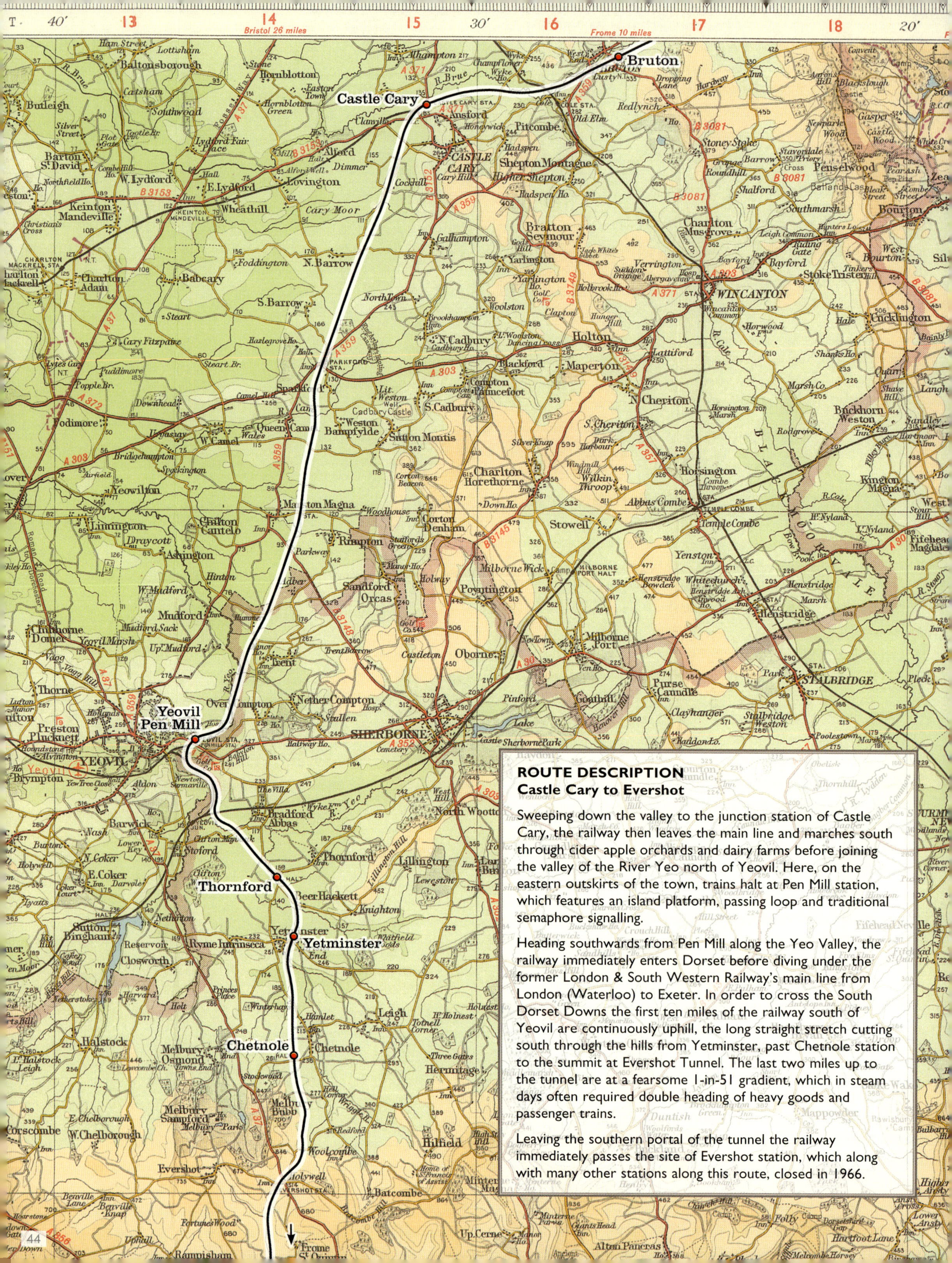

ROUTE DESCRIPTION
Castle Cary to Evershot

Sweeping down the valley to the junction station of Castle Cary, the railway then leaves the main line and marches south through cider apple orchards and dairy farms before joining the valley of the River Yeo north of Yeovil. Here, on the eastern outskirts of the town, trains halt at Pen Mill station, which features an island platform, passing loop and traditional semaphore signalling.

Heading southwards from Pen Mill along the Yeo Valley, the railway immediately enters Dorset before diving under the former London & South Western Railway's main line from London (Waterloo) to Exeter. In order to cross the South Dorset Downs the first ten miles of the railway south of Yeovil are continuously uphill, the long straight stretch cutting south through the hills from Yetminster, past Chetnole station to the summit at Evershot Tunnel. The last two miles up to the tunnel are at a fearsome 1-in-51 gradient, which in steam days often required double heading of heavy goods and passenger trains.

Leaving the southern portal of the tunnel the railway immediately passes the site of Evershot station, which along with many other stations along this route, closed in 1966.

Ex-GWR 'Castle' Class 4-6-0 No. 5029 'Nunney Castle' heads the 'Weymouth Seaside Express' towards Bristol at Castle Cary in August 2011.

Maiden Newton
Dorchester West
Upwey
Weymouth
DORCHESTER
WEYMOUTH
WEYMOUTH BAY
PORTLAND HARBOUR
WEST BAY
Chesil Bank
Portland

ROUTE DESCRIPTION
Evershot to Weymouth

From Evershot the railways heads downhill into the valley of the River Frome, cutting its path through the South Dorset hills before reaching Maiden Newton station and its passing loop. This former junction was once the starting point for branch line trains to Bridport, which were axed in 1975 when the line became one of the last victims of the 'Beeching Report'.

From Maiden Newton the railway continues its descent down the valley towards the county town of Dorchester, criss-crossing the meandering River Frome at three points. Relics of the area's ancient settlers such as tumuli and the earthworks of a Roman road and aqueduct are scattered on the hillside to the south of the railway on the approaches to the town. The single-track section ends at Dorchester West station and half-a-mile to the south at Dorchester Junction the railway is joined by the electrified line from Bournemouth (see page 60).

Heading south from the junction, the railway climbs steadily for two miles towards Bincombe Tunnel, the mighty Iron Age hill fort of Maiden Castle overlooking the line from the west. The tunnel is the beginning of the 1-in-50 descent to Upwey – for northbound steam trains this was a challenging obstacle with trains often double-headed out of Weymouth. The final approach to the seaside town is made along the eastern shore of Radipole Lake with trains ending their long journey from Bristol at the unprepossessing modern terminus station. The mothballed Weymouth Quay Tramway, which was once used by the Channel Islands boat trains, diverges southwards from the mainline on the approach to the station.

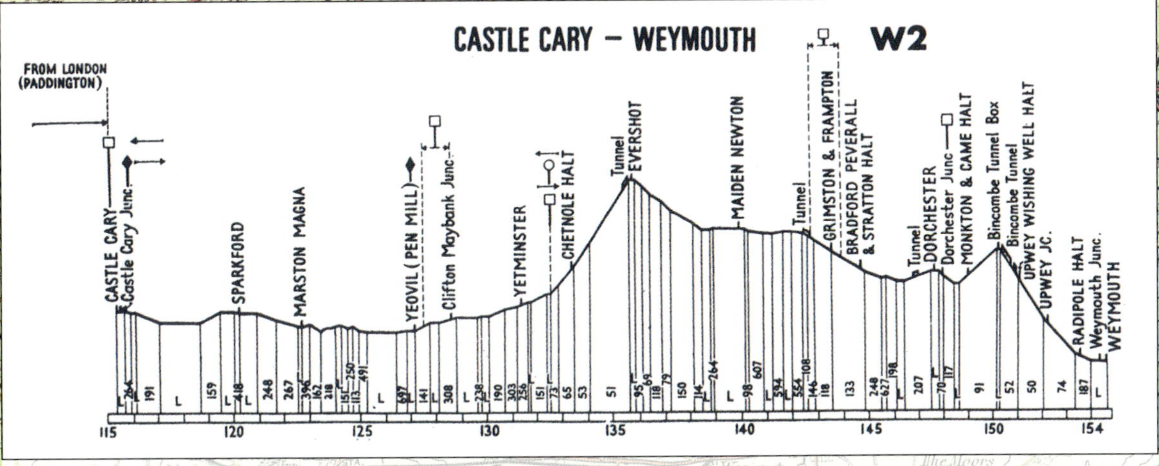

The gradient profile of the steeply graded route over the South Dorset Downs between Castle Cary and Weymouth.

Ex-SR 'Battle of Britain' Class 4-6-2 No. 34067 'Tangmere' storms up the 1-in-50 gradient from Upwey towards Bincombe Tunnel with a charter train from Weymouth in August 2009.

Headed by 'M7' 0-4-4T No. 30053, a steam train on the Swanage Railway gently weaves its way through the lush Purbeck landscape near Corfe Castle.

SOUTHERN ENGLAND

TONBRIDGE TO HASTINGS

HASTINGS LINE

With its winding route across the sandstone hills of the High Weald characterized by tunnels and steep gradients, the Hastings line continues to serve the historic towns and villages along its thirty-three-mile stretch to the sea.

Railways first reached the town of Tonbridge in 1842 when the South Eastern Railway (SER) opened the first section of its London (London Bridge) to Dover main line. Running via Ashford and Folkestone, the full line was opened for traffic in 1844. Intent on reaching the resort of Hastings, the SER initially achieved this with a roundabout route via Ashford and across Romney Marsh on a line that was opened in 1851. With a station at West Marina, Hastings had already been reached from the west by the Brighton, Lewes & Hastings Railway in 1846, which was taken over in 1847 by the London, Brighton & South Coast Railway (LBSCR). The SER's arrival in the resort led to a bitter price war with the LBSCR, which was only resolved with the opening of the SER's Tonbridge to Hastings direct line.

What was to become the SER's route to Hastings across the High Weald started life in 1845 with the opening of a five-and-three-quarter-mile steeply graded branch line from Tonbridge through Somerhill Tunnel to the spa town of Tunbridge Wells. The line was extended through the half-mile Wells Tunnel to Tunbridge Wells Central station a year later. Tunbridge Wells had been developed in a grand style in the eighteenth century and was frequented by royalty through the early nineteenth century, but its popularity had begun to wane due to competition from more accessible coastal resorts. The arrival of the railway arrested the town's decline.

Continuing southwards from Tunbridge Wells, the SER's route via Robertsbridge and Battle involved the building of seven tunnels through the sandstone of the High Weald and was opened in 1852. At Hastings it joined the LBSCR line from Brighton at Bopeep Junction. The ongoing price war was eventually settled by an amicable sharing of receipts, although LBSCR trains heading for Hastings station were not allowed to stop at the intermediate St Leonards Warrior Square station until 1870.

The Hastings Line as it is now known was left with one curious legacy that was not resolved until the late twentieth century. The original contractors who built the tunnels cut costs by lining them with four layers of bricks instead of the specified six. This was only discovered in 1862 when part of the 1,205-yd Wadhurst Tunnel collapsed and the only solution was for the SER to add two layers of bricks to all of the tunnels, making them narrower in the process. From then until the track was singled through the tunnels during electrification in 1986, trains operating on the line had to be specially built with a narrower profile – the most famous were the Hastings diesel-electric multiple units which took over from steam in 1957.

Passenger train services between Tonbridge and Hastings are operated by Southeastern. Freight traffic is still generated from the gypsum mine at Brightling, beginning its journey on a four-mile conveyor belt to Mountfield, south of Robertsbridge, before being despatched by rail to various destinations throughout England.

Set in an idyllic Wealden landscape of forested hills and small river valleys, the Tonbridge to Hastings line rates as one of the most scenic railway journeys in southern England.

A pair of Class 375 electric multiple units leave Etchingham with a Hastings to Charing Cross train in March 2012.

ROUTE DESCRIPTION
Tonbridge to Robertsbridge

For most of its thirty-three-mile route the line from Tonbridge to Hastings winds through and burrows under a region of hard sandstone hills known as the High Weald. Located at a strategically important crossing of the River Medway, the town of Tonbridge (population 30,000) has historic connections dating back to Anglo-Saxon and Norman times. Its eleventh-century castle, destroyed in a siege, was later rebuilt and became an official residence of King Edward II. During the eighteenth century transport links were much improved with the opening of the Medway Navigation, which remained a profitable concern until the arrival of the railway in 1842.

Immediately south of Tonbridge the railway climbs steadily through Somerhill Tunnel, across Southborough Viaduct and through Wells Tunnel to arrive at Tunbridge Wells Central station. Famous as a Georgian spa with its elegant Pantiles promenade, this town of 56,500 people is located at the head of the valley of the River Teise, a tributary of the Medway. Leaving Tunbridge Wells the railway continues southwards to climb through Grove Tunnel and Strawberry Hill Tunnel before emerging into a lush High Wealden landscape of small valleys and wooded hillsides. Hugging the contours the railway makes a sweeping approach to the small town of Wadhurst, the uppermost point of the line at the summit of the high, wooded ridge of the Weald.

Once surrounded by forests, Wadhurst was granted a royal charter in 1253 to hold a market each Saturday and went on to become an important centre for the Wealden iron industry from the sixteenth to the eighteenth century. Leaving the town through the long Wadhurst Tunnel the railway weaves its way down through the patchwork of wooded hillsides to descend into the valley of the River Rother, following it as far as the village of Robertsbridge. Named after the twelfth-century founder of a Cistercian abbey, Robert de Martin, this picturesque village is home to many fine medieval buildings, its peace and tranquillity preserved by a modern bypass to the east. It was also once the junction of the delightfully eccentric Kent & East Sussex Light Railway (see page 52), which ran along the Rother Valley to Tenterden and Headcorn.

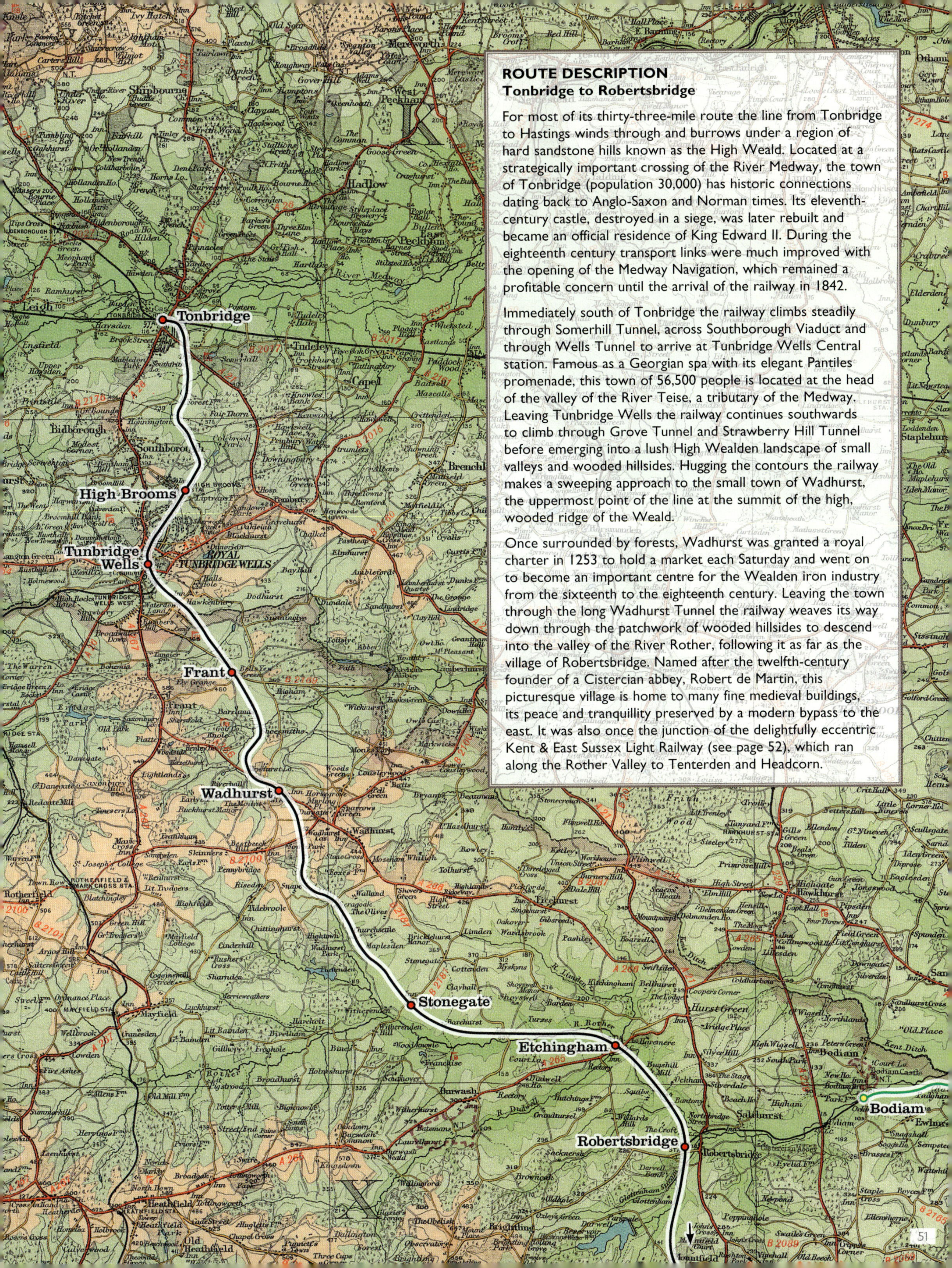

A pair of Class 375 electric multiple units emerge from Mountfield Tunnel south of Robertsbridge with a Hastings to Charing Cross train.

ROUTE DESCRIPTION
Robertsbridge to Hastings

Leaving Robertsbridge behind, the Hastings Line climbs away from the valley, winding its way up through Mountfield Tunnel and once more into the wooded hills before reaching the line's second summit at Battle. This small town of 6,100 people is famous as the site of the 1066 Battle of Hastings when the Norman invader, William the Conqueror, defeated the incumbent King Harold II.

From Battle it makes a meandering descent through wooded hills and past the village of Crowhurst to join the coastal route from Brighton at Bopeep Junction. The final one and a half miles to the modern Hastings station bores through the sandstone beds beneath the town, passing the delightful Victorian St Leonards Warrior Square station, which sits sandwiched between the 1,318-yd Bopeep Tunnel and 788-yd Hastings Tunnel, en route. Fuelled by claims that its seawater has health-enhancing properties, Hastings grew into a fashionable seaside resort following the arrival of the railway and now has a population of 87,000.

KENT & EAST SUSSEX RAILWAY
Tenterden Town to Bodiam

Built and owned by Colonel Holman F Stephens – the leading advocate of low cost rural light railways in Britain – the Kent & East Sussex Railway (K&ESR) opened in its entirety from Robertsbridge to Tenterden and Headcorn in 1905. Employing second-hand locomotives and rolling stock the railway led an impoverished existence until 1948 when it was nationalized. Surprisingly, the line survived a further thirteen years until its closure in 1961. A preservation group moved into Tenterden Town station following the closure, but it was not until 1974 that trains started to run again, with the present extent westwards to Bodiam being opened in 2000. Plans are currently in hand to extend the line further to Robertsbridge where it will join the Rother Valley Railway and the national rail network.

K&ESR trains start their ten-and-a-half-mile journey to Bodiam at the delightful Tenterden Town station. The town is set on a low hill overlooking the Rother Valley, with numerous pubs dotted along its wide tree-lined High Street. Beginning its descent in a westerly direction, the railway turns southwards at Rolvenden station, site of one of the four level crossings along the route that are so characteristic of light railways, to follow the Newmill Channel to Wittersham Road where it joins the lush low-lying meadows of the Rother Valley. The railway currently terminates at Bodiam, famous for its magnificent fourteenth-century moated castle, which was restored to its former glory in the nineteenth century and is now owned by the National Trust.

Ex-SECR 'C' Class 0-6-0 No. 592 heads a vintage train on the Kent & East Sussex Railway.

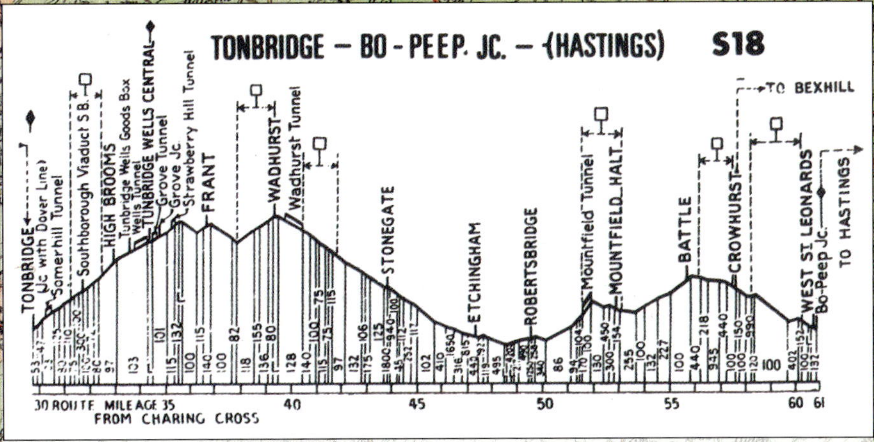

The Tonbridge to Hastings line climbs over and burrows through the High Weald on its way down to the valley of the River Rother at Robertsbridge.

HYTHE TO DUNGENESS

ROMNEY, HYTHE & DYMCHURCH RAILWAY

Billed as the 'World's Smallest Public Railway', the thirteen-and-a-half-mile Romney, Hythe & Dymchurch Railway has served the Kentish ribbon development of chalets, bungalows and holiday camps between Romney Marsh and the sea for the last eighty-five years.

During the 1920s Count Louis Zborowski and Captain Jack Howey were not only wealthy racing drivers but also shared a passion for miniature railways. Their first attempt to build a 15-in-gauge passenger-carrying railway along the narrow gauge Ravenglass & Eskdale Railway (see page 204) in Cumbria fell through, but not before they had ordered two miniature LNER Pacific locomotives from railway engineers Davey Paxman of Colchester. Tragically, before a new home for their railway could be found, Zborowski was killed in a motor racing accident at Monza in Italy. Howey then teamed up with miniature locomotive designer Henry Greenly and chose a level eight-mile stretch of coast between Hythe and New Romney in Kent, which was already becoming popular with holidaymakers.

Opened between Hythe and New Romney in 1927 by the Duke of York (later King George VI), the Romney, Hythe & Dymchurch Railway (RH&DR) is a one-third scale fully signalled double-track main line with express trains operating at up to twenty-five mph, hauled by scale versions of LNER and Canadian steam locomotives. A five-and-a-half-mile single-track extension was opened in 1928 from New Romney to Dungeness where the line makes a 360-degree loop.

Following a period of popularity during the 1930s, the onset of war placed the RH&DR on the front line. Train services were suspended while the railway, operated by the Somerset Light Infantry, played its part by employing an armoured train and transporting materials for the D-Day landings. The railway was re-opened by Laurel and Hardy in 1947, but declining passenger numbers and Howey's death in 1963 led to its sale in 1968. The new owners struggled to make ends meet and the railway was only saved from extinction when it was purchased and then restored by a consortium led by Sir William McAlpine in 1973. Today, while goods traffic such as shingle is no longer carried, the railway remains popular with local residents and tourists and operates a train for school children during term time. This unique railway also continues to run a postal and parcels service and issues its own stamps.

Operated by a small band of permanent staff and a large group of volunteers, train services (mainly steam hauled) on the Romney Hythe & Dymchurch Railway run daily between April and October and on most weekends except January.

Bounded to the west by the vast, lonely expanse of Romney Marsh and to the east by enormous coastal shingle banks, this miniature public railway offers a unique travelling experience that is unrivalled in Britain.

Built in 1931, North American-style 4-6-2 No. 10 'Dr Syn' heads a train through the shingle banks near Dungeness on the Romney, Hythe & Dymchurch Railway.

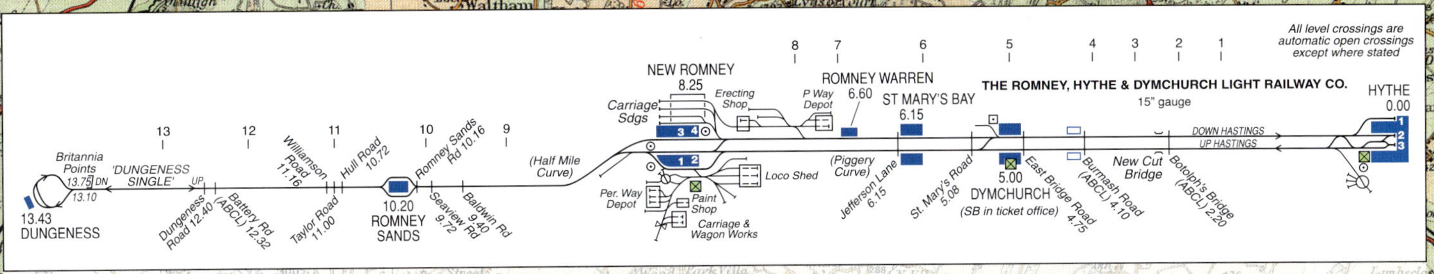

The diagrammatic track plan of the Romney, Hythe & Dymchurch Railway clearly shows the return loop at Dungeness.

ROUTE DESCRIPTION

Trains leave the three-platform overall-roofed terminus at Hythe, taking a southwesterly route between the vast expanse of Romney Marsh and the coastline. Reclaimed from the sea between the thirteenth and sixteenth centuries by the building of a maze of drainage channels and sluices, the 100 square miles of sparsely populated marsh is bordered far to the west by the Royal Military Canal. Completed in 1809, the twenty-eight-mile canal was built below the cliffs of the former coastline during the Napoleonic Wars. Once a haven for smugglers, the marsh also supports sheep farming – famed for its fleece, the medieval Romney Sheep has been exported around the world for over 200 years. To the east of the railway are vast coastal shingle banks, tossed there by storms over the centuries, punctuated by man-made seawalls and hundreds of groynes which stretch along the coastline to protect against longshore drift. Between the railway and the sea is the twentieth-century ribbon development of chalets, bungalows and holiday camps. The railway's workshops and engine sheds are located at the site of the line's headquarters at New Romney.

In addition to the canal, the area abounds with reminders of former military importance including Martello Towers, army ranges and the remains of huge Second World War concrete acoustic mirrors, built before the introduction of radar to detect enemy aircraft.

With its route punctuated by numerous level crossings, the railway makes its approach to Dungeness through a strange landscape of vast shingle banks dominated by two lighthouses and the massive bulk of two nuclear power stations. Perversely, the latter are now within a Site of Special Scientific Interest as the waste hot water pumped into the sea enriches the sea bed and attracts marine birds in vast numbers. The area as a whole is one of the largest shingle banks in the world and supports an enormous number of rare plant, bird and insect species.

SOUTHAMPTON TO WEYMOUTH

THE NEW FOREST AND THE DORSET COAST

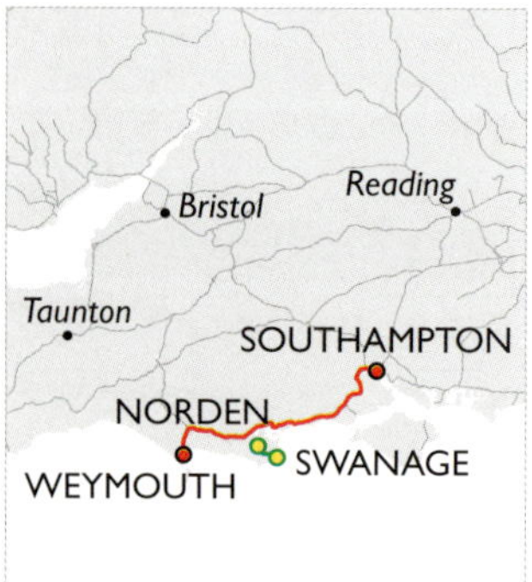

Once seen as a railway route to the West of England, the sixty-three-and-a-half-mile railway between Southampton and Weymouth was only completed in 1893, but its arrival along the Hampshire coastline led to the rapid growth of once small villages into major seaside resorts.

The London & South Western Railway (LSWR) completed its main line from Nine Elms in London to Southampton in 1840. To the west the Southampton & Dorchester Railway had been promoted by a Wimborne solicitor, Mr Castleman, and the meandering route, which opened in 1847 via Brockenhurst, Ringwood, Wimborne Minster and Broadstone became known as Castleman's Corkscrew. The railway was taken over by the LSWR in 1848. As the coastal resorts of Christchurch and Bournemouth grew, the need for a railway to serve them led to the opening of the single-track Ringwood, Christchurch & Bournemouth Railway in 1870, with a further branch from Broadstone to Bournemouth (West) added in 1874. Neither of these new lines was ideal for travellers to the two resorts as they involved either a lengthy roundabout journey or a change of train at Ringwood. This was only resolved in 1888 when the LSWR opened a double-track main line from Brockenhurst to Christchurch and Bournemouth (Central).

Meanwhile, in the west, Castleman's Corkscrew had already finished at a terminus in Dorchester (East) in 1847, with a planned westward extension along the Dorset and Devon coast to Exeter failing to materialize. The Dorchester (West) to Weymouth line was opened by the broad-gauge Great Western Railway in 1857 – the line was laid to a mixed gauge, which allowed LSWR trains from Southampton to reach Weymouth via a connection at Dorchester.

With the opening of the main line from Brockenhurst to Bournemouth in 1888 and a new line between Branksome and Hamworthy Junction in 1893 the LSWR main line from Southampton to Weymouth was complete. This inevitably led to the original Ringwood Loop between Brockenhurst and Broadstone being relegated to a secondary status. From then until its closure in 1964 it was used only by local stopping trains and trains from Waterloo to Swanage and Weymouth on summer Saturdays.

Notable expresses on the main line were the all-Pullman 'Bournemouth Belle' and the 'Channel Island Boat Train', the latter completing its journey via the streets of Weymouth on the harbour tramway. Trains continued to be steam hauled until July 1967 when third-rail electric trains were introduced from Waterloo to Bournemouth. The Bournemouth to Weymouth section was electrified in 1988 and a frequent electric train service is operated between Southampton and Weymouth by South West Trains.

As it traces the south coast from the major seaport of Southampton, through the New Forest, past the Hampshire coastal resorts and round Poole Harbour to the South Dorset Downs and Weymouth's Jurassic Coast, this railway witnesses a gentle unravelling of the many layers of scenic English countryside.

A South West Trains electric multiple unit crosses the causeway at Holes Bay near Poole with a Weymouth to Waterloo express.

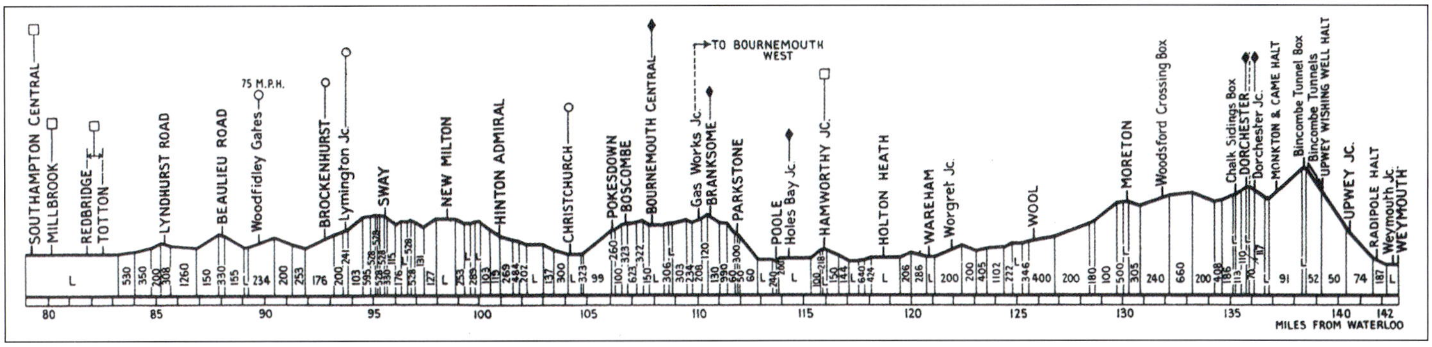

The gradient profile of the Southampton to Weymouth route through the New Forest, Bournemouth and Wareham.

ROUTE DESCRIPTION
Southampton to Poole

Trains for Weymouth depart from Southampton Central station, close to the centre of this large city (population 234,600). Located at the head of Southampton Water and on the east bank of the River Test the city has been a seagoing port for several centuries. However, the arrival of the railway in 1840 and the subsequent major expansion and modernization of dock facilities by the LSWR in 1892 transformed the city, making it the most important seaport in Britain. Playing a key role during both World Wars, Southampton Docks was also the terminal for ocean-going liners from all over the world, with passengers conveyed by rail to and from London Waterloo via the purpose-built Ocean Liner Terminal. Today, Southampton Docks is one of the largest commercial ports in Europe and still plays an important role as a terminus for enormous container ships and cruise liners. Long trains carrying containers leave two Freightliner terminals in and adjacent to Southampton Western Docks every day, heading for destinations around Britain.

Soon after leaving Southampton and its docks behind, the railway to Weymouth crosses the River Test at Totton and heads off in a southerly direction to enter the New Forest National Park at Ashurst (New Forest) station. For the next ten miles the line winds through forest and heathland via Beaulieu Road and Brockenhurst to the town of Sway. Created as a royal hunting ground by William I in 1079 and a source of timber for the Royal Navy in the eighteenth century, the 219 square miles of the New Forest centred on Lyndhurst became a National Park in 2005. Rich in rare wildlife and with miles of footpaths, cycle tracks and bridleways, the forest now attracts walkers, cyclists and horse riders in their hundreds of thousands every year.

Leaving the forest behind at Sway, the railway heads off in a westerly direction at New Milton where it enters the continuous built-up ribbon development that runs along the coast for seventeen miles through Christchurch and Bournemouth to Poole. Now one of Britain's major seaside resorts, with a population of 163,000, Bournemouth was no more than a tiny village until the coming of the railway in the nineteenth century. Famous for its nine miles of sandy beaches and pine trees that stretch from Hengistbury Head to Sandbanks along with its golf courses, conference centre, hotels and English language schools, this sprawling town was once the destination of the 'Pines Express' which ran daily from Manchester via the now-closed and much-lamented Somerset & Dorset Joint Railway.

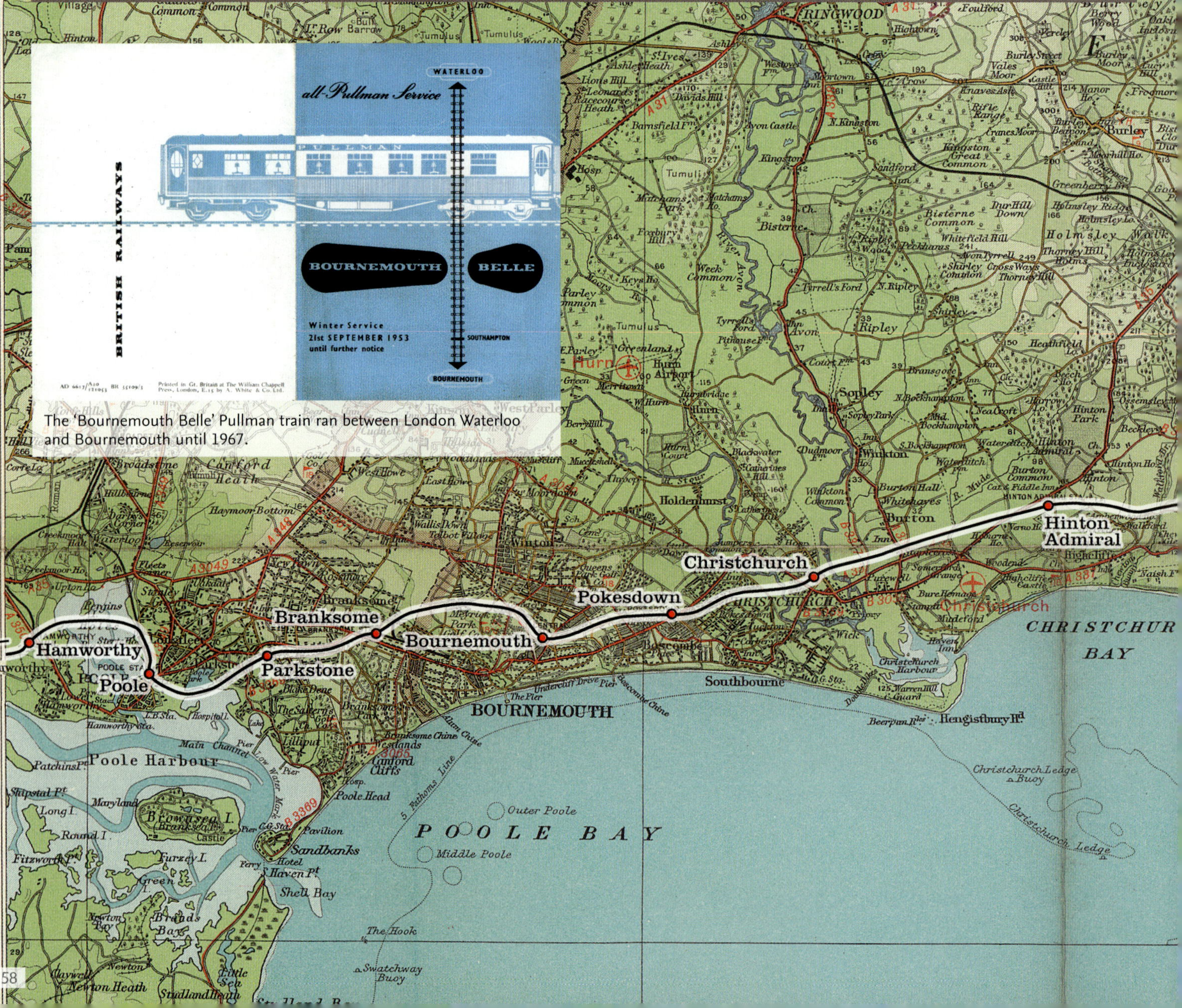

The 'Bournemouth Belle' Pullman train ran between London Waterloo and Bournemouth until 1967.

A South West Trains electric multiple unit approaches Beaulieu Road station in the midst of the New Forest.

ROUTE DESCRIPTION
Poole to Weymouth

From Poole station the railway crosses the marshy saltings of Holes Bay on an embankment and the built-up Hamworthy peninsula before reaching the western shore of Poole Harbour. Apart from a tiny entrance from the sea where two sand spits nearly meet at Sandbanks, the harbour is landlocked, its southern shores and islands a curious mixture of internationally important nature reserves and well-camouflaged oil wells.

Leaving Poole Harbour behind, the railway enters Wareham, famous for the remains of its ninth-century town walls and as one of the towns where the infamous Judge Jeffreys held the Bloody Assize in the aftermath of the Battle of Sedgemoor in 1685. Until 1972 the station also served the branch line that ran down the Isle of Purbeck peninsula to Corfe Castle and Swanage. Having fallen victim to the 'Beeching Report', the line has since been restored by the Swanage Railway (see opposite). Leading west from Wareham the main line trails the broad valley of the meandering River Frome as far as Dorchester. To the south are the South Dorset Downs and the coastline, famous for its dramatic cliffs and unusual geological features such as stacks, arches and a horseshoe cove as well as for the army tank ranges around Lulworth. Spreading out to the north of Wool station lie large tracts of mainly coniferous woodland, interspersed by the army tank training area around Bovington Camp. Alongside the railway to the west of Wool lies the decommissioned nuclear power station at Winfrith, its gaunt modern features at odds with the surrounding heathland.

Weaving through the lush countryside immortalized by the author Thomas Hardy, the railway reaches Dorchester where it makes a ninety-degree turn south to join the former GWR line from Castle Cary to Weymouth (for route details see page 47). With a population of 20,000 the county town of Dorchester has many historic connections dating back to the Iron Age and the Roman occupation. During the seventeenth century it was at the centre of the Puritans' migration to New England, while in 1833 the leaders of the Tolpuddle Martyrs were tried in the Shire Hall and sentenced to transportation to Australia.

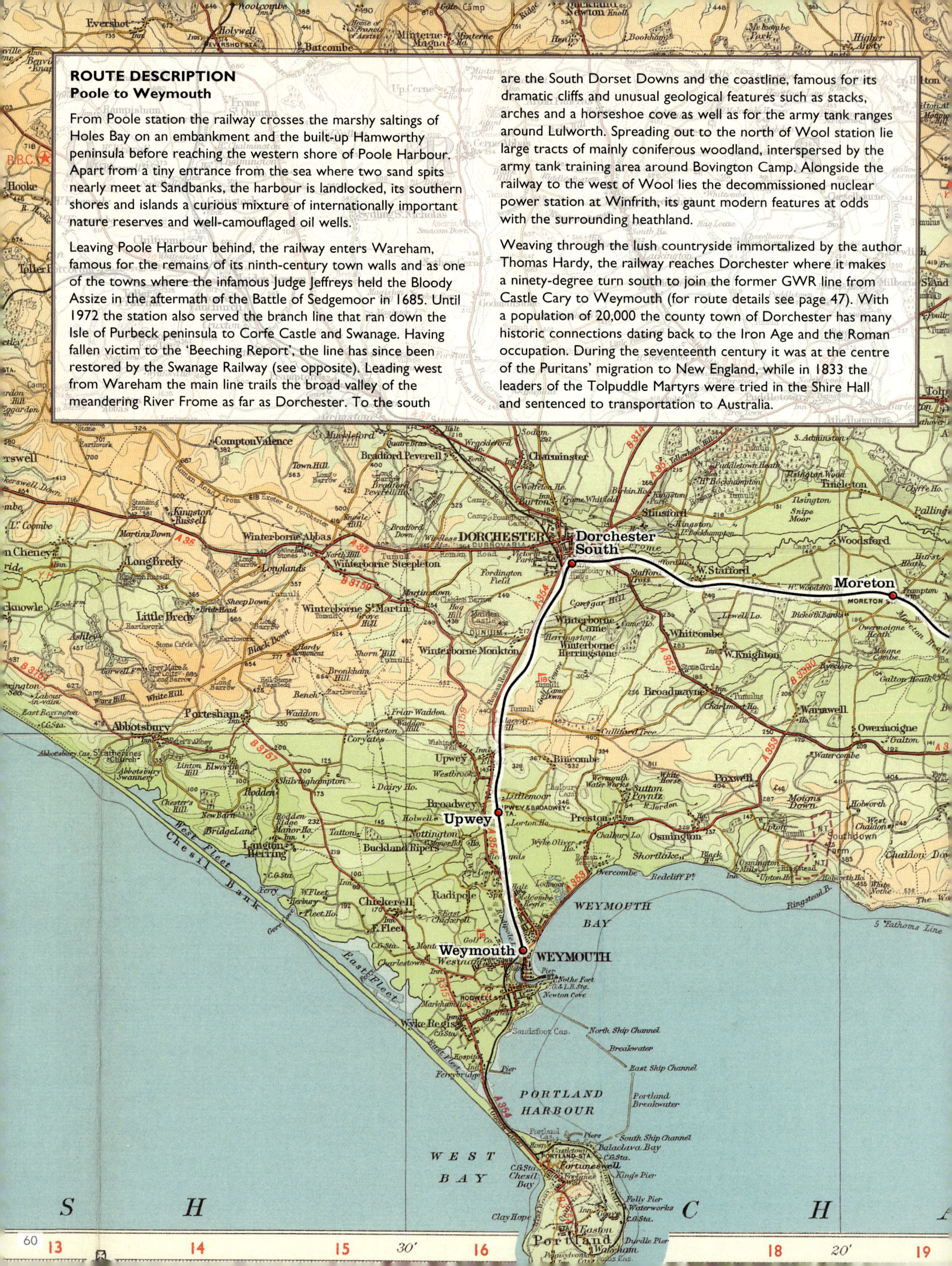

SWANAGE RAILWAY

Norden to Swanage

The ten-and-a-quarter-mile branch line across the Isle of Purbeck from Worgret Junction, one mile west of Wareham, to the seaside resort of Swanage was opened by the LSWR in 1885. Busy with holidaymakers during the summer months, the branch also saw clay traffic from Furzebrook and, in later years, oil trains from the Wytch Farm oil terminal. Despite not being listed for closure in the 'Beeching Report' the railway was closed in 1972. The section from Swanage to Corfe Castle and a new station at Norden were reopened as a steam heritage railway in 1995. A physical connection with the national rail network still exists at Worgret Junction, although it is currently only used by charter trains from other parts of the country.

The railway is operated by a small team of permanent staff supported by a large team of volunteers. Trains, mainly steam hauled, operate between Norden and Swanage most weekends and daily from April to October.

The Isle of Purbeck is in fact a peninsula formed of two distinctly different geological regions. The northern half is low-lying partially forested heathland with important reserves of shale oil lying beneath, while the southern half rises to a high ridge of limestone hills, in places over 600 ft high, before ending sharply at the coast as dramatic cliffs. The coastline forms the eastern end of the Jurassic Coast, a World Heritage Site. Here the oil-bearing shale surfaces at Kimmeridge Bay, site of the oldest working oil well in the world. While the railway crosses both these features, the section that is currently open from Swanage to Norden lies on the lower slopes of the Purbeck Hills, populated by small villages and farms. Located strategically in a gap in the Purbeck Hills, Corfe Castle station is overlooked by the famous ruins of the eleventh-century castle. The railway ends at the delightful terminus close to the sea front at Swanage, a town of 10,000 people that was once a fishing village and small port before the railway arrived. Vast amounts of locally quarried Purbeck Marble, shipped out from Swanage, were used in the rebuilding of London after the Great Fire of 1666.

Ex-SR 'Battle of Britain' Class 4-6-2 No. 34067 'Tangmere' leaves Corfe Castle station with a train on the Swanage Railway.

RYDE TO SHANKLIN

THE ISLE OF WIGHT

Marketed as the 'Island Line', this short route was saved from the 'Beeching Axe' in the 1960s and, along with ferries from the mainland and island buses, still plays a key role in the Isle of Wight's integrated transport system.

Following the opening of the Cowes & Newport Railway (later to form part of the Isle of Wight Central Railway) in 1862, the completely separate railway from Ryde (St John's Road) to Shanklin and Ventnor became the second railway to open on the Isle of Wight. Operated by the Isle of Wight (Eastern Section) Railway it was opened to Shanklin in 1864, but the building of a three-quarter-mile tunnel under St Boniface Down (at 791 ft the highest point on the island) held up the Ventnor extension for another two years. The company then changed its name to the Isle of Wight Railway and, in 1880, extended its line northwards from St John's Road to run out along Ryde Pier. Here, at Pier Head station, trains met ferries to and from Portsmouth, a service that is still provided by today's electric railway.

By 1900 the Isle of Wight possessed over forty-five miles of railways and their coming certainly boosted the island's popularity as a holiday destination. However, by the 1950s increased competition from island buses coupled with worn-out steam locomotives and rolling stock saw the closure of three island lines. The remaining lines (Ryde to Ventnor and Ryde to Cowes) struggled on until 1963 when they were listed for closure in the 'Beeching Report'. However, a vociferous campaign by islanders brought a reprieve for the eight-and-a-half-mile Ryde Pier Head to Shanklin line – the four-mile section running south to Ventnor was closed in April 1966 due to the high cost of maintaining the tunnel under St Boniface Down. The Ryde to Cowes line (see 'Isle of Wight Steam Railway' on page 64) had already closed in February of that year.

Steam continued to operate services between Ryde and Shanklin until the end of 1966 when the line was temporarily closed for the installation of third-rail electrification. Services then recommenced in March 1967 using redundant London Underground trains, a tradition that still continues today. Train services on the Island Line between Ryde Pier Head and Shanklin are operated as a subsidiary of South West Trains.

All that remains of a once-extensive island railway system, the unique electric railway between Ryde and Shanklin and the steam heritage railway between Smallbrook Junction and Wootton still form a historical and physical link between the island's Victorian seaside resorts, chalk downland and picturesque villages.

Ex-LSWR '02' Class 0-4-4T No. 24 'Calbourne' hauls a train of vintage carriages on the Isle of Wight Steam Railway.

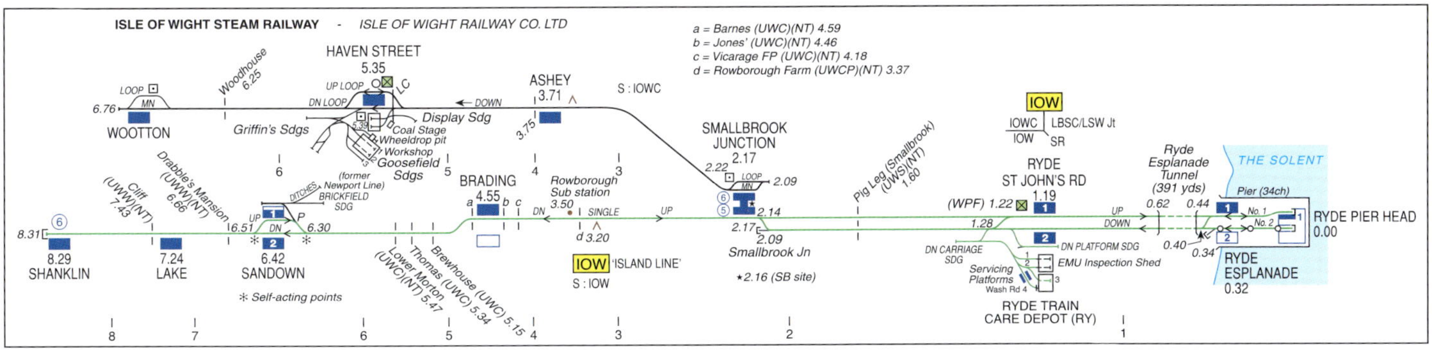

The route diagram of the two railways – one electric and one steam – on the Isle of Wight.

An ex-London Underground third-rail electric train on the Island Line near Shanklin.

ROUTE DESCRIPTION
Ryde to Shanklin

Offering extensive views across Spithead to the mainland, Ryde Pier was originally built as a timber structure in 1814 to enable the island's rail passengers to board mainland ferries without getting their feet wet. The pier was extended to its current length of 2,234 ft in 1833 and widened with a second pier to carry horse-drawn trams in 1864. By then the island was already popular with Queen Victoria and well-heeled Victorian tourists, and the building of railways on the island only fuelled its dramatic growth as a holiday destination. A third pier was added alongside the others in 1880 to carry the main railway line to Ventnor and Cowes, a function it still serves for today's electric trains to Shanklin. With a population of 30,000, Ryde is the largest town on the island – from the long pier and esplanade to grand seafront architecture and Gilbert Scott's lofty-spired All Saint's Church, its Victorian influences are visible at every turn.

Beyond Ryde the railway heads off in a southerly direction, first passing through gently undulating farmland before skirting around the central belt of chalk downland at the small town of Brading – world famous for the preserved remains of an impressive first-century Roman villa nearby.

South of Brading the railway approaches the island's east coast and the ribbon development of the resorts of Sandown and Shanklin. Sandown is the start of a footpath and cycleway along the track bed of the long-closed railway to Merstone, Newport and Cowes. With a combined population of 13,000, Sandown and Shanklin boast long stretches of sandy beaches, guest houses, holiday apartments, amusement piers and other twentieth-century tourist attractions which combine to make this a popular destination for family holidays. After skirting the coast near Sandown Bay, the railway ends at the small Victorian station of Shanklin, an interchange for bus services, including the summer-only Island Explorer, which runs in both directions around the forty-six miles of coastline.

ISLE OF WIGHT STEAM RAILWAY
Smallbrook Junction to Wootton

Opened in 1875 by the Ryde & Newport Railway, the ten-mile line between Ryde and Newport became part of the Isle of Wight Central Railway in 1887. This rural route across the northeastern part of the island met the Ryde to Shanklin and Ventnor line at Smallbrook Junction from where it ran on a parallel track up to Ryde. Serving only small villages and farming communities along its length, this was the route used by trains from Ryde to Cowes via Newport. Steam hauled to the end, the line closed in February 1966. Five years later the newly formed Isle of Wight Steam Railway made its headquarters at Havenstreet station and since then the line has been reopened westwards to Wootton and eastwards to Smallbrook Junction where there is an interchange station with the Island Line.

Operated by a small number of permanent staff and an enthusiastic group of volunteers, steam train services operate daily between Smallbrook Junction and Wootton from June to September. A reduced service also operates on certain days throughout the rest of the year apart from January.

With its beautifully restored stations, vintage carriages and steam engines, the Isle of Wight Steam Railway faithfully relives its Victorian heritage while passing through a delightful rural setting, its patchwork of rolling farmland and woodland still unspoilt by modern development.

Ex-LSWR '02' Class 0-4-4T No. 24 'Calbourne' emerges from the early morning mist with a train for Havenstreet on the Isle of Wight Steam Railway.

Ryde Pier Head
Ryde Esplanade
Ryde St John's Road
Smallbrook Junction
Brading
Sandown
Lake
Shanklin
Wootton
Havenstreet
Ashey
THE SOLENT
Spithead
Portsmouth Harbour
COWES
E. COWES
NEWPORT
RYDE
SANDOWN
SHANKLIN
VENTNOR
Sandown Bay
ISLE OF WIGHT
Stokes Bay
Mother Bank
Bramble Bank
Chillerton Down
St CATHERINE'S POINT
HEIGHT OF LAND IN FEET
Low Tide 0 High Tide 100 200 300 400 600 800
ENGLISH MILES
KILOMETRES
Railways
Canals
County Boundaries
Overhead Electrical High Power Lines
The figures thus 190 represent heights in feet
BY APPOINTMENT TO THE LATE KING GEORGE V.

ROUTE DESCRIPTION

The railway's eastern terminus lies in the town of Alton, its 17,000 population within daily commuting distance of London. Recorded in the Domesday Book of 1086 as having the most valuable market in the land, the town also witnessed the signing of the Treaty of Alton between Robert, Duke of Normandy, and King Henry I in 1101. Granted a royal charter in 1320, the town and the surrounding area bore witness to the seventeenth-century Battle of Alton and boast many historic sites including the homes of Jane Austen in Chawton and Gilbert White in Selborne and the grave of eight-year-old 'Sweet Fanny Adams' who was murdered in 1867. Surrounded by fields of hops and barley, Alton has also been an important centre for brewing since 1763, a practice which is continued in the town today by the multi-national giant Coors Brewers.

Threading its way through Alton's southern suburbs, the railway passes the site of Butts Junction, once the starting point of railways to Basingstoke and Fareham. Here, the once rail-connected Lord Mayor Treloar Hospital first opened its doors in 1908 for the treatment of children crippled by polio and tuberculosis. With the Forestry Commission's Chawton Park Wood to the north, the single-track railway embarks on its steep climb away from the valley of the River Wey towards its 652-ft summit near Medstead & Four Marks station, surrounded by hills rich with arable farming.

From the watershed at Medstead & Four Marks, the highest railway station in the south of England at 644 ft above sea level, the line descends through Ropley station, famous for its nineteenth-century topiary, to meet the River Arle, a tributary of the slow-flowing, meandering Itchen. The railway ends at the attractive village of New Alresford, population 5,000, which is famed as a centre for the growing of watercress in the chalk streams that abound in the area.

Visiting ex-LNWR 'G2' Class 0-8-0 No. 49395 enters Ropley station on the Mid-Hants Railway's Watercress Line.

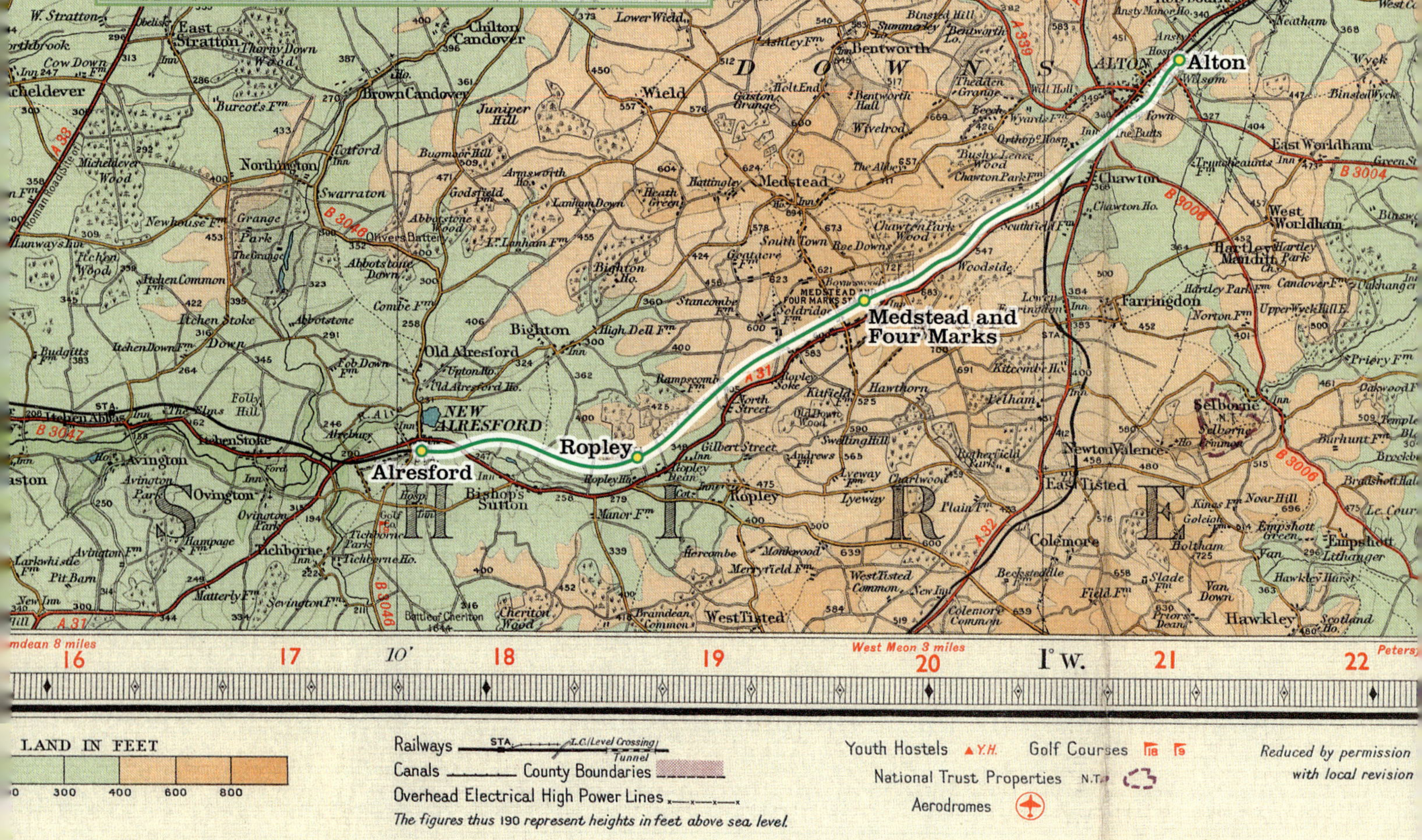

ALTON TO ALRESFORD

WATERCRESS LINE

Named in modern times as the 'Watercress Line', after the locally grown produce, the ten-and-a-quarter-mile Mid-Hants Railway reopened as a steam-hauled heritage line over thirty-five years ago.

The seventeen-mile Mid-Hants Railway opened between the historic market town of Alton and the cathedral city of Winchester in 1865. Connecting with the London & South Western Railway (LSWR) at both ends, the LSWR operated the railway from the outset and took it over in 1884. Besides linking villages along the Itchen Valley and the Hampshire hills with the outside world and conveying locally grown fresh watercress to market, the line was also useful as a diversionary route when engineering work closed the LSWR main line. With its connections eastwards to Aldershot and southwards to Southampton Docks, the railway was heavily used during both World Wars. The climb to the line's highest point near Four Marks, 652 ft above sea level, often necessitated double heading of locomotives in both directions, leading this steeply graded section to become fondly known as 'The Alps'.

While third-rail electrification reached Alton from Waterloo as early as 1937, the Mid-Hants remained steam hauled until the introduction of diesel electric multiple units in the late 1950s. Despite a regular interval service being introduced, the line was listed for closure in the 1963 'Beeching Report' and although strong local objections secured a ten-year reprieve, it finally closed in February 1973.

In 1975 the ten and a quarter miles of track bed between Alton and Alresford were bought by the Winchester & Alton Railway Company and the railway opened in stages between 1977 and 1985, when it once again connected to the national rail network at Alton. With this all-important cross-platform link and direct connection with London, the Mid-Hants Railway is today one of the most-visited steam-hauled heritage lines in Britain.

Operated by a nucleus of permanent staff and a large band of volunteers, the Mid-Hants Railway runs mainly steam-hauled trains on most weekends and school holidays with a more intensive service between April and September.

The comparatively short Mid-Hants Railway weaves its way up through the rolling Hampshire hills dotted with picturesque villages before descending to the broad valley of the headwaters of that very English of rivers, the meandering Itchen.

Ex-SR 'U' Class 2-6-0 No. 31806 hauls a passenger train bound for Alton across 'The Alps'.

DOCKLANDS LIGHT RAILWAY

Providing multiple interchange points with the London Underground and the capital's bus and overground rail systems, the automated Docklands Light Railway has played a pivotal role in the redevelopment of former rundown areas of East London and is now used by around 200,000 passengers each day.

Greatly expanded during the nineteenth century, London's docklands area, known as the Port of London, was by the early twentieth century the largest and busiest port in the world. Seriously damaged during the Blitz in the Second World War, the docks had lost their importance by the 1970s when new and larger container ships favoured the deepwater ports down river at Tilbury and at Felixstowe on the east coast. By 1980 all of London's docks had closed leaving an eight-square-mile area of derelict land in the east of the city. The Conservative government of the time stepped in and created the London Docklands Development Corporation (LDDC), with wide-ranging powers to redevelop the Docklands area.

A Lewisham to Tower Gateway train crosses over regenerated docklands as it approaches Heron Quays station on the Docklands Light Railway.

One of the first problems to be addressed by the LDDC was the provision of good transport links with the rest of London – an extension of the Underground Jubilee Line was initially turned down due to the high construction costs. Instead LDDC proposed a low-cost electrified light railway, running for much of its length on disused railway lines and derelict land – thus the Docklands Light Railway (DLR) was born.

Opened in 1987, the first seven-and-a-half-mile phase of the DLR consisted of three branches radiating out from a triangular junction at Poplar – to the west a branch to Tower Gateway station in the City of London (close to Fenchurch Street station), to the south a branch to Island Gardens on the north shore of the Thames and to the north a branch to Stratford via Bow. Two-thirds of the railway was built along disused or little-used railways including the London & Blackwall Railway's viaduct of 1839 between Limehouse and West India Docks. The southern terminus at Island Gardens was built on the site of the old North Greenwich station, which was once the terminus of the Millwall Extension Railway.

Despite teething problems, the automated lightweight railway was a great success and certainly contributed to the rapid redevelopment of the Docklands area. To cope with the increased traffic, stations and trains were lengthened and a more conveniently located western terminus was opened in 1991 at Bank in the City. Here there is an interchange with the Underground system. In the same year Canary Wharf was reopened as an enlarged station with six platforms and an overall glass roof. A fourth branch was opened eastwards from Poplar to Beckton via Canning Town in 1994 – Beckton is now the main stabling and maintenance facility on the DLR. In 1999, an extension was opened under the Thames from Island Gardens to Lewisham.

The DLR further expanded in 2005 with the opening of another eastbound branch to King George V Docks via London City Airport – this being extended under the Thames to Woolwich Arsenal in 2009. To coincide with the Olympic Games coming to Stratford in 2012, the DLR has been upgraded to three two-car trains and lengthened platforms, with a sixth branch, from Canning Town to Stratford International, being opened in 2011 along the route of the former Great Eastern Railway line from Stratford to North Woolwich.

Today the DLR's six branches total twenty-one route miles serving forty-five stations – proposals to extend to Dagenham Dock, Victoria, Euston and Catford are also being considered. With a top speed of 62 mph and controlled remotely from the operations centre at Poplar, electric multiple unit trains can operate in either direction and are driverless, although the onboard guard (known as the Passenger Service Agent) is able to take over the controls in an emergency.

In London's East End, the Docklands Light Railway slips magically through the canyons of glass and steel of Canary Wharf.

ROUTE DESCRIPTION
Bank to Woolwich Arsenal

Travelling on a DLR train is great fun – the two-car trains (run in multiples of two or three) have a swaying motion as they round sharp curves or rattle down steep gradients and over flyovers, and sitting at the front (there are no drivers) gives a grandstand view of the route ahead.

Trains for Woolwich Arsenal leave the underground interchange station at Bank and travel through a long tunnel before emerging alongside the Fenchurch Street to Shoeburyness main line to arrive at Shadwell. Set high above the streets on a Victorian railway viaduct, the railway offers a unique view of the regenerated East End of London, the remnants of a former world-wide trading empire contrasting vividly with the gleaming new twenty-first-century towers of international commerce. Old blocks of flats with washing hanging out to dry on their verandas, pubs, rows of roofs and chimney pots, an old Second World War bomb site and a city allotment give way to the boat marina at Limehouse Basin where the DLR crosses the Regent's Canal on an historic viaduct. From here the vista opens up through West Ferry, the towering modern glass buildings of Canary Wharf and the distant O_2 Arena contrasting with a nineteenth-century Old Dockmaster's House (now an Indian restaurant).

A flyunder and sharp curves lead to the important interchange station at Poplar, which is also the main hub and control centre of the DLR. From here the scenery continues to enthral – modern Billingsgate Fish Market, views of the Thames and the O_2 Arena again, modern flats, a cross-river cable car, derelict docks, the old Tate & Lyle building near Canning Town, the Thames Barrage at West Silvertown, the derelict Millennium Mills (built in 1905 to make imported grain into flour) and dockside wasteland awaiting redevelopment, the enormous modern Tate & Lyle sugar factory near Pontoon Dock, London City Airport (best viewed from the Beckton branch, see opposite) and the roofs of nineteenth-century terraced houses all combine to paint a vivid picture before the train dives under the River Thames in a long curving tunnel. The railway ends its journey from Bank at the underground station at Woolwich Arsenal where there is an interchange with the national rail network.

Surrounded by sky-rise office blocks, Heron Quays station is on the Lewisham branch of the Docklands Light Railway.

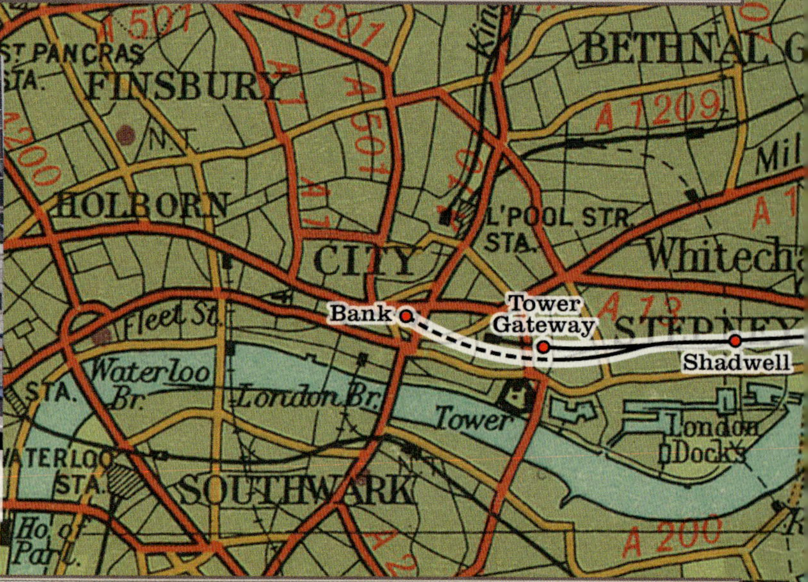

ROUTE DESCRIPTION
Stratford to Lewisham

Overlooked from the north by the ultra-modern 2012 Olympic Games' complex built on the site of the Great Eastern Railway's locomotive works, DLR trains leave the busy interchange station at Stratford before paralleling the electrified main line into Liverpool Street through Pudding Mill Lane station. Here, Olympic hopes give way to the massive former Bryant & May match factory (now flats) before the railway (initially single line) heads south along the route of the old North London Railway's line through Bow Church, Devons Road, Langdon Park and All Saints to the DLR interchange station and control centre at Poplar, where, to the south, the glass-fronted high-rise offices of Canary Wharf dwarf the modern Billingsgate Fish Market below.

With wheels squealing on the sharp curves, DLR trains leave Poplar behind to climb the steep flyover to West India Quay before entering canyons of glass and the glass-roofed interchange station at Canary Wharf – towering over the station is the landmark 770-ft-high iconic pyramid pinnacle of One Canada Square. From here the DLR continues its amazing journey suspended over the former West India Docks through Heron Quays, snaking on sharp curves between high-rise offices, apartments and hotels with the diminutive red trains reflected in their glass walls. From South Quay, where the station is set high above the water, there are glimpses of preserved dockside cranes, waterfront bars, restaurants and the 'Christina O', a large private yacht once owned by Aristotle Onassis and now available for private functions. Leaving the modern waterfront behind, the DLR heads through Crossharbour before passing the forty-acre City Farm at Mudchute that was famously saved from high-rise development by local residents in 1977.

Still heading south, the DLR dives underground for a while, first calling at Island Gardens station before burrowing in a long tunnel under the Thames to Cutty Sark for Maritime Greenwich station and then emerging into daylight again at Greenwich. Seen from high above the streets, the highlights of the next part of this journey to Deptford Bridge include rotting Thames barges and tugs left high and dry in an old dock basin and the splendidly preserved nineteenth-century flour mill of S. P. Mumford & Co ('1790–1897'). The final leg of this route runs through a mixture of Edwardian tree-lined suburbia and modern flats before ending at Lewisham station where there is an interchange with buses and the national rail network.

ROUTE DESCRIPTION

Stratford International to Beckton

Surrounded by the trappings of the 2012 Olympic Games, the DLR station at Stratford International is conveniently located adjacent to the modern station serving Eurostar trains between St Pancras and Paris. From here, DLR trains head south along the former route of the Great Eastern Railway's (GER) line to North Woolwich and Beckton through Stratford, Stratford High Street, Abbey Road, West Ham and Star Lane before reaching the interchange station at Canning Town. The Jubilee Line that operates between Stratford and Stanmore parallels the entire length of this section. The DLR station at Canning Town is on two levels and is also served by the Jubilee Line and an adjacent bus station.

From Canning Town the railway climbs on a flyover above the Woolwich Arsenal branch (see opposite), its route eastwards closely paralleled to the south by a row of enormous electric pylons through Royal Victoria and Custom House. Here, the modern ExCel Centre immediately to the south contrasts with the derelict platforms of the old Custom House station on the ex-GER's line to Beckton to the north – the trackbed of which is now being rebuilt here as a road. Continuing eastwards through Prince Regent and Beckton Park, the scenic highlights are all to the south of this route and include a hotel built of multi-coloured glass, derelict docklands and London City Airport with its commuter jets taking off from a waterside runway. After emerging from a cutting at Cyprus the railway passes the brightly coloured modern buildings of the University of East London, its circular students' quarters set on the waterside, each painted in a different colour of the spectrum giving them the appearance of enormous ship's funnels. The final leg through Gallions Reach and its multi-coloured warehouse seemingly ends in open countryside at Beckton although passengers should not get too relaxed on leaving the station as they are greeted by a dual carriageway, a bus interchange station and a massive supermarket!

The Docklands Light Railway splits into the Beckton and Woolwich Arsenal branches at Canning Town, just across the Thames from the O_2 Arena.

Heading a freight train on the Oxford to Worcester route, Colas Rail Class 47 diesel No. 47739 interrupts the peace and tranquility of these narrow boats moored on the River Avon at Evesham.

EASTERN & CENTRAL ENGLAND

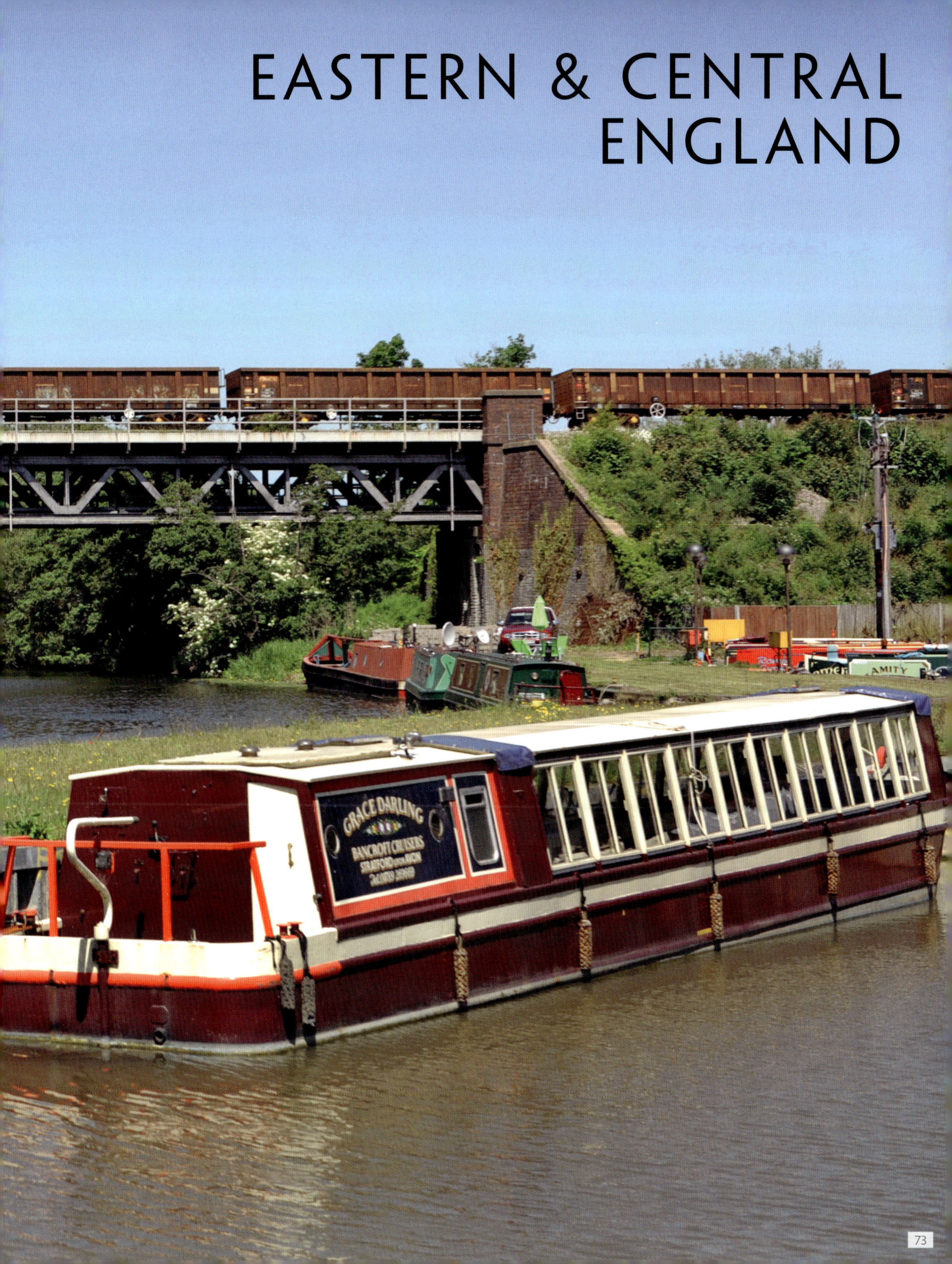

NORWICH TO GREAT YARMOUTH

VIA BERNEY ARMS

Marketed as one of the Wherry Lines, this community railway that runs along the valley of the River Yare was one of the first lines to be built in Norfolk and still features England's only station without road access.

Originally seen as the first phase of a major east-west trunk route, the twenty-and-a-half-mile Yarmouth & Norwich Railway was authorized in 1842. Surveyed by George and Robert Stephenson, the level route tracks the meandering Yare Valley before straddling reclaimed marshland to the coast, its only major engineering feature being a half-mile diversion of the river to the east of Norwich, which was considered cheaper than building two bridges.

Opened in 1844, the line was amalgamated with the Norwich & Brandon Railway in 1845 to form the Norfolk Railway. The arrival of the railway in Great Yarmouth soon breathed life into the ailing port and by the late nineteenth century the enlarged and modernized harbour was home to 1,000 trawlers, their vast catches of herring and other fish being rapidly transported by train to distant markets. With rail connections from London, the Midlands and the North in place, the town and its long sandy beaches were to become one of the most popular holiday destinations in Britain. The line became so congested that a new relief railway was opened in 1883 between Brundall and Breydon Junction, one mile west of Great Yarmouth, via the town of Acle.

Despite a closure threat in the 1980s, the railway remains open today and also serves the three intermediate villages of Brundall, Cantley and Reedham – the junction for the eleven-and-a-half-mile line to Lowestoft, which opened in 1847 – before striking off in a northeasterly direction on a five-mile dead level and straight route across the marshes to Great Yarmouth. Halfway along this stretch is tiny Berney Arms station. Now a request stop during daylight hours only, it is over three miles from the nearest road and is accessible only by train, by boat or on foot. Berney Arms takes its name from the nineteenth-century landowner, Thomas Berney, who sold the land to the railway company on condition that it built a station. Used by five passengers each day on average, it is East Anglia's least-used but perhaps most fascinating station.

Today's route is double track from Norwich to Reedham and thence single track via Berney Arms to Great Yarmouth. Diesel train services between Norwich and Great Yarmouth are operated by Greater Anglia – services alternate with those via the single-track line through Acle.

Popular with walkers and birdwatchers, this scenic railway provides the perfect meeting point between the wide horizons of the Norfolk Broads and the scenic beauty of the area's reedbeds, marshes, nature reserves and windmills.

A two-car Sprinter diesel multiple unit approaches remote Berney Arms station with a Great Yarmouth to Norwich train.

Class 47 diesel-electric No. 47818 heads away from Reedham station with a Norwich to Great Yarmouth train.

ROUTE DESCRIPTION

This scenic route begins its journey across the southern section of the Norfolk Broads in the cathedral city of Norwich. With a population of 135,000, the city has historic links dating back to Roman, Anglo-Saxon, Viking and Norman times, the latter represented by its formidable twelfth-century keep and market place. First produced in 1814, Norwich's most famous product, Colman's Mustard, is still exported around the world from its original factory in the city.

The railway strikes east from Norwich's grand nineteenth-century Thorpe terminus and immediately follows the north bank of the River Yare – for the entire route the meandering river is always to the south of the railway. Gradually broadening out as the river flows slowly eastwards, the marshes and reed beds along the valley floor are criss-crossed by a myriad of drainage channels that, by the end of the journey, stretch for miles to the north and south. At Brundall, famous as a Norfolk Broads boat-building centre, are the Mid-Yare National Nature Reserve and the Royal Society for the Protection of Birds' Surlingham Church Marsh reserve; other RSPB reserves along the railway's route are at Strumpshaw Fen, Berney Marshes, Halvergate Marshes and Breydon Water. In the small village of Cantley an enormous sugar beet factory – one of only four left in Britain but no longer rail-connected – is sandwiched between railway and river.

Just over two miles beyond Cantley lie the riverside village and railway junction of Reedham, famous as the crossing point for both road and rail over the Yare. The chain-operated Reedham Ferry is the only road crossing of the river between Norwich and Great Yarmouth, while the swing bridge on the Lowestoft branch is one of only a few of this type still in operation on Britain's railways.

For the next five miles the railway runs on a dead-straight and level route, the eerily quiet reedbeds and marshland reaching away to the north and south. Along this stretch is the lonely Berney Arms station and the nearby riverside pub of the same name, a favourite haunt of birdwatchers, sailors and walkers making their along the Weavers' Way long-distance path. The windmill here, one of several along this stretch of the river, is the tallest in Norfolk and has been restored to working order by English Heritage. To the south, the site of the Roman fort of Gariannonum stands on a small strategic prominence at Burgh Castle overlooking what was once a wide estuary. Beyond Berney Arms the river widens out into Breydon Water, where at low tide in winter the mud and sand flats teem with thousands of wading birds and wildfowl.

The railway ends at Great Yarmouth station, named Vauxhall when there were two other stations in the town. The town lies on a long and narrow spit of land between the sea and the River Yare and was for centuries a major fishing port. The port experienced a boom time in the late nineteenth and early twentieth centuries, during the height of the herring fishing industry, but today the fishing boats are long gone, replaced by companies servicing offshore oil and gas rigs. The coming of the railways brought hundreds of thousands of holidaymakers each summer and by the early twentieth century Great Yarmouth, with its long sandy beaches, had become one of the most popular seaside resorts in Britain.

Popular with walkers and birdwatchers, remote Berney Arms station is inaccessible by road and is only served by two trains a day in each direction.

NORWICH TO SHERINGHAM

BITTERN LINE

Now marketed as the Bittern Line, in recognition of the threatened bird that makes its home in the local wetlands, the thirty-and-a-half-mile Norwich to Sheringham railway is an unusual survivor of north Norfolk's once-extensive railway network. Linked to two 'heritage' branch lines, not only does it provide an essential public transport link, it also offers an attractive access route to the Norfolk coast that is becoming increasingly popular with walkers, birdwatchers and beach lovers.

For many years after the arrival of the railway in Norwich the villages of the Norfolk Broads and the fishing villages of Cromer and Sheringham remained cut off from the outside world. It was only in 1864 that the East Norfolk Railway (ENR) was authorized to build a line from Norwich to the small market town of North Walsham. The line eventually opened in 1874 – followed by an extension to Cromer, which opened in 1877. The Great Eastern Railway (GER) operated the ENR from the outset and took it over in 1881.

Cromer was also reached from the west in 1887 over a fifteen-and-a-quarter-mile branch from Melton Constable via Sheringham, part of the sprawling rural Midland & Great Northern Joint Railway (M&GNJR) network which was incorporated in 1893.

By the early twentieth century through trains from London Liverpool Street to Cromer and Sheringham were being operated by the GER, serving the burgeoning coastal holiday market that was opened up by the railway. Services with romantic-sounding names such as the 'Norfolk Coast Express' and the 'Broadsman' ran until the end of steam traction in 1962. Neither Cromer nor Sheringham were listed for closure in the 'Beeching Report', but all the other rural lines in this area had succumbed by 1964. At Sheringham, in order to eliminate a level crossing, British Rail withdrew from the ex M&GNJR station and built a simple single-platform terminus on the other side of the road. Fortunately, the old station was taken over by the volunteer-led North Norfolk Railway, which re-opened the six and a quarter miles of line from Sheringham to Holt as a heritage route in 1989 (see page 80).

Like all surviving rural branch lines, the Sheringham line has seen its share of economies, but the first nine miles of the line to Hoveton & Wroxham are still double track. At North Walsham,

BURE VALLEY RAILWAY
Hoveton and Wroxham to Aylsham South

This 15-in-gauge miniature railway runs for nine miles along the pretty valley of the River Bure. Opened in 1990, the line follows the trackbed of the former East Norfolk Railway's branch line from Wroxham to County School. Train services are usually steam hauled and take passengers to Coltishall village before crossing the river to reach the village of Buxton. The railway passes Brampton village before ending its journey in the historic market town of Aylsham, which was founded by the Anglo-Saxons about 1,500 years ago and later became an important centre for the linen and weaving industries.

A steam train approaches Buxton station on the fifteen-inch-gauge Bure Valley Railway.

ROUTE DESCRIPTION
Norwich to Gunton

Leaving the grand Norwich terminus (formerly Norwich Thorpe), the Bittern Line soon turns north away from the Yare Valley on a level and dead-straight route across open countryside to the small village of Salhouse, its station a jumping-off point for walks to Salhouse Broad. The railway then crosses the wet woodland banks of the River Bure to reach Hoveton and Wroxham station. The village of Wroxham is known as the 'Capital of the Norfolk Broads' and is a popular spot for sailing. This station is also the interchange point for the Bure Valley Railway (see above) and the Bure Valley Path to Aylsham.

Beyond Wroxham the railway reaches the village of Worstead before arriving at the prosperous market town of North Walsham, by far the busiest intermediate station. With a population of 18,000, it was once an important and wealthy centre of the Flemish weaving industry. Today, the semi-derelict North Walsham & Dilham Canal is the reminder of a transport link that predated the railway by nearly fifty years. Making use of long-closed railway lines, the fifty-six-mile long-distance footpath known as the Weavers' Way passes through the town.

The line then passes through Gunton station, built for the convenience of Lord Suffield of nearby Gunton Hall and used by royalty when visiting his lordship in the nineteenth century.

Gunton
North Walsham
Worstead
Hoveton and Wroxham
Salhouse
Norwich
Aylsham South
Brampton
Buxton
Coltishall
Coltishall
Horsham St. Faith
AYLSHAM
NORWICH
Mundesley

NORTH NORFOLK RAILWAY
Sheringham to Holt

With its delightfully restored stations, the standard-gauge North Norfolk Railway (also known as the Poppy Line) runs for five and a quarter miles from Sheringham to Holt. Originally part of the extensive Midland & Great Northern Joint Railway network, this mainly steam-hauled railway offers glimpses of the sea to the west of Sheringham before heading southwest through Weybourne, skirting typical Norfolk heath land and wooded country before ending its journey at the historic town of Holt, with its many attractive Georgian buildings.

Visiting ex-GWR 'City' Class 4-4-0 'City of Truro' hauls a train near Kelling Heath Park on the North Norfolk Railway.

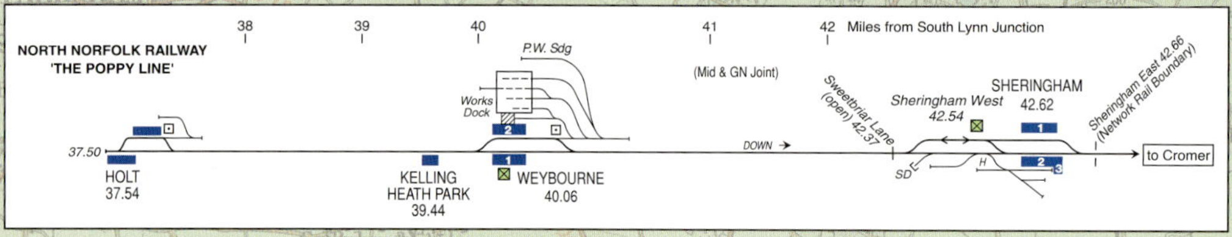

The track diagram of the North Norfolk Railway between Sheringham and Holt.

unusually for a line of this type, freight traffic is still handled – servicing a gas condensates plant linked by pipeline to the nearby North Sea.

The closure of the former Cromer High station in the 1950s now requires trains to take a wide sweep into the town and then reverse out of the intermediate terminus station at Cromer Beach en route to Sheringham. There, at the end of the Network Rail line, the railway has been reinstated across the level crossing displaced in the 1960s, so steam-hauled charter trains can run direct from the rest of the network on to the popular North Norfolk Railway.

One of the most scenic railways in East Anglia, the Bittern Line passes through a variety of attractive landscapes to reach a coastline that is otherwise inaccessible by train. With the added attractions of two connecting heritage railways – and now promoted by a 'Community Rail Partnership' – this unusual branch line makes for an outstanding day out in and beyond the Norfolk Broads.

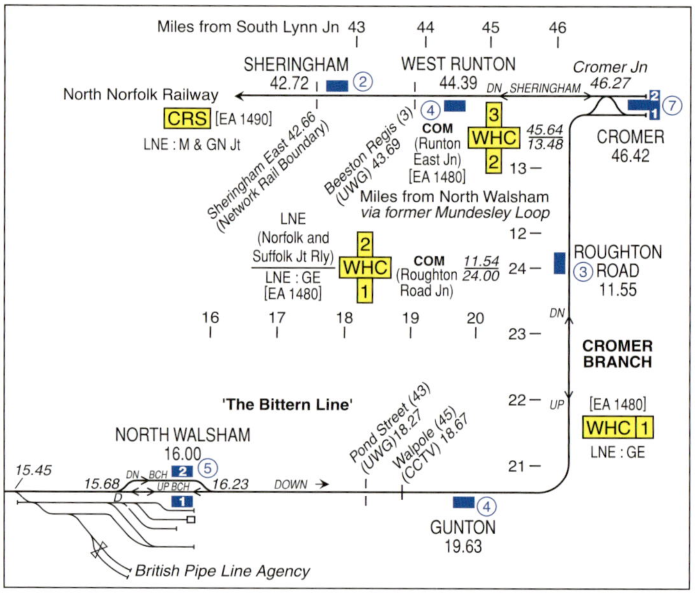

The track diagram of the 'Bittern Line' between North Walsham and Sheringham.

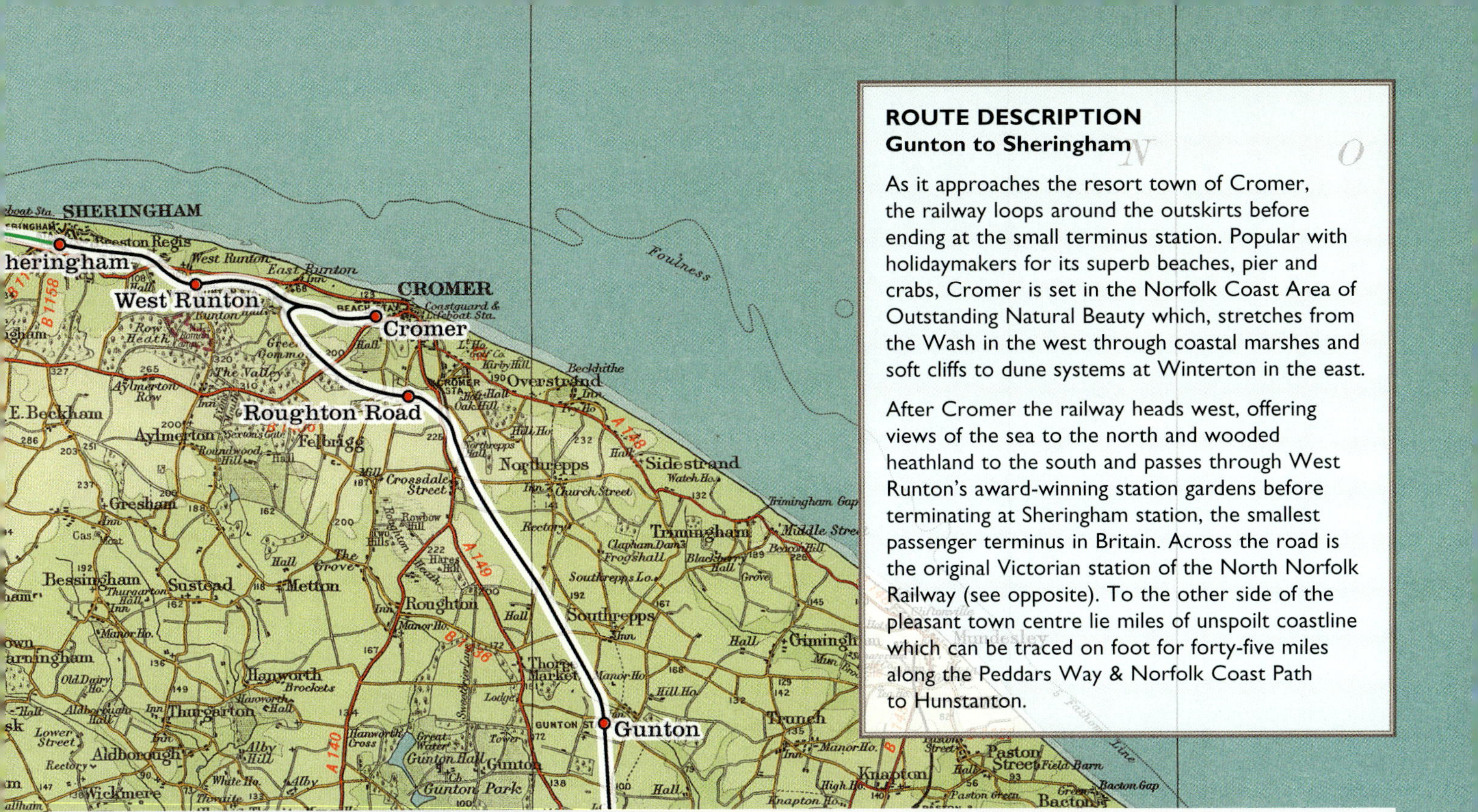

ROUTE DESCRIPTION

Gunton to Sheringham

As it approaches the resort town of Cromer, the railway loops around the outskirts before ending at the small terminus station. Popular with holidaymakers for its superb beaches, pier and crabs, Cromer is set in the Norfolk Coast Area of Outstanding Natural Beauty which, stretches from the Wash in the west through coastal marshes and soft cliffs to dune systems at Winterton in the east.

After Cromer the railway heads west, offering views of the sea to the north and wooded heathland to the south and passes through West Runton's award-winning station gardens before terminating at Sheringham station, the smallest passenger terminus in Britain. Across the road is the original Victorian station of the North Norfolk Railway (see opposite). To the other side of the pleasant town centre lie miles of unspoilt coastline which can be traced on foot for forty-five miles along the Peddars Way & Norfolk Coast Path to Hunstanton.

With the coast never far away, a two-car diesel unit heads past the golf links at West Runton between Sheringham and Cromer.

ROUTE DESCRIPTION

The Gainsborough Line leaves Marks Tey station in a northerly direction with much of its route to Chappel & Wakes Colne slicing through undulating farmland in a series of long cuttings before crossing the Colne Valley on the magnificent viaduct at Chappel. Flowing into the sea south of Colchester, the River Colne to the west of Chappel was once followed by the Colne Valley & Halstead Railway, which closed in 1965. The twin villages of Chappel and Wakes Colne are served by Chappel & Wakes Colne station, also home to the East Anglian Railway Museum. Climbing northwards, the railway enters a long cutting before joining the valley of the River Stour at Bures. The intervening countryside, a patchwork of fields and orchards, still carries reminders of the Second World War, with concrete pillboxes and the remains of a military airfield.

The village of Bures straddles the River Stour, itself the boundary between Suffolk and Essex. Navigation along the river, famously portrayed in paintings by local artists such as Thomas Gainsborough and John Constable, was much improved in the early eighteenth century with the construction of locks and cuts.

North of Bures the railway follows the meandering Stour and its water meadows, crossing the river into Suffolk some two miles north of the village. To the west, the gently rolling farmland is crossed by the Stour Valley Path, a sixty-mile long-distance path that runs from Newmarket to Cattawade through some of the most attractive countryside in East Anglia. Following the east bank of this delightful river, the railway ends its journey in Sudbury.

With a population of 12,000, this riverside town has historic links dating back to the Saxons and an eleventh-century market. Settled by Flemish weavers in the fourteenth century, the town had become a wealthy and important centre for the wool and silk industries by the late Middle Ages. A corn exchange, riverside quays and warehouses still stand as a testimony to the town's former importance on the navigable River Stour.

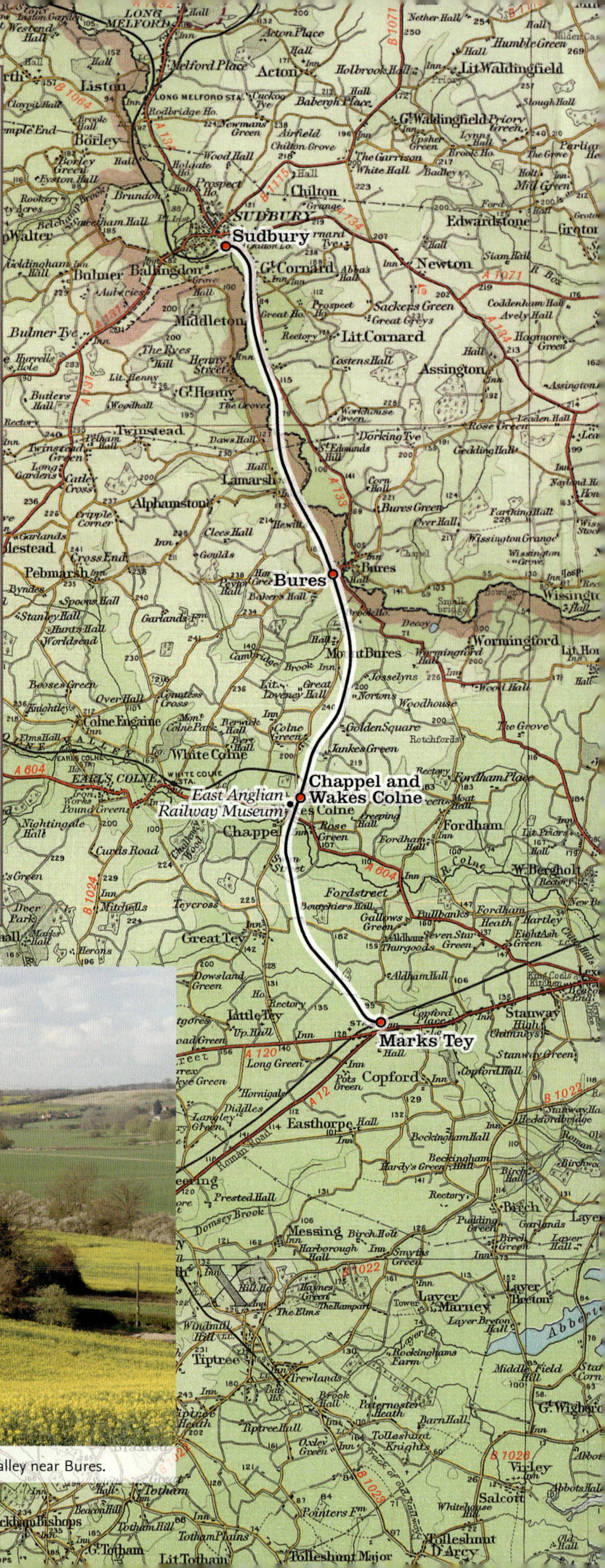

A two-car Sprinter diesel multiple unit heads along the picturesque Stour Valley near Bures.

MARKS TEY TO SUDBURY

GAINSBOROUGH LINE

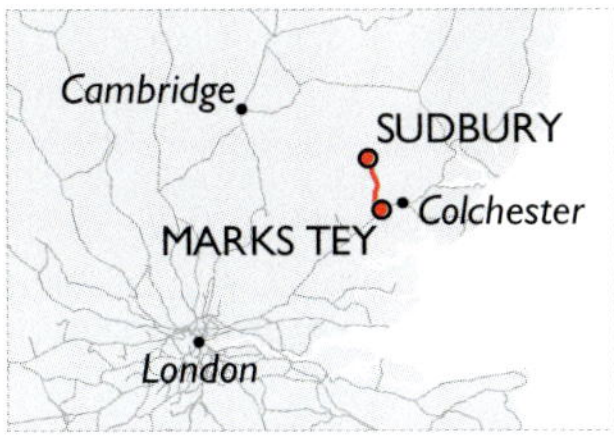

Once forming part of a rural cross-country route and now marketed as the Gainsborough Line, this single-track branch line fortunately escaped the 'Beeching Axe'.

The grandly titled Colchester, Stour Valley, Sudbury & Halstead Railway (CSVS&HR) began operation in 1849 between Marks Tey, on the Eastern Counties Railway main line from London to Colchester (opened in 1843), and the market and silk-weaving town of Sudbury. Although the eleven-and-three-quarter-mile railway was in direct competition with the recently modernized Stour Navigation, both forms of transport lived in harmony until the 1890s, by which time most freight traffic had been transferred to rail. Originally leased to the Eastern Union Railway, the CSVS&HR was taken over by the Great Eastern Railway in 1898.

The building of the Sudbury branch involved the construction of a massive viaduct across the Colne Valley at Chappel & Wakes Colne. Originally planned as a timber structure on brick piers, the discovery of brick clay on the site led to it being entirely brick built. With seven million used in its construction it is reputed to be the second-largest brick structure in Britain and is 1,066 ft long with thirty-two arches at a height of 75 ft. North of the viaduct a two-mile cutting also had to be excavated to take the line to its summit at Mount Bures.

The railway was extended northwards to Long Melford and west to Clare and Haverhill (for Cambridge) in 1865. A branch line from Long Melford to Bury St Edmunds also opened in the same year. Passenger services consisted mainly of through trains running between Haverhill or Cambridge and Colchester while the backbone of the goods traffic was the transport of cattle, grain and farm produce. The railway was particularly busy during the Second World War delivering fuel and bombs to RAF bases in the area. Steam traction was replaced by diesel multiple units in 1959 but steadily declining passenger numbers and freight traffic saw the line listed for closure in the 'Beeching Report'. The Marks Tey to Sudbury section was reprieved due to the town's future growth prospects but the line northwards closed in 1967. Named after the artist Thomas Gainsborough, who was born in Sudbury in 1727, the Gainsborough Line became a Community Railway in 2006.

With diesel train services operated by Greater Anglia, this community railway is single track throughout with no passing loops. Marks Tey is also served by trains on the Liverpool Street to Norwich main line.

A Marks Tey to Sudbury train crosses the 32-arch viaduct over the Colne Valley on the approach to Chappel & Wakes Colne station.

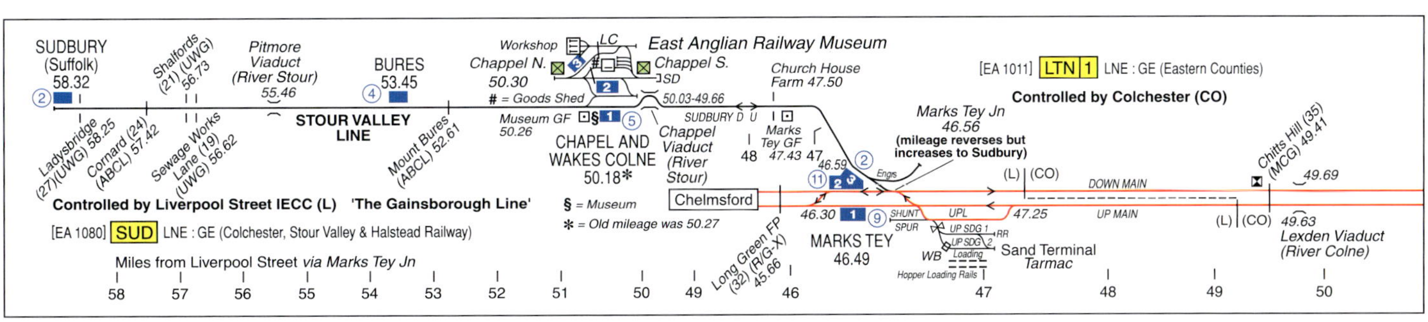

The diagrammatic track plan of the Marks Tey to Sudbury branch line also shows the layout of the East Anglian Railway Museum at Chappel and Wakes Colne.

KINGSLEY & FROGHALL TO LEEKBROOK JUNCTION

CHURNET VALLEY RAILWAY

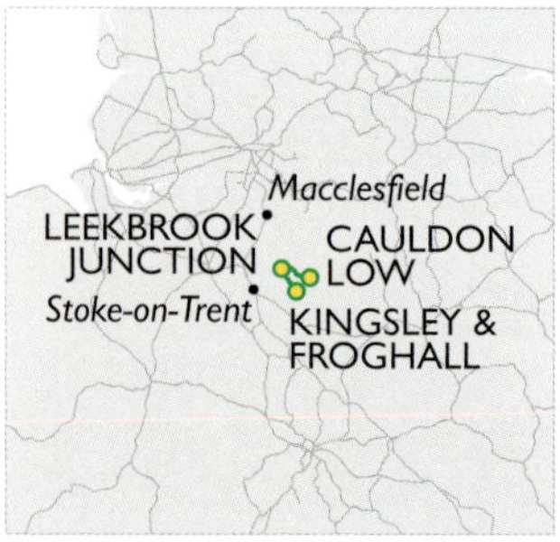

Punctuated by a series of ornate Victorian stations, the five-and-a-quarter-mile Churnet Valley Railway offers a scenic journey, keeping company with river and canal through a delightful hidden valley rich in industrial archaeology. It is now the starting point for an exciting re-established rail journey across moorlands to the limestone hills of Staffordshire.

By the late eighteenth century Stoke-on-Trent and the surrounding area were at the heart of the Industrial Revolution. Rich in coal, ironstone and limestone deposits, the area was a heavy mix of mines, quarries, blast furnaces and ironworks. All of these industries depended on a transport network to move raw materials and finished goods – the main artery being the ninety-three-and-a-half-mile Trent & Mersey Canal, which as its name implies, linked two of Britain's major rivers through the industrial heartland of Staffordshire. Completed in 1777, its construction included the building of seventy locks and five tunnels. An eighteen-mile branch, known as the Caldon Canal, was opened between Etruria and Froghall via the Churnet Valley

Stanier Class '5' 4-6-0 No. 44767 hauls a photographers' chartered parcels train along the Churnet Valley Railway near Cheddleton.

in 1779 – with branches to Leek and Uttoxeter it was linked to limestone quarries at Caldon by a horse-drawn tramway.

However, the canals' monopoly was short lived with the opening of the North Staffordshire Railway (NSR) from Macclesfield southwards to a junction with the Grand Junction Railway at Norton Bridge in 1845. By 1852, with its headquarters in Stoke, the railway, nicknamed the 'Knotty' after the company's Staffordshire Knot trademark, owned over 200 miles of railway and the Trent & Mersey Canal – all serving the area's growing industries.

Despite its industrial heartland, the highly profitable NSR also owned several routes that were later widely promoted for their scenic beauty. One of these, opened in 1849, was the twenty-seven-and-three-quarter-mile double-track line along the scenic Churnet Valley from North Rode, south of Macclesfield, to Uttoxeter on the company's line from Crewe to Derby. Running parallel with the Caldon Canal from Leek to Froghall, the railway was built over the course of the Uttoxeter branch of the canal to the south. With its ornate stations and practicality as an alternative route for through goods trains avoiding Stoke, the line also carried locally generated agricultural and livestock traffic, daytrippers from the industrial Potteries keen to visit the scenic delights along the line and spectators to Uttoxeter racecourse.

The Churnet Valley line ended as a through route in 1960 with the closure of the section north of Leek. The remainder of the line from Leek to Uttoxeter struggled on with a few workmen's trains until they were withdrawn in 1965. All that remained were industrial sand trains from Oakamoor northwards to Leekbrook Junction and on to Stoke, which finally ceased in 1988.

In the meantime, the Cheddleton Railway Centre had opened its doors to the public at Cheddleton station in the 1970s. The purchase of seven miles of the mothballed line between Oakamoor and Leekbrook Junction followed in 1994 with public services beginning a few years later. The railway trackbed south of Oakamoor through Alton Towers station is now a footpath and cycleway forming part of National Cycle Network Route 54. North of Leekbrook Junction the railway awaits reopening through Birchall Tunnel to the town of Leek.

Operated almost entirely by a large band of volunteers, the scenic Churnet Valley Railway operates both steam and diesel trains between Kingsley & Froghall and Cheddleton stations. North of here trains continue on to Leekbrook Junction (no alighting here as there is no access other than by train), with some services continuing eastwards to Cauldon Low on the Moorlands & City Railway (see page 87). Trains occasionally use the line south of Kingsley & Froghall to Oakamoor although, again, there is no public access at this point.

Popular with steam enthusiasts, walkers and real ale aficionados, the Churnet Valley Railway blends into its peaceful rural setting only a short distance from the sprawling conurbation of the Potteries.

Ex-LNER Class 'V2' 2-6-2 No. 4771 'Green Arrow' crosses the Caldon Canal near the Black Lion pub at Consall on the Churnet Valley Railway.

ROUTE DESCRIPTION

Kingsley and Froghall to Leekbrook Junction

Along much of its route the Churnet Valley Railway closely follows the Caldon Canal along the wooded Churnet Valley. Trains start or terminate their journeys at Kingsley & Froghall station, although some diesel multiple units occasionally make a return journey for two and a half miles along the valley to Oakamoor – here the railway awaits reopening through the 497-yd Oakamoor Tunnel to Oakamoor station and beyond to Alton Towers.

The small village of Froghall was once a hive of industry, serving the Thomas Bolton copper factory and acting as a junction for the tramway to the limestone quarries at Caldon Low. Before the coming of the railway, limestone was loaded into barges at the canal basin, now a tranquil spot favoured by walkers or pleasure boats, or crushed for burning in the lime kilns that still dominate the valley. Still in operation, the factory once made transatlantic cables and wiring for Spitfire aircraft. Winding northwards in company with canal and river along the wooded valley, the railway reaches Consall, a popular spot for walkers and for real ale aficionados visiting the charming and picturesque Black Lion pub, which is completely inaccessible by road.

Leaving Consall, the railway accompanies the Staffordshire Way long distance path up the steep gradient of the valley walls, the wooded sides thinning out and giving way to moorland in the east. At Cheddleton, the ornate Grade II listed station building, saved from demolition in the 1970s, is now the headquarters of the railway. With a population of 5,000, this large village is home to a restored flint mill that once supplied ground flint for use in the nearby pottery industry. From Cheddleton the railway continues northwards, briefly leaving the valley and burrowing through the 531-yd Cheddleton Tunnel to terminate at Leekbrook Junction. Although this is currently the northerly limit of the railway there is no platform or road access – trains retrace their route down the valley to Cheddleton. Some trains continue eastwards along the eight-mile freight line to Cauldon Low that has recently been reopened by Moorlands & City Railway (see opposite).

Ex-LNWR Class 'G2' 0-8-0 No. 49395 hauls a short engineers' train through Consall Wood on the Churnet Valley Railway.

MOORLANDS & CITY RAILWAY
Leekbrook Junction to Cauldon Low

Seeking to return commercial traffic to the railways, the Moorlands & City Railway is reopening the mothballed freight-only line from Stoke-on-Trent to the limestone quarries at Cauldon Low. The eight-mile steeply graded section from Leekbrook Junction, where it connects with the Churnet Valley Railway (CVR), to Cauldon Low was opened for heritage trains in 2010. With an award from the Government's Regional Growth Fund, the remaining section from the national rail network at Stoke-on-Trent eastwards to Leekbrook Junction is currently being restored. Plans to extend the Churnet Valley Line southwards to Alton Towers and northwards to Leek are also being considered. On selected weekends, steam trains originating at Kingsley & Froghall on the CVR now make a twenty-seven-mile round trip to Cauldon Low taking in gradients as steep as 1 in 40 across the Staffordshire moorlands. A recent successful share issue by the CVR will cement the relationship between the two companies and create the largest operating network of any UK heritage railway.

ROUTE DESCRIPTION
Oxford to Adlestrop

Known as the 'city of dreaming spires', Oxford has long been an important railway crossroads with destinations to the north, south, east and west. It was here that inter-regional trains once changed steam locomotives before resuming their long cross-country journeys. Today, with a population of 165,000, this world-famous university city has a diverse economy which includes car manufacturing and science and technology industries. Leaving the unprepossessing railway station on the west of the city, the railway to Worcester first heads north before branching off at Wolvercote Junction and heading west into open countryside. Populated by wealthy landowners, this region of the eastern Cotswolds contains a number of large country estates that the railway builders were forced to avoid along the valley of the River Evenlode: north of the line at Hanborough is the birthplace of Winston Churchill, eighteenth-century Blenheim Palace whose lake and grounds were landscaped by Capability Brown; to the south lies Eynsham Hall and Park; and at Charlbury, parliamentary constituency of David Cameron, the line skirts around Cornbury Deer Park and Wychwood Forest.

Continuing along the Evenlode Valley, the railway heads northwards at Shipton-under-Wychwood through pretty Cotswold villages such as Kingham and Adlestrop. Kingham was once an important cross-country railway junction while Adlestrop was immortalized in Edward Thomas's famous poem of the same name although, sadly, the station was closed in 1966.

Ex-GWR 'Castle' Class 4-6-0 No. 5043 'Earl of Mount Edgcumbe' storms through Honeybourne station with the 'Cathedrals Express' in September 2011.

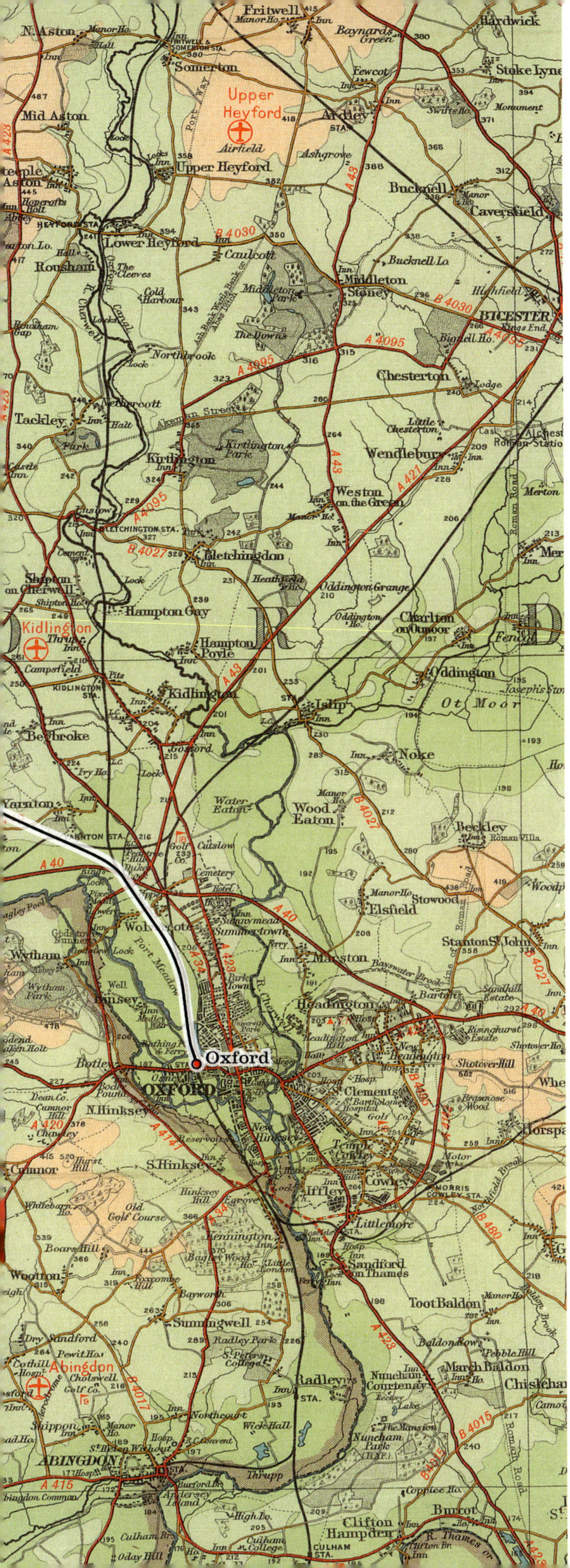

OXFORD TO WORCESTER AND HEREFORD

COTSWOLD LINE

This eighty-six-and-a-quarter-mile railway serves a fascinating mix of pretty Cotswold towns and villages, two cathedral cities and a Victorian spa resort. After years in the doldrums as a run-down cross-country route, the service is now greatly improved thanks to extensive recent upgrades.

While the broad-gauge Great Western Railway (GWR) had already reached Oxford from Didcot in 1844, the authorized route westwards across the Cotswolds to Worcester was dogged by numerous problems including escalating construction costs, a disgruntled contractor and a change of gauge. With initial support from the GWR, who saw it as part of a through route to Birmingham, and with Brunel as engineer, the line began life as the broad-gauge Oxford, Worcester & Wolverhampton Railway (OW&WR). This would soon change when the GWR switched sides and took over a new and shorter route from Oxford to Birmingham via Banbury. Faced with a dilemma, the OW&WR directors decided to reduce construction costs by building their railway to the narrower standard gauge and by 1852 it had opened from Evesham to Stourbridge via Worcester. To the east of Evesham lay the Cotswolds, the line's summit and the 875-yd Chipping Campden Tunnel – here Brunel was faced with a sacked contractor who refused to leave because of non-payment of his bill. The 'Battle of Chipping Campden' was only resolved when Brunel brought in several thousand navvies and threatened the use of government soldiers to break the impasse. Costing over double its original estimate, the railway eventually opened throughout in 1853 and the impecunious company soon earned the nickname of 'Old Worse and Worse' for its dire train service.

To the west of Worcester, the Worcester & Hereford Railway (W&HR) had just begun the construction of its line between the two cities. The most significant engineering feature was a long single-bore tunnel beneath the Malvern Hills at Colwall and it was here that the original contractors came unstuck – dug completely by hand through very hard rock and constantly flooded, it was only completed in 1861 with the assistance of Welsh coal miners. By 1863 the OW&WR and the W&HR had become absorbed by the GWR but Colwall Tunnel, with its steep gradient, continued to be a problem into the twentieth century. Following a collapse of the tunnel in 1907 the GWR decided to build a new one with a more workable gradient. This finally

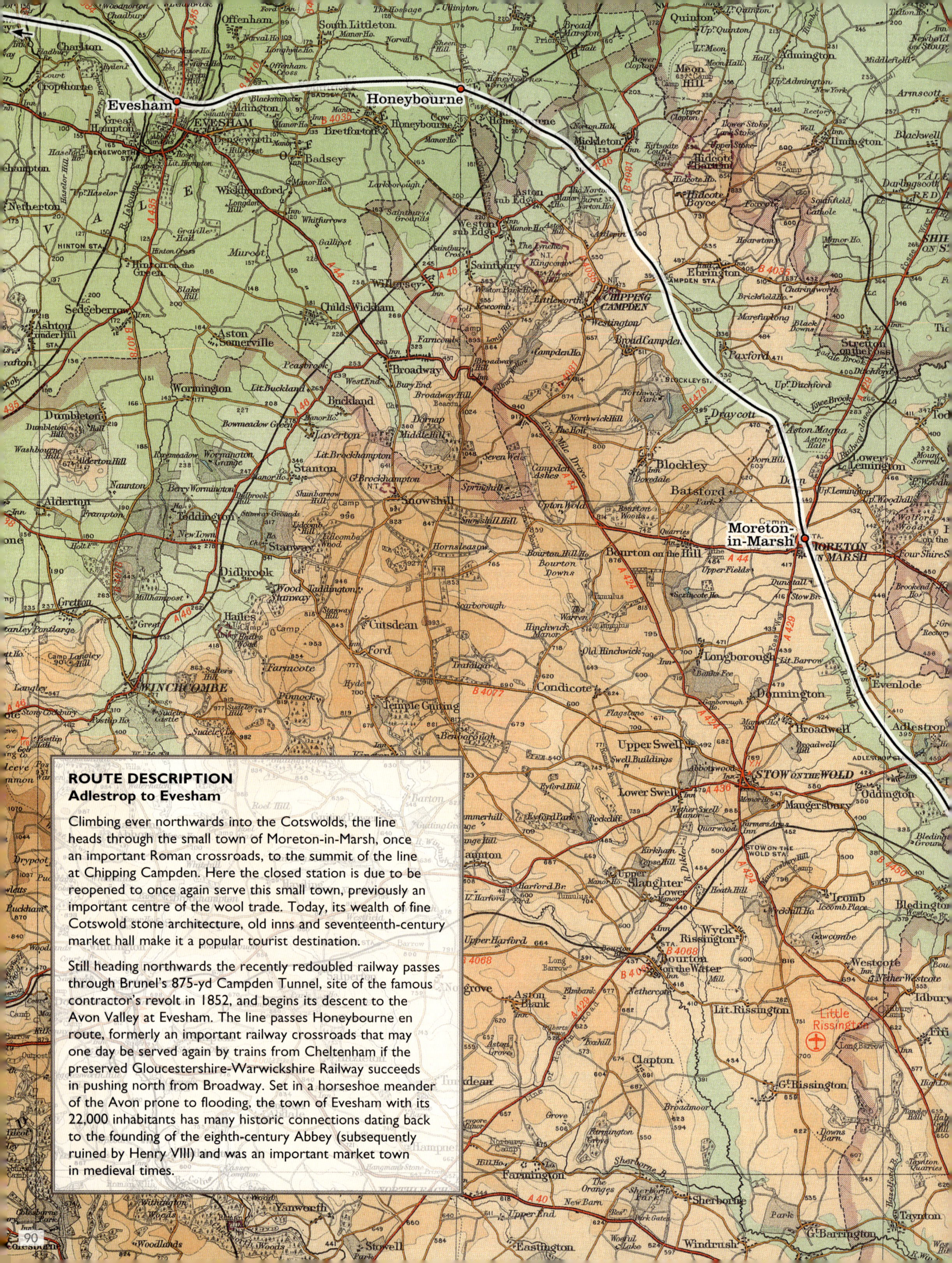

ROUTE DESCRIPTION
Adlestrop to Evesham

Climbing ever northwards into the Cotswolds, the line heads through the small town of Moreton-in-Marsh, once an important Roman crossroads, to the summit of the line at Chipping Campden. Here the closed station is due to be reopened to once again serve this small town, previously an important centre of the wool trade. Today, its wealth of fine Cotswold stone architecture, old inns and seventeenth-century market hall make it a popular tourist destination.

Still heading northwards the recently redoubled railway passes through Brunel's 875-yd Campden Tunnel, site of the famous contractor's revolt in 1852, and begins its descent to the Avon Valley at Evesham. The line passes Honeybourne en route, formerly an important railway crossroads that may one day be served again by trains from Cheltenham if the preserved Gloucestershire-Warwickshire Railway succeeds in pushing north from Broadway. Set in a horseshoe meander of the Avon prone to flooding, the town of Evesham with its 22,000 inhabitants has many historic connections dating back to the founding of the eighth-century Abbey (subsequently ruined by Henry VIII) and was an important market town in medieval times.

opened in 1926 and after a stint storing torpedoes during the Second World War the old tunnel was finally bricked up.

Patronized by well-heeled Londoners seeking health cures, the mineral water springs at Malvern were once well served by express dining car trains from Paddington, the last of these was the 'Cathedrals Express', which remained steam hauled until the mid 1960s. There then followed years of retrenchment: through freight trains were diverted away from the route, many intermediate stations were closed and much of the line was singled.

By the 1970s this once-important cross-country route had fallen on hard times and it is only in recent years that the 'Cotswold Line', as it is now marketed, has seen a change in its fortunes with the reinstatement of double track between Charlbury East and Ascott and between Moreton-in-Marsh and Evesham. First Great Western operates diesel train services between Oxford, Worcester and Hereford while additional services between Worcester and Hereford are provided by London Midland.

From the world-renowned university city of Oxford this rejuvenated cross-country route follows river valleys across the Cotswold Hills, crossing the Avon and Severn and burrowing beneath the iconic Malvern Hills before ending its journey in the historic border city of Hereford.

The working manual signal box at Moreton-in-Marsh is a reminder of a bygone age.

Given a clear road ahead by the ex-GWR semaphore signal, a 1970s-vintage High Speed Train rattles through Moreton-in-Marsh in 2011.

A London Midland Class 170 diesel unit heads away from the Malvern Hills with a train for Worcester and Birmingham.

ROUTE DESCRIPTION

Evesham to Worcester

Setting off westwards across the Avon the railway enters the Vale of Evesham, its fertile soil making this an important area for commercial market gardening. Also served by the railway, the small town of Pershore is famous for its Anglo-Saxon abbey and its delicious local plums. Here the railway leaves the River Avon and heads westwards across the watershed to meet the River Severn at Worcester. Straddling the river, the cathedral city (population 94,000) is the site of the final battle of the English Civil War and birthplace of composer Edward Elgar. Also known for Royal Worcester porcelain and glove making, the city is still home to the Lea & Perrins factory that makes the world-famous Worcestershire Sauce.

Worcester to Malvern

Trains for Hereford leave Shrub Hill station before heading west across the Worcester & Birmingham Canal to Foregate Street station. From here, high above the city, the railway continues over the River Severn close to the city's famous racecourse before heading off southwards towards the Malvern Hills. Crossing the Teme Valley on a series of bridges the railway soon reaches Malvern, a picturesque Victorian spa town of 29,000 people set on the eastern slopes of the Malvern Hills. Founded in the eleventh century by Benedictine monks, the town rose to prominence in the seventeenth century with the discovery of the health-giving qualities of its spring waters. The arrival of the railway in 1859 brought visitors in their thousands, in search of cures, rest and recreation. Patronized by many notable Victorians such as Charles Darwin, Florence Nightingale and Lord Tennyson the hillside town was soon bursting at the seams with hotels and boarding houses.

A London Midland Class 170 diesel unit crosses the River Severn to the west of Worcester's Foregate Street station.

The gradient profile of the Worcester to Hereford railway clearly shows the climb up through the Malverns to Colwall.

ROUTE DESCRIPTION
Malvern to Hereford

South of Great Malvern station the railway tunnels under the Malvern Hills through the 'new' 1,589-yd Colwall Tunnel opened in 1926. Formed in the pre-Cambrian era some 675 million years ago, this iconic eight-mile-long range of hills with a high point of 1,394 ft on Worcestershire Beacon is now a designated Area of Outstanding Natural Beauty. Emerging into daylight from the long dark tunnel, the railway soon approaches the market town of Ledbury through a single-bore 1,323-yd tunnel. With a population of 10,000 this lively market town contains many medieval timber buildings and remains the centre of an important soft-fruit growing and cider-making industry.

Departing westwards from Ledbury station, perched high above the town, the railway crosses the valley of the River Leadon on a 330-yd viaduct. Constructed from five million locally made bricks, its thirty arches dominate the surrounding farmland. After skirting Bunker's Hill the railway crosses the valleys of the River Frome (confusingly, there are four other River Fromes in England) and then the River Lugg to end its long journey at Hereford.

Set astride the River Wye, this cathedral city of 56,000 residents has historic connections dating back to the Anglo-Saxons. Located close to the Welsh borderlands, the city was destroyed by a combination of Welsh and Viking warriors in 1056.
For centuries an important centre for farming and agriculture, the city has recently seen a new cattle market opened on its outskirts. Undoubtedly the most famous employer in the city is Bulmers Cider. Founded by Percy Bulmer in 1887 and awarded a Royal Warrant in 1911, this refreshing local nectar is now enjoying booming worldwide sales.

The approach to Colwall Tunnel under the Malvern Hills. Replaced by the present tunnel in 1926, the old bricked-up tunnel is now inhabited by a colony of bats.

Opened by the Worcester & Hereford Railway in 1861, Ledbury Viaduct was built of five million locally made bricks.

GLOUCESTER TO SWINDON

GOLDEN VALLEY LINE

Originally forming part of the GWR's main line to South Wales, this scenic journey up the Stroud Valley and through the Cotswolds takes in the picturesque patchwork of honey-coloured mills and villages that once lay at the heart of a thriving eighteenth-century wool-weaving industry.

By the late eighteenth century the spa town of Cheltenham had become a popular destination for wealthy Georgians seeking health cures from its mineral waters. Graced by a royal visit in 1788, the town continued to expand with the building of fine promenades, hotels and elegant town houses – by 1815 it also had a racecourse. Eager to serve this popular destination, both the London & Birmingham Railway (L&BR) and the Great Western Railway (GWR) had set their eyes on reaching the town from London. In the end the GWR won the prize but only because it was the first important step for the company to reach South Wales via what was then the lowest crossing point of the River Severn at Gloucester.

Preserved Hastings-line diesel-electric multiple unit No. 1001 works a railtour along the Golden Valley at Frampton Mansell in 2010.

ROUTE DESCRIPTION
Gloucester to Kemble

Once a Roman walled garrison town strategically located at the lowest crossing point of the River Severn, the historic cathedral city of Gloucester became an important railway crossroads in the mid-nineteenth century. Along with its canal and river transport links the city soon became home to numerous industries including mills, engineering, railway carriage and wagon construction and, from 1926, the aviation industry – the latter witnessing the flight of Britain's first jet aircraft in 1941. Unfortunately, since the closure of the former Midland Railway's Birmingham to Bristol main line through Eastgate station, the city's 118,000 population now has a rail service that is markedly inferior to that operating in the 1960s – today most north-south trains avoid reversal at the former Central station by bypassing the city without stopping.

Leaving the city outskirts behind, the railway sets off in a southerly direction down the Severn Vale – to the east lie the Cotswold Hills, an Area of Outstanding Natural Beauty that stretches ninety miles from Warwickshire to the River Avon at Bath. At Standish Junction the railway diverges from the route to Bristol, sweeping eastwards up the valley of the River Frome to begin its climb up the Stroud Valley and over the Cotswold Ridge to Kemble.

Keeping company with road and canal the railway cuts up the valley to the town of Stroud. With a population of 32,000, Stroud marks the meeting point of five valleys and was an important centre for the woollen industry during the Industrial Revolution. With a workforce of immigrant weavers from northern France, numerous water-powered mills along the valley once wove quality cloth made from Cotswold sheep – dotted alongside the railway many of these mills still stand today, characterized by their distinctive honey-coloured Cotswold stone. Marking the same course as the railway is the Stroudwater Navigation which opened in 1779 from the River Severn to Stroud – beyond here the valley is also accompanied as far as Daneway by the Thames & Severn Canal which opened in 1789. Featuring a 3,817-yd tunnel beneath the Cotswolds, it is now the subject of a major restoration scheme.

Beyond Stroud the railway begins to climb the valley in earnest, through a charming patchwork of Cotswold stone houses, fields and woodland and past Brimscombe and Chalford before plunging into the depths of Sapperton Tunnel. Here the summit of the line is reached at the halfway point, while above are the extensive woodlands of the 14,000-acre Bathurst Estate. All trains stop at the pretty Cotswold stone station of Kemble, once the junction for branch lines to Tetbury and Cirencester. A reminder of its importance during steam days is the surviving large water tower that still graces the west end of the down platform. While the line south of Kemble to Swindon was singled as an economy measure in 1968, an upsurge in traffic will soon see it restored to double track.

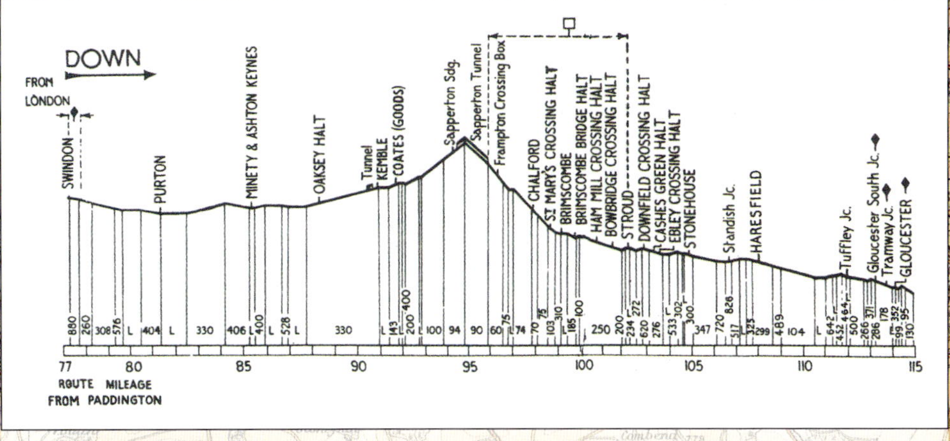

The gradient profile of the Gloucester to Swindon route shows the steep southbound climb up to the summit of the line in Sapperton Tunnel.

Ex-GWR 'Hall' Class 4-6-0 No. 4965 'Rood Ashton Hall' and 'Castle' Class 4-6-0 No. 5043 'Earl of Mount Edgcumbe' meet the former Midland Railway main line at Standish Junction with a Didcot to Tyseley charter train in April 2011.

With opposition from the Thames & Severn Canal swept away, the broad-gauge railway opened from Swindon to Cirencester in 1841. Completion of the rest of the railway from the Cotswolds down the Stroud Valley was much slower due to the construction of the 1,864-yd Sapperton Tunnel and nine viaducts. Involving gradients as steep as 1 in 60, the three-mile climb to the summit of the line inside the tunnel required the banking of heavy eastbound freight trains until the end of steam in the 1960s. The final section opened in 1845, leaving Cirencester at the end of a branch line from a new junction at Kemble.

North of Gloucester the railway to Cheltenham had two owners – the Birmingham & Gloucester Railway, which had reached the cathedral city in 1840, and the GWR, which terminated its trains from Paddington at the St James' terminus close to the centre of Cheltenham.

Converted to standard gauge in 1872, the Gloucester to Swindon line remained the GWR's through route to South Wales until the opening of the Severn Tunnel in 1886. Route of the record-breaking 'Cheltenham Flyer' in the 1930s, the section between Chalford and Stonehouse saw the introduction of a steam rail-motor service in 1903. Serving a number of destinations along the Stroud Valley, it was an immediate success amongst the workers commuting to the numerous mills, foundries and breweries scattered along the valley. The service was extended to Gloucester in the 1920s and remained in operation until 1964.

Steam gave way to diesel in the early 1960s, and there is still a regular interval service of six through trains each weekday between Cheltenham, Gloucester (where trains reverse direction) and Paddington – the line is also well served by additional trains between Gloucester and Swindon. First Great Western operates diesel train services between Gloucester and Swindon while occasional steam charter trains can still be seen tackling the climb up Sapperton Bank. The line is also used as a diversionary route when the Severn Tunnel is closed for routine maintenance.

Linking the historic cathedral city of Gloucester with the former railway town of Swindon, this demanding railway route retraces the footsteps of the world-famous 'Cheltenham Flyer' and an innovative steam rail-motor service up through the Cotswold Hills.

Seen at Standish Junction, double-headed Class 37s Nos. 37401 and 37417 head off up the Golden Valley to Stroud with a charter train for Swindon.

Hauled by 'Castle' Class locomotives between Gloucester and Paddington until the early 1960s, the 'Cheltenham Spa Express' was British Railways' successor to the GWR's record-breaking 'Cheltenham Flyer'.

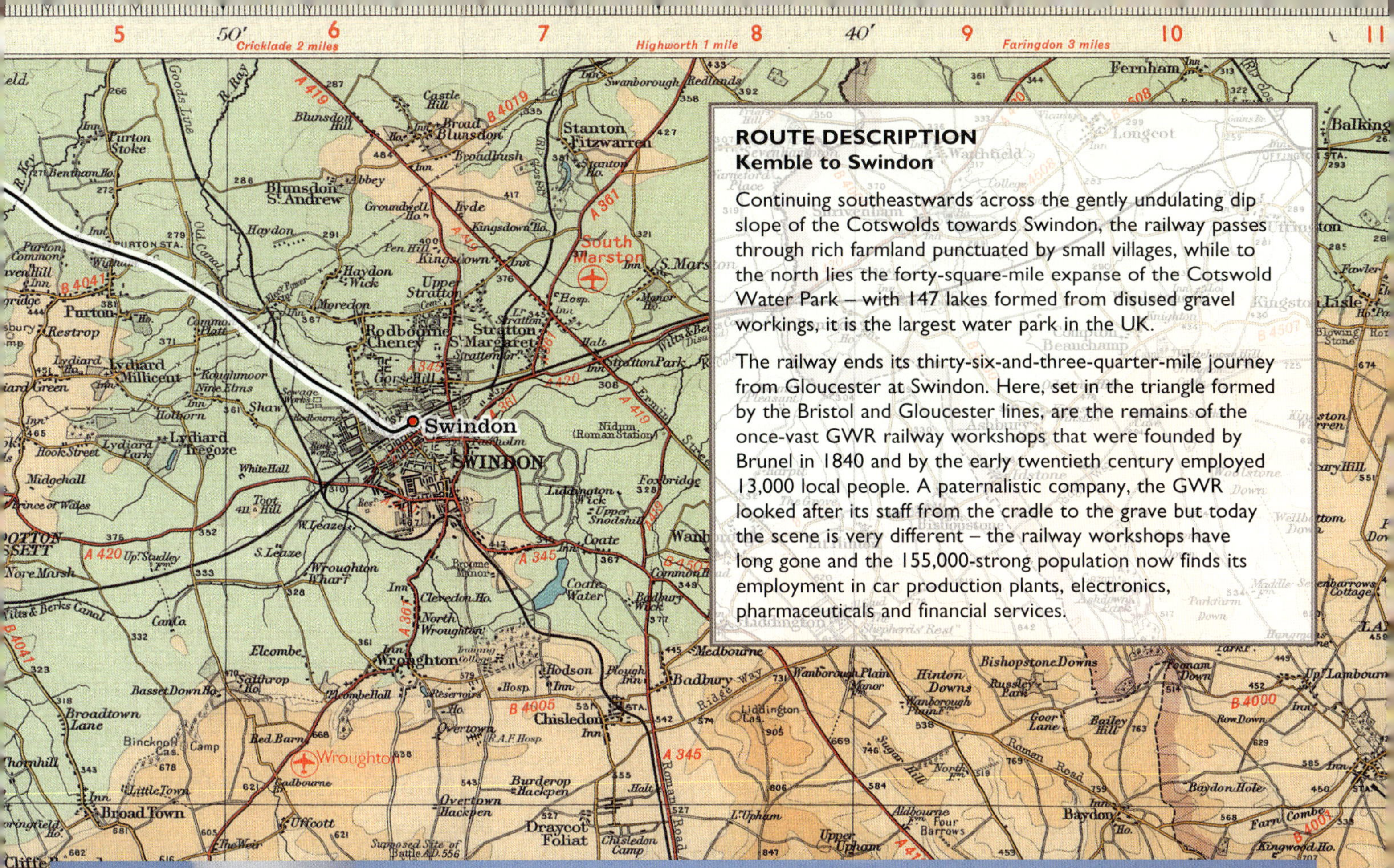

ROUTE DESCRIPTION

Kemble to Swindon

Continuing southeastwards across the gently undulating dip slope of the Cotswolds towards Swindon, the railway passes through rich farmland punctuated by small villages, while to the north lies the forty-square-mile expanse of the Cotswold Water Park – with 147 lakes formed from disused gravel workings, it is the largest water park in the UK.

The railway ends its thirty-six-and-three-quarter-mile journey from Gloucester at Swindon. Here, set in the triangle formed by the Bristol and Gloucester lines, are the remains of the once-vast GWR railway workshops that were founded by Brunel in 1840 and by the early twentieth century employed 13,000 local people. A paternalistic company, the GWR looked after its staff from the cradle to the grave but today the scene is very different – the railway workshops have long gone and the 155,000-strong population now finds its employment in car production plants, electronics, pharmaceuticals and financial services.

Seen at Frampton Mansell between Chalford and Kemble, ex-GWR 'Hall' Class 4-6-0 No. 4965 'Rood Ashton Hall' and 'Castle' Class 4-6-0 No. 5043 'Earl of Mount Edgcumbe' storm up the 1-in-60 gradient towards Sapperton Tunnel.

KIDDERMINSTER TO BRIDGNORTH

SEVERN VALLEY RAILWAY

Once a quiet country railway backwater, the scenic Severn Valley Railway has been reborn as one of Britain's leading steam heritage lines – its easy accessibility from the heavily populated West Midlands conurbation ensuring its continuing success.

The forty-and-three-quarter-mile Severn Valley Railway (SVR) between Shrewsbury and Hartlebury, north of Droitwich, was incorporated in 1853. As it name suggests, the railway was built along the valley of the River Severn, closely following the west bank of the river for much of its route and crossing it only once, at Victoria Bridge north of Bewdley. Leased from its opening in 1862 by the West Midland Railway, the SVR was eventually taken over by the Great Western Railway (GWR) in 1872. A three-mile connection from Bewdley to Kidderminster was opened by the GWR in 1878.

Traffic was never heavy on this long rural line – goods consisted mainly of coal from mines at Highley and Alveley, along with some agricultural traffic. Up to the Second World War, passenger services consisted of some five return journeys each weekday between Shrewsbury and Worcester – travel times for the fifty-two-mile journey were around two and three-quarter hours at a leisurely pace of 19 mph. With increasing competition from road transport, traffic declined considerably in the post-war years – this was certainly not helped by British Railways' timetabling which forced passengers to change trains at Bridgnorth on most through journeys. Inevitably, closure between Shrewsbury and Bewdley followed in 1963, and between Bewdley and Kidderminster/Hartlebury in 1970.

Since then the sixteen miles of the Severn Valley Railway south of Bridgnorth has been reborn as a heritage railway. A preservation group ran its first train to Hampton Loade in 1970 and the Severn Valley Railway now operates a mainly steam-hauled service throughout the year between Bridgnorth and Kidderminster, where it connects with the national rail network. With its six beautifully restored stations, historic steam locomotives, vintage rolling stock, well-equipped workshops and Engine House Museum, the railway is today one of the most attractive heritage lines in Britain.

The mainly steam-hauled services along the Severn Valley Railway are operated by a team of full-time staff and a large band of volunteers. The railway is open on most weekends, during school holidays and on most days from Easter to September. National rail network trains to and from Birmingham and Worcester to Kidderminster are operated by London Midland.

Linking historic towns and unspoilt villages along a picturesque river valley, the Severn Valley Railway is deservedly one of the most popular steam-operated heritage railways in Britain.

Ex-GWR 'Manor' Class 4-6-0 No. 7812 'Erlestoke Manor' steams along the Severn Valley near Arley with a train of vintage carriages for Bridgnorth in the autumn of 2011.

Ex-GWR 2-6-2T No. 5164 crosses graceful Victoria Bridge over the River Severn near Arley with a Kidderminster to Bridgnorth train.

ROUTE DESCRIPTION

The Severn Valley Railway (SVR) starts at the town of Kidderminster, famous for its carpet making since the late eighteenth century. Straddling the River Stour and the Staffordshire & Worcestershire Canal and with a population of 55,000, the town has historic connections dating back to the Anglo-Saxons, Vikings and Normans. Today, in addition to carpet making, it is also home to several other important industries such as yacht building and missile manufacturing.

The railway leaves the 1984-built Victorian-style town station terminus in a southerly direction before heading west across the river and canal to approach the town of Bewdley through the 480-yd Foley Park Tunnel. Bewdley (population 9,000) straddles the River Severn, which is crossed by a road bridge built by Thomas Telford in 1798. Striking north from the town the railway hugs the wooded eastern bank of the Severn, while from the far bank some 6,500 acres of Wyre Forest stretch away towards the west. One of the largest areas of semi-natural woodland still remaining in Britain, the forest features the trackbed of the long-closed branch line to Tenbury Wells which is now a footpath and cycleway.

The SVR makes its only crossing of the Severn at Victoria Bridge, three and a half miles north of Bewdley. Built of cast iron in 1861, this graceful bridge carries steam trains high above the river on its single 200-ft span. When built of Coalbrookdale iron, it was the largest cast-iron arch in Britain. Soon crossing from Worcestershire into Shropshire, the railway then heads north along the west bank of the river, passing the village of Arley and on through Highley and Hampton Loade, both of which operate passenger ferries to real ale pubs on the opposite shore. Set in a rural farming area, all three villages depended on the river as their main form of transport until the railway arrived. Across the river from Highley, the former coal-mining area around the village of Alveley has now been landscaped into the Severn Valley Country Park. Taking the train in one direction, walkers can return along the Severn Way long-distance footpath which accompanies the line throughout its route from Bewdley to Bridgnorth – this 220-mile path runs from Severn Beach to the river's source at Plynlimon.

Without straying far from the waters of the Severn, the railway heads north through pastoral scenery to terminate at the hilltop town of Bridgnorth. Named after an ancient river bridge, the town and its 12,000 population have many historic connections including the leaning ruins of a twelfth-century castle, a thirteenth-century town gate, a sixteenth-century Bishop's House and a seventeenth-century timbered town hall. Caves cut into the sandstone banks of the river to the east of the town were once home to a tenth-century hermitage, witches and in later years the town's poor. Separated by the River Severn, Low Town is linked to High Town via a Victorian cliff railway, its lower terminus close to the SVR station. With its 1950s' passenger cars, this is now the only inland cliff railway still in operation in Britain. To the west of Bridgnorth lie the Shropshire Hills, their 310 square miles designated as an Area of Outstanding Natural Beauty.

Ex-GWR outside-frame 'Dukedog' Class 4-4-0 No. 9017 'Earl of Berkeley' hauls a vintage train along the Severn Valley Railway near Rock Cutting, south of Eardington.

Ex-LMS Stanier Class '6P5F' 2-6-0 No. 42968 crosses Oldbury Viaduct just south of Bridgnorth with a train for Kidderminster.

DERBY TO MATLOCK

DERWENT VALLEY LINE

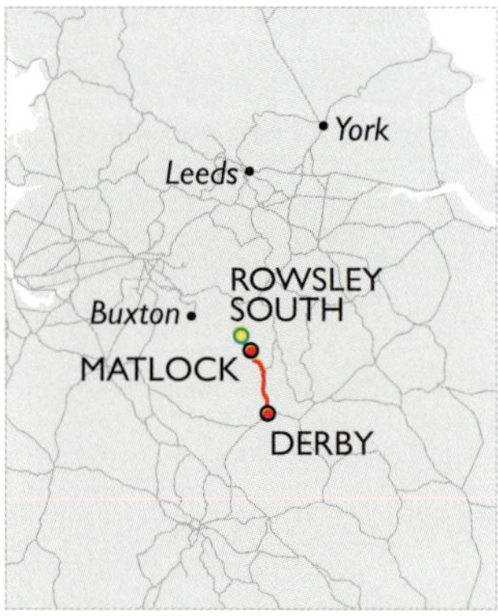

Once forming part of the Midland Railway's main line from St Pancras to Manchester, the Derwent Valley Line has survived as a truncated branch railway serving the Victorian spa resort of Matlock and historic towns and villages that were once at the heart of Britain's Industrial Revolution.

The Derby to Matlock route is a story of two railways. Linking the cities and coalfields of the north Midlands and south Yorkshire, the seventy-two-mile North Midland Railway (NMR) between Derby and Leeds was surveyed by George Stephenson in 1835 and authorized a year later. Meeting the Sheffield & Rotherham Railway at Masborough and George Hudson's York & North Midland railway at Leeds, the line was expensive to build – tracing major river valleys for much of its route, it required seven tunnels, 200 bridges and a canal aqueduct as well as miles of embankments and cuttings.

Opening in 1840, the railway soon ran into financial difficulties and by 1842 a group of shareholders including George Hudson had forced severe economies including the closing of stations, a reduction in services and the sacking of many key staff. Later disgraced for his underhand machinations and financial irregularities, Hudson effectively took control of the NMR and in 1844 amalgamated it with the Birmingham & Derby Junction Railway and the Midland Counties Railway to form the Midland Railway (MR). With its headquarters at Derby, the MR expanded and by 1876 owned a railway empire that stretched from London (St Pancras) to Carlisle.

The NMR's route had taken it through the village of Ambergate, ten and a quarter miles north of Derby. It was from here that the Manchester, Buxton, Matlock & Midlands Railway (MBM&MR) sought to build a railway through the heart of the Peak District to the spa town of Buxton in 1846. Involving extensive engineering works including many viaducts and tunnels, the railway was completed only as far as Rowsley (for Chatsworth House) in 1849 when funds ran out. This eleven-and-a-quarter-mile branch line was then leased jointly by the MR and the London & North Western Railway (LNWR) and eventually extended to Buxton in 1863, albeit with much dissension from the local landed gentry and the poet John Ruskin. By 1867 railway connections had been opened north of Buxton to Manchester and, with the LNWR relinquishing its lease, the MBM&MR was taken over by the MR in 1871. The latter finally had its own route from London to Manchester via Derby and Matlock.

While the railway from Derby to Leeds remains open for business today, the MR's former main line through the Peak District was not so lucky. Although the 'Beeching Report' of 1963 only recommended the withdrawal of stopping trains and the

A two-car Class 153 diesel multiple unit heads south down the Derwent Valley near Duffield with a Matlock to Derby train.

ROUTE DESCRIPTION

With a population of 236,000, the city of Derby still maintains links with the railway industry that go back to 1840 when the North Midland Railway first set up its headquarters and workshops on a green-field site. At the centre of Britain's Industrial Revolution in the eighteenth and nineteenth centuries, the city went on to experience an industrial boom in the early twentieth century with much-enlarged railway workshops and the opening of the Rolls-Royce car and aviation factory. Today, while the railway industry is greatly reduced in size, the city is still an important producer of cars and aircraft engines.

North of Derby the railway clings to the valley of the meandering River Derwent, crossing one of its tributaries, the River Ecclesbourne, at the village of Duffield before plunging into the 855-yd Milford Tunnel. Crossing the Derwent, the railway reaches the town of Belper, an important centre for medieval iron working and one of the first mill towns in Britain in the eighteenth century. North of Belper the railway crosses the Derwent four times before reaching Ambergate, the junction for Matlock.

While the main line sets off in a northeasterly direction up the Amber Valley, the Matlock branch continues up the narrowing, wooded valley of the Derwent, keeping company with the A6 trunk road and the Cromford Canal along its route as far as Cromford. The first station on the branch is Whatstandwell, jumping-off point for the National Tramway Museum at Crich. Along this five-mile stretch the railway crosses the river four times and passes through two tunnels – at Leawood, where the 1849 pump house has been restored to working order, the canal crosses the railway and river on aqueducts. A short distance along the valley lies High Peak Junction with its canal wharf and preserved railway works, built in 1830 for the Cromford & High Peak Railway.

Set in a region of limestone hills with deep valleys and fast-flowing streams, Cromford was one of the most important sites in the early development of the Industrial Revolution – it was here that Richard Arkwright built his first cotton mill in 1771 and for this reason the valley has been designated as the Derwent Valley Mills World Heritage Site. Opened in 1794 along the 300-ft contour of the valley from Ambergate, the fourteen-and-a-half-mile Cromford Canal terminates at Cromford Wharf. Leaving Cromford, the railway immediately passes through the 764-yd Willersley Tunnel before entering Matlock Bath station.

Located in the steep-sided valley, Matlock Bath sprang to prominence as a spa resort in 1831 when it was visited by Princess Victoria – the coming of the railway in 1849 only served to increase its popularity with Victorians seeking health cures from its warm springs. Once served by direct train services from London and marketed for its scenic beauty by the Midland Railway as 'Little Switzerland', Matlock Bath and its northerly neighbour of Matlock were soon linked by a ribbon development of hotels, guest houses and health hydros along the valley. After passing through the two High Tor Tunnels the railway finally terminates at Matlock station.

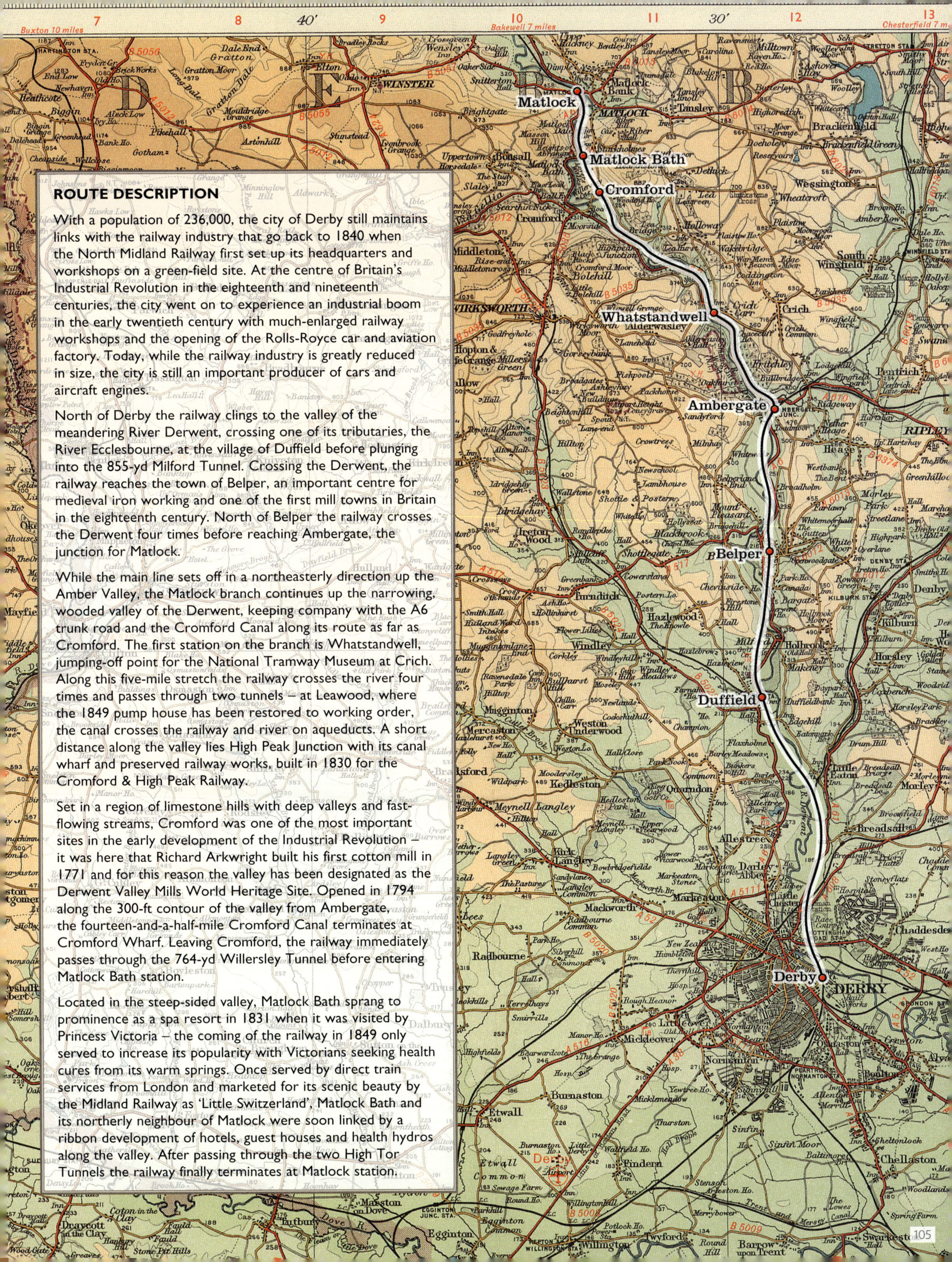

Framed by a graceful Midland Railway footbridge, a Class 153 diesel unit enters Whatstandwell station on the single-track line between Ambergate South Junction and Matlock.

complete closure of the former Midland Railway Sheffield to Manchester route via the Hope Valley to the north, in fact the opposite happened. Closure north of Matlock came in 1968 and since then the spa town has been at the end of a branch line served by trains from Derby and Nottingham. Since 1991, the five miles from Matlock to Rowsley have been re-opened by Peak Rail (see page 107) as a heritage railway. National rail network train services between Derby and Matlock are operated by East Midlands Trains. The branch line from Ambergate Junction is single track through to the run-round loop at Matlock.

Following the Derwent Valley deep into the limestone hills of the Derbyshire Dales, this scenic railway passes through a World Heritage Site into a region once marketed as 'Little Switzerland' by a publicity-conscious railway company.

Peak Rail's preserved 0-6-0ST WD150 gets ready to leave the former Midland Railway station at Darley Dale.

A steam-hauled train rattles along Peak Rail's heritage line in the scenic Derwent Valley between Matlock and Rowsley South.

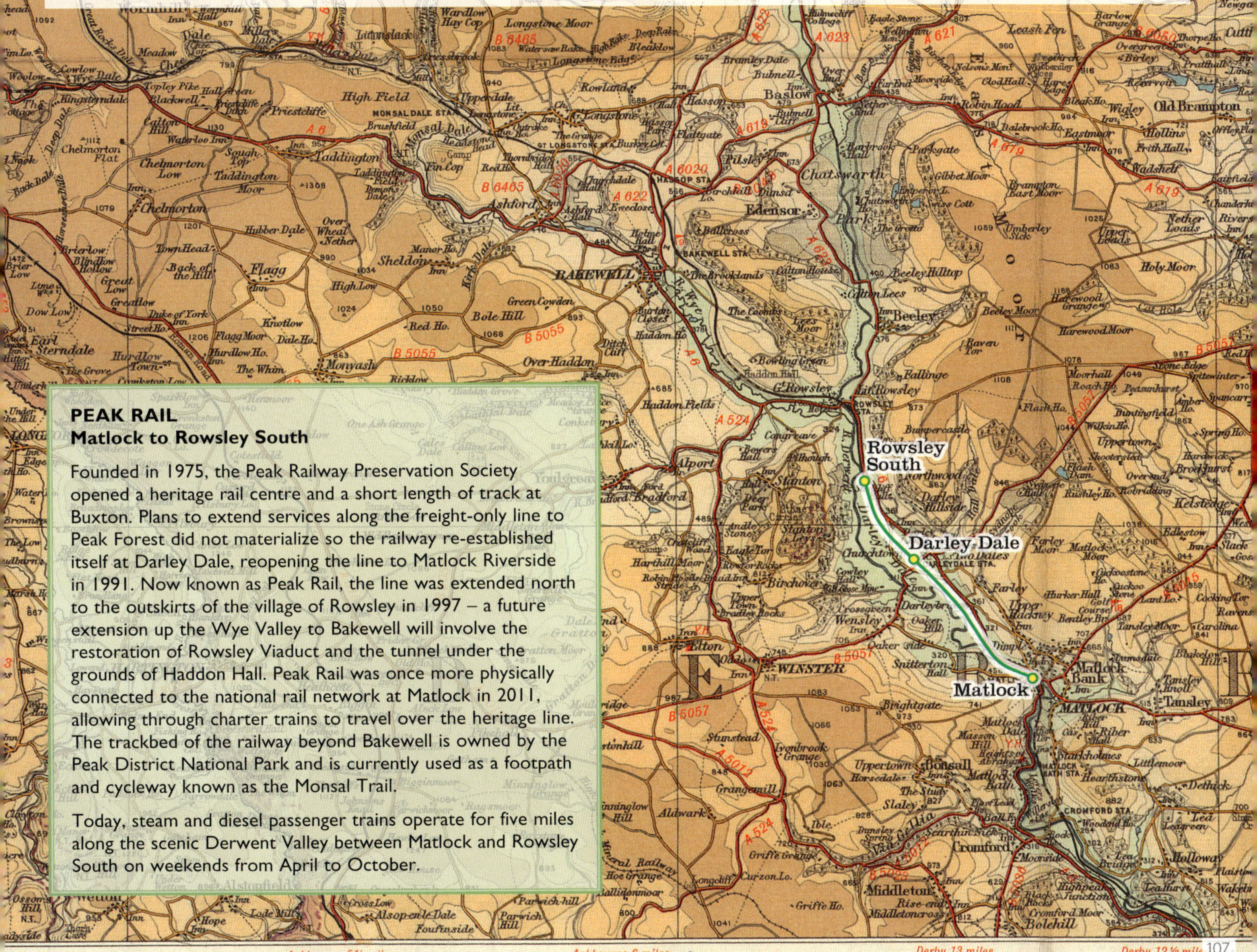

PEAK RAIL
Matlock to Rowsley South

Founded in 1975, the Peak Railway Preservation Society opened a heritage rail centre and a short length of track at Buxton. Plans to extend services along the freight-only line to Peak Forest did not materialize so the railway re-established itself at Darley Dale, reopening the line to Matlock Riverside in 1991. Now known as Peak Rail, the line was extended north to the outskirts of the village of Rowsley in 1997 – a future extension up the Wye Valley to Bakewell will involve the restoration of Rowsley Viaduct and the tunnel under the grounds of Haddon Hall. Peak Rail was once more physically connected to the national rail network at Matlock in 2011, allowing through charter trains to travel over the heritage line. The trackbed of the railway beyond Bakewell is owned by the Peak District National Park and is currently used as a footpath and cycleway known as the Monsal Trail.

Today, steam and diesel passenger trains operate for five miles along the scenic Derwent Valley between Matlock and Rowsley South on weekends from April to October.

WALES

A southbound steam charter train on the Cambrian Coast line heads out across the 990-yd, 121-span Barmouth Bridge over the Mawddach Estuary.

SHREWSBURY TO LLANELLI

HEART OF WALES LINE

This long and sinuous single-track line across the backbone of Wales has defied two threats of closure and today remains one of the most scenic railway backwaters in Britain. Recent service improvements and enhanced promotion of the line have attracted an increasing number of leisure visitors to this remarkable survivor.

The Central Wales Line was born out of a desire by the London & North Western Railway (LNWR), with its headquarters at Euston, to tap into the lucrative docks and coal traffic at Swansea. However, the 110-mile route taken by today's passenger trains between Shrewsbury and Llanelli took many years to be completed. First on the scene at the northern end was the Shrewsbury & Hereford Railway (S&HR), which opened throughout in 1853 – it was double track through Church Stretton and Craven Arms as far as Ludlow and single track thereafter through 1,056-yd Dinmore Tunnel to Hereford.

A highly profitable operation, it was leased jointly to the Great Western Railway (GWR), LNWR and West Midland Railway in 1862. In 1870 it was taken over jointly by the GWR and LNWR and, with the opening of the Severn Tunnel in 1886, became an important route for North-to-West expresses avoiding Birmingham.

Down in South Wales the Llanelly Railway (LR), primarily an anthracite-carrying operation, opened its line from Llanelli to Llandeilo via Pontarddulais in 1857 – Ammanford had already been reached as early as 1841. Operations were extended northwards along the leased Vale of Towy Railway to Llandovery a year later. The LR further expanded in 1865 by opening a line westwards from Llandeilo to Carmarthen in 1865 and southwards from Pontarddulais across the Gower Peninsula to Swansea two years later.

Meanwhile, up north at Craven Arms the Knighton Railway (KR) had opened westwards from its junction with the S&HR and across the Welsh border to the town of Knighton in 1861. With the LNWR keeping a close eye on events from distant Euston, the KR merged with the uncompleted Central Wales Railway (CWR) in 1863 – the latter opening from Knighton across castellated Knucklas Viaduct, through tunnels at Llangynllo

New-build Class 'A1' 4-6-2 No. 60163 'Tornado' heads north along the Welsh Marches line near Ludlow with a Christmas Market special from Paddington to Shrewsbury in November 2011.

and Pen-y-Bont, to the spa town of Llandrindod Wells in 1865. The Central Wales Extension Railway (CWER) completed the intervening gap south to Llandovery in 1868. By far the most heavily engineered section of the Central Wales line, construction included the curving Cynghordy Viaduct and the single-bore Sugar Loaf Tunnel, approached on long gradients as steep as 1 in 60 from both directions.

The LNWR moved fast, absorbing both the CWR and the CWER less than four months before the latter had even opened. Despite this, Euston's goal of reaching Swansea was still some years away – in 1873 the LNWR was granted running powers over the Swansea & Carmarthen Railway's (formerly Llanelly Railway) lines from Llandovery to Swansea and from Llandeilo to Carmarthen. The situation became more complicated in 1891 when the LNWR absorbed both the Llandeilo to Carmarthen and the Pontarddulais to Swansea lines – the intervening gap between Llandeilo and Pontarddulais was then operated jointly with the GWR.

The LNWR now had a through route from Euston to Swansea Victoria, albeit 278 miles against the GWR's more direct

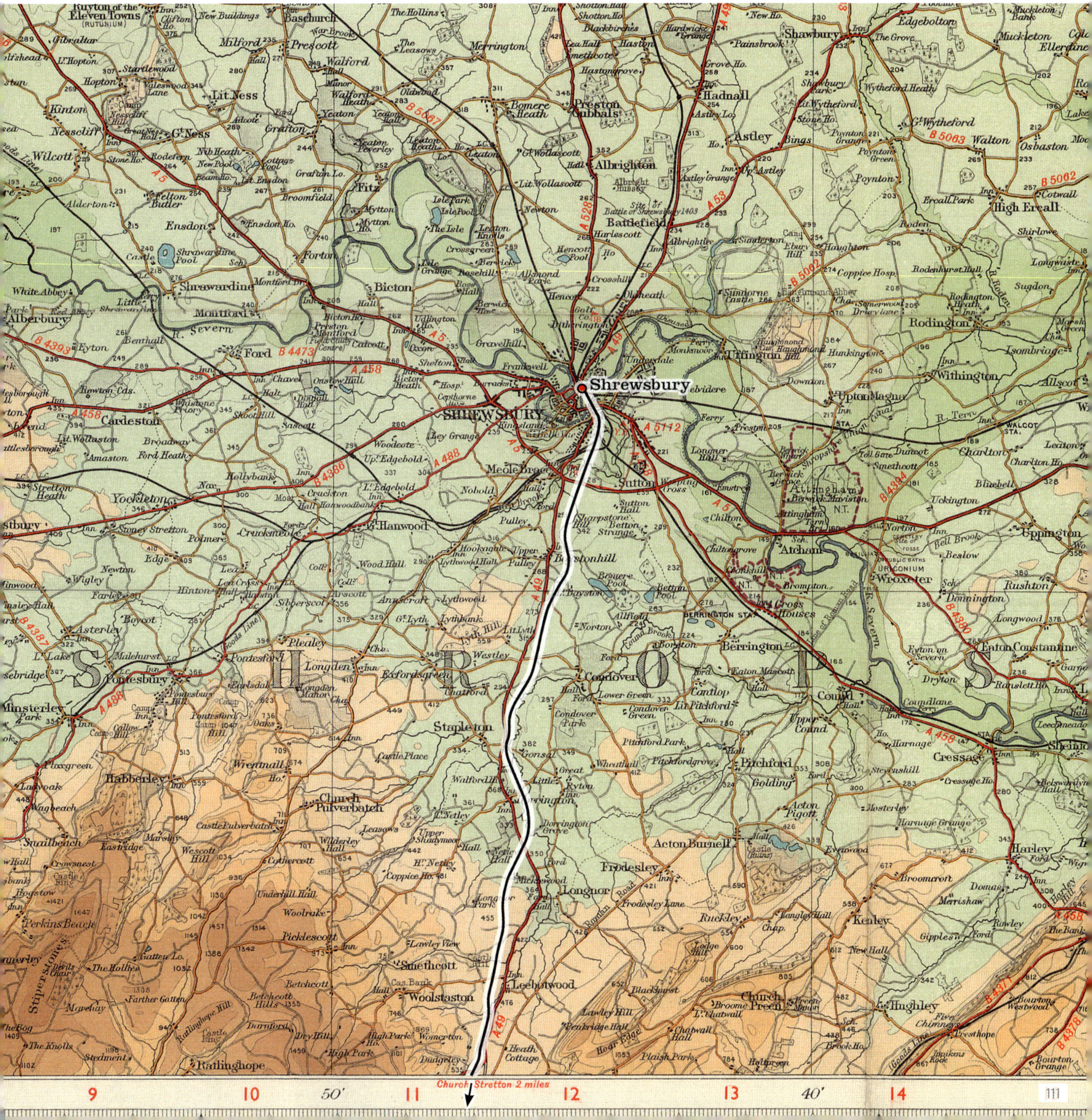

191-mile route from Paddington. The line was soon being served by through coaches not only from Euston but also from York, Birmingham, Liverpool and Manchester but journey times along the heavily graded single-track route were of necessity rather slow. Even so, the small spa resorts of Llandrindod Wells, Builth Wells, Llangammarch Wells and Llanwrtyd Wells all benefitted from an influx of Victorian visitors seeking the health-curing properties of the local mineral spring waters.

Freight traffic was also heavy, with long trains of anthracite headed for the North and the Midlands labouring up to the choking confines of Sugar Loaf Tunnel. One through working that survived until the end of steam haulage in 1964 was the weekday overnight mail train between York and Swansea Victoria, which also conveyed passenger coaches and took over nine hours to complete its long journey.

Although listed for closure in the 1963 'Beeching Report', most of the Central Wales Line precariously clung on to life following a 1964 reprieve by the new Labour Government – the fact that the route passed through three marginal constituencies was the critical political factor. However, the section southwards from Pontarddulais to Swansea Victoria did not escape and was closed completely in the summer of 1964, with trains to and from Shrewsbury diverted to Llanelli. Apart from remaining coal traffic south from Pantyffynnon, all goods traffic was diverted away from the line, which settled down to an infrequent passenger service operated by diesel multiple units.

Paring running costs to the bone, the line has been operating under a Light Railway Order since 1972 – apart from Llandrindod, all intermediate stations also became unstaffed with fares being collected by a conductor guard on the train. Surviving another closure threat in 1969, the line saw the introduction of a new signalling system in 1986 – known as No Signalman Token with Remote Token Stations, it replaced surviving signal boxes at passing loops with just one operator at Pantyffynnon. Today, apart from a short section from Llanelli to Morlais Junction, the entire route is single track with five passing loops: Knighton, Llandrindod, Llanwrtyd, Llandovery and Llandeilo, although only two (Llandrindod and Llanwrtyd) are regularly in use.

Serving twenty-eight intermediate stations and halts, trains are normally operated by single-car diesel units, strengthened to two cars during the peak summer months. With a journey time of around three and a half hours there are four return services operated by Arriva Trains Wales on weekdays with two on Sundays – a more frequent service is now possible following recent upgrades of passing loops. Marketed as the Heart of Wales Line, the route is also very occasionally used by diverted freight trains and charter trains.

Linking faded Victorian spa resorts and isolated farming villages, travellers on this marathon rail journey are rewarded by inspiring scenery through the remote and unspoilt hills of Central Wales. This is a journey to savour slowly and remember at leisure.

Stanier Class '5' 4-6-0 No. 45407 and BR Standard Class 4 2-6-0 No. 76079 double-head a charter train on the Heart of Wales Line near Knighton.

ROUTE DESCRIPTION
Shrewsbury to Knighton

Heart of Wales Line trains to Llanelli head southwards from Shrewsbury's grand Tudor-style station set on the west bank of the River Severn, taking the Welsh Marches Line (the former GWR/LNWR joint line) for the first twenty miles as far as Craven Arms. At Church Stretton the railway's Victorian builders took advantage of a narrow gap in the Shropshire Hills. To the west lies the high heathland plateau of the Long Mynd with its 1,693-ft peak of Pole Bank, while to the east is Caer Caradoc Hill, its 1,506-ft summit crowned by an Iron Age hill fort and, beyond, the limestone escarpment of Wenlock Edge. Nestled in the Shropshire Hills Area of Outstanding Natural Beauty, the small market town of Church Stretton was once an important textile centre, which later developed as a spa resort during the nineteenth century. The Victorian and Edwardian tourists who flocked to the town by train gave it the name of 'Little Switzerland' for its numerous streets and houses that cling to the hillsides.

Heading south away from Church Stretton station and its award-winning gardens, the railway winds its way down through the hills past Acton Scott and its working Victorian Farm Museum to enter the valley of the River Onny near Craven Arms. Until the coming of the railway, Craven Arms was just a small village but its importance grew as a railway junction following the opening of the Knighton Railway in 1861.

Turning westwards away from the Welsh Marches Line at Craven Arms, the Heart of Wales Line heads into the border hills through Broome before bearing south along the Clun Valley through Hoptonheath to join the valley of the River Teme at Bucknell. Turning west up the narrowing valley the railway hugs the England-Wales border to the historic market town of Knighton – here the station and its passing loop lie in England while across the river the town lies in Wales. Set astride the earthworks of Offa's Dyke and on the route of two former drovers' roads, this border town, with its two ruined Norman castles, has seen its fair share of mayhem – in the early fifteenth century it was besieged and destroyed by Owain Glyndwr during his long crusade against the English oppressors.

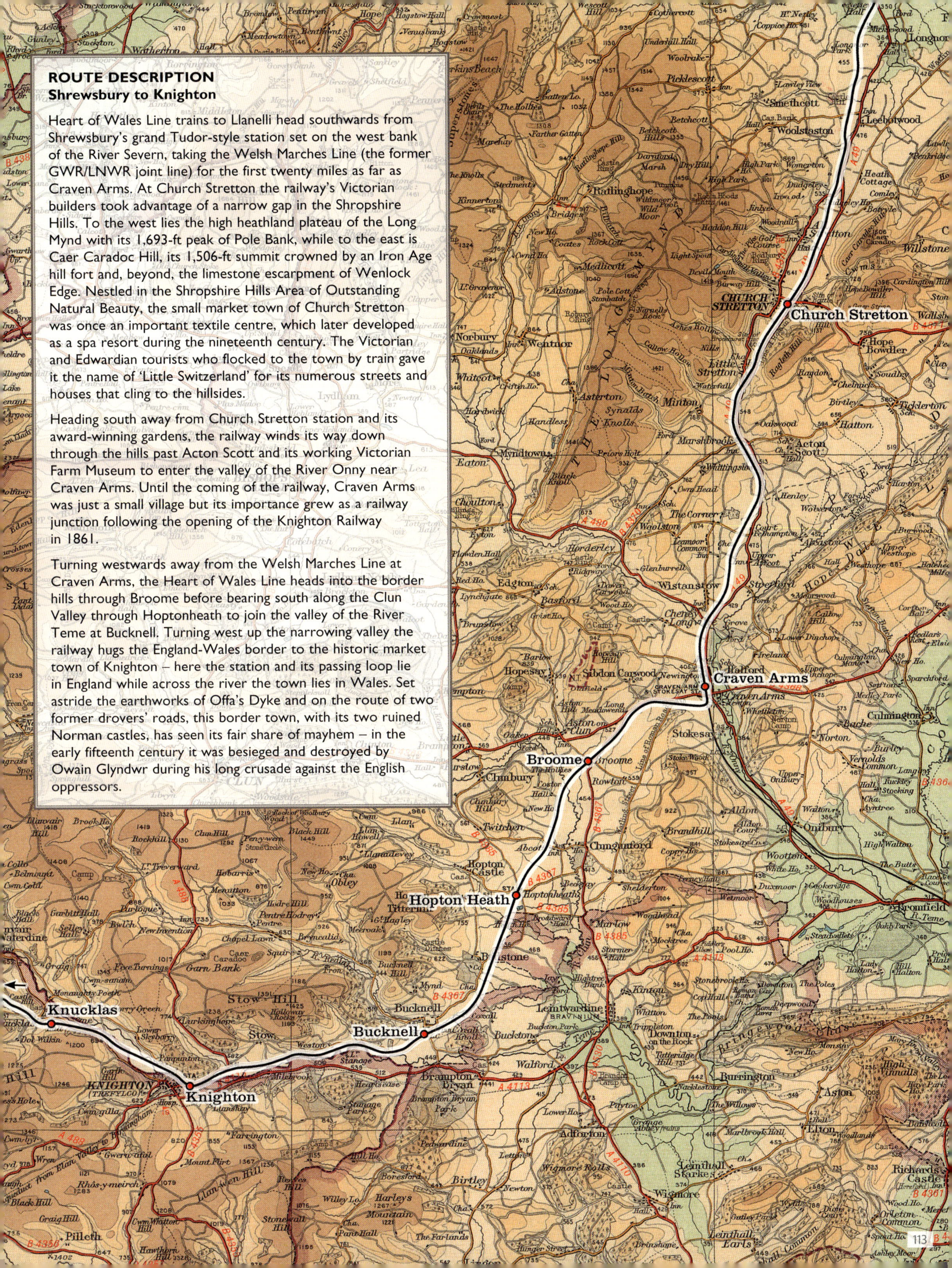

ROUTE DESCRIPTION
Knighton to Llangammarch Wells

Still heading up the Teme Valley from Knighton, the railway crosses into Wales before starting its long climb into the hills at Knucklas. Here, the fine castellated thirteen-arch railway viaduct overlooks the remains of the thirteenth-century Norman castle, which was also destroyed by Owain Glyndwr. Climbing steadily up through curving Llangynllo Tunnel the railway reaches the summit of the line just to the north of Llangynllo, 980 ft above sea level, before heading south down the valley of the River Lugg. Closely following the contours, the railway soon turns west again, climbing to the watershed at Llanbister Road station before descending once more, through the valley of the River Aran through Dolau station. Set amidst rolling hill-farming countryside, this remote station is famous for its beautifully tended award-winning garden. To the south lies the featureless rock dome known as Radnor Forest, dominated by the 2,165-ft peak of Rhos Fawr (Great Rhos).

Leaving Dolau, the railway crosses the Ithon Valley for the first time before taking a short cut through Pen-y-Bont Tunnel and crossing the valley again at Crossgates. Following the meandering Ithon, the railway reaches the town of Llandrindod Wells. The busiest intermediate station on the line, it is the only one with a ticket office and now boasts a museum in the restored signal box. The town developed as a spa resort in the eighteenth century but the coming of the railway in the nineteenth century resulted in massive expansion with the building of grand hotels, boarding houses, pump rooms and an ornamental park and lake. With visitors arriving on direct railway carriages from London, Birmingham and Manchester, the town experienced a boom period that lasted until the First World War. In the 1970s it became a magnet for those seeking an alternative lifestyle and evolved as a local government administrative centre for the reformed county of Powys.

Heading south from Llandrindod Wells, the railway climbs away from the Ithon Valley to Builth Road station. Once an important railway crossroads that was also served by the lower-level Mid-Wales Line (closed at the end of 1962), the station lies some two miles north of the former spa town of Builth Wells and sees additional trains during the Royal Welsh Show which is held every July. Cutting across the valleys of the Wye and the Chwefri rivers, the railway heads through Cilmeri to follow the winding valley of the river Irfon through Garth, crossing over the river twice before reaching the village of Llangammarch Wells – one of several former Victorian spa resorts still served by the railway.

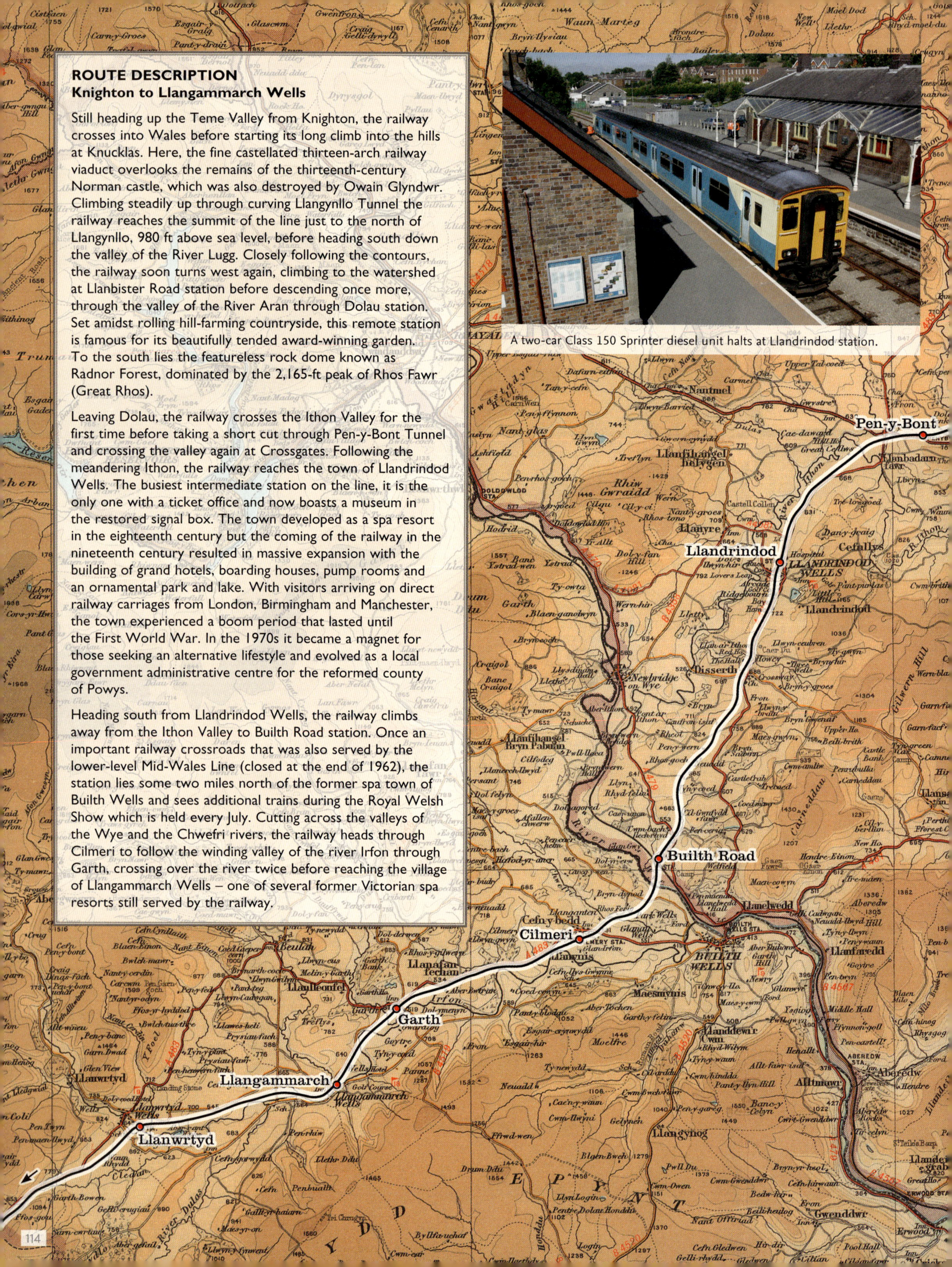

A two-car Class 150 Sprinter diesel unit halts at Llandrindod station.

Stanier Class '5' 4-6-0 No. 45407 and BR Standard Class 4 2-6-0 No. 76079 cross Knucklas Viaduct with a southbound charter train on the Heart of Wales Line.

A two-car Class 150 diesel multiple unit slowly crosses the curving Cynghordy Viaduct southwest of Sugar Loaf Tunnel.

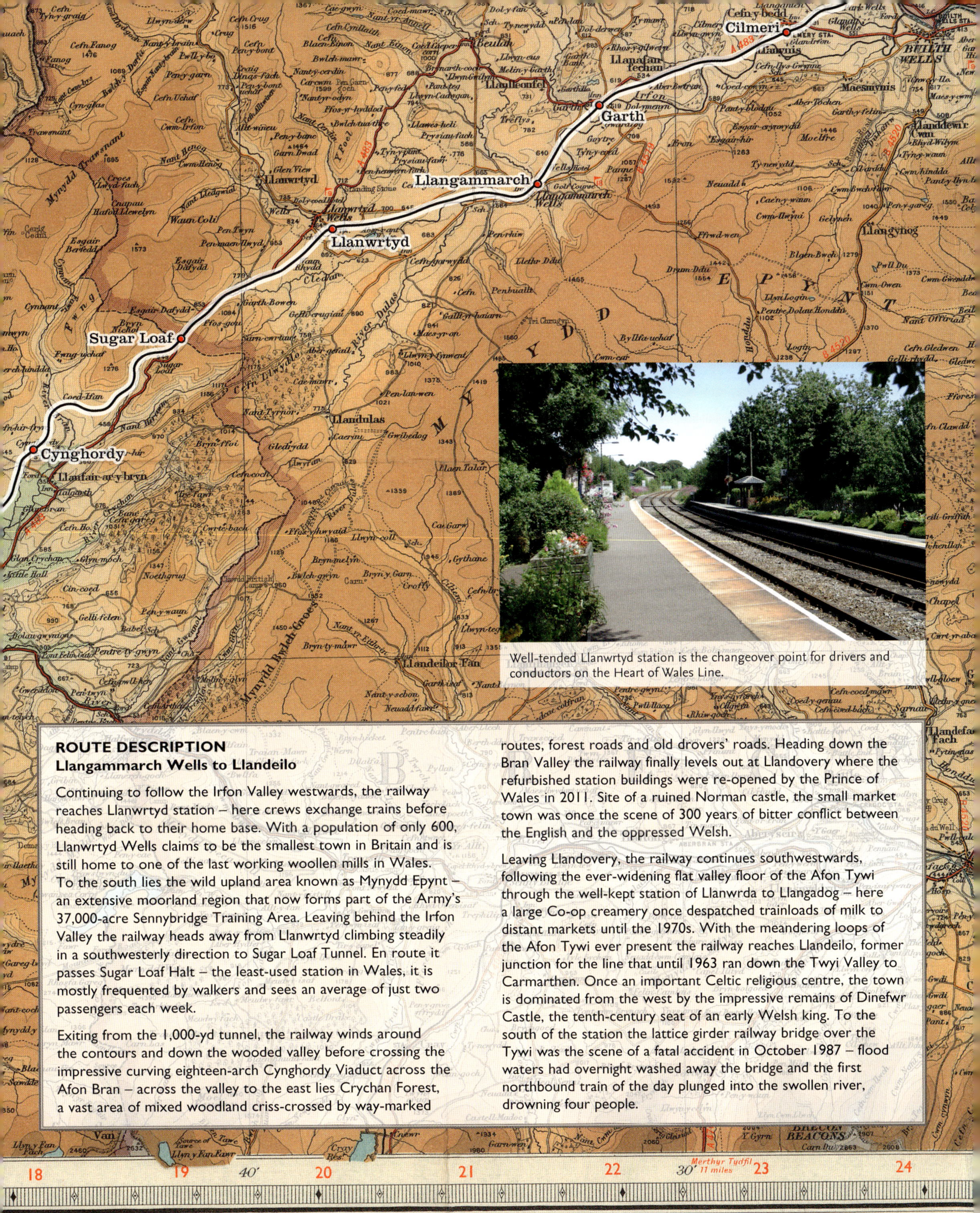

Well-tended Llanwrtyd station is the changeover point for drivers and conductors on the Heart of Wales Line.

ROUTE DESCRIPTION
Llangammarch Wells to Llandeilo

Continuing to follow the Irfon Valley westwards, the railway reaches Llanwrtyd station – here crews exchange trains before heading back to their home base. With a population of only 600, Llanwrtyd Wells claims to be the smallest town in Britain and is still home to one of the last working woollen mills in Wales. To the south lies the wild upland area known as Mynydd Epynt – an extensive moorland region that now forms part of the Army's 37,000-acre Sennybridge Training Area. Leaving behind the Irfon Valley the railway heads away from Llanwrtyd climbing steadily in a southwesterly direction to Sugar Loaf Tunnel. En route it passes Sugar Loaf Halt – the least-used station in Wales, it is mostly frequented by walkers and sees an average of just two passengers each week.

Exiting from the 1,000-yd tunnel, the railway winds around the contours and down the wooded valley before crossing the impressive curving eighteen-arch Cynghordy Viaduct across the Afon Bran – across the valley to the east lies Crychan Forest, a vast area of mixed woodland criss-crossed by way-marked routes, forest roads and old drovers' roads. Heading down the Bran Valley the railway finally levels out at Llandovery where the refurbished station buildings were re-opened by the Prince of Wales in 2011. Site of a ruined Norman castle, the small market town was once the scene of 300 years of bitter conflict between the English and the oppressed Welsh.

Leaving Llandovery, the railway continues southwestwards, following the ever-widening flat valley floor of the Afon Tywi through the well-kept station of Llanwrda to Llangadog – here a large Co-op creamery once despatched trainloads of milk to distant markets until the 1970s. With the meandering loops of the Afon Tywi ever present the railway reaches Llandeilo, former junction for the line that until 1963 ran down the Twyi Valley to Carmarthen. Once an important Celtic religious centre, the town is dominated from the west by the impressive remains of Dinefwr Castle, the tenth-century seat of an early Welsh king. To the south of the station the lattice girder railway bridge over the Tywi was the scene of a fatal accident in October 1987 – flood waters had overnight washed away the bridge and the first northbound train of the day plunged into the swollen river, drowning four people.

18 19 40′ 20 21 22 30′ 23 24

Railways — Canals — County Boundaries — Overhead Electrical High Power Lines — Youth Hostels Y.H. — Golf Courses — National Trust Properties N.T.

The figures thus 190 represent heights in feet above sea level.

2250 2500 2750

1960

The publishers record their appreciation ... services rendered in the past by map users ... maintain the accuracy of this series and are ... to acknowledge any corrections brought to ...

ROUTE DESCRIPTION
Llandeilo to Llanelli

The Heart of Wales Line heads south away from the Tywi Valley at Llandeilo, first climbing up the valley of the Afon Cennen close to the western boundary of the Black Mountains in the Brecon Beacons National Park before crossing the watershed north of Llandybie. Heading down to Ammanford the railway joins the Loughor Valley, which it follows for the rest of its route to Llanelli. At Ammanford the original station buildings have long since been demolished, only to be replaced by one of the functional 'bus' shelters that are now common at most stations along the route. Large amounts of anthracite were once conveyed by rail from local mines in the area to fuel blast furnaces in South Wales and in the Midlands – opencast mined coal is still carried by rail from Gwaun-Cae-Gurwen and down the Amman Valley to the next station on the line at Pantyffynnon and thence for distribution in South Wales. At Pantyffynnon the only remaining working signal box controls the entire Heart of Wales Line north to Craven Arms.

Leaving Pantyffynnon behind, the railway heads south down the tranquil and unspoilt Loughor Valley to the former tinplate-making town of Pontarddulais – the tinplate works closed soon after the Second World War and were replaced by a new plant at Llanelli. One mile southwest of Pontarddulais the Heart of Wales Line joins the double-track Swansea District Line at Morlais Junction and is soon heading along the west shore of the tidal Loughor Estuary through Llangennech and Bynea. Meeting Brunel's South Wales Railway at Llandeilo Junction, where the northern landscape is dominated by the rail-connected Corus tinplate works at Trostre, trains from Shrewsbury rattle into Llanelli station where they reverse direction and continue their journey eastward to Swansea.

A double-headed steam charter train passes Sugar Loaf Halt, the least-used station in Wales, as it heads south towards Sugar Loaf Tunnel.

SHREWSBURY TO ABERYSTWYTH AND PWLLHELI

CAMBRIAN LINE

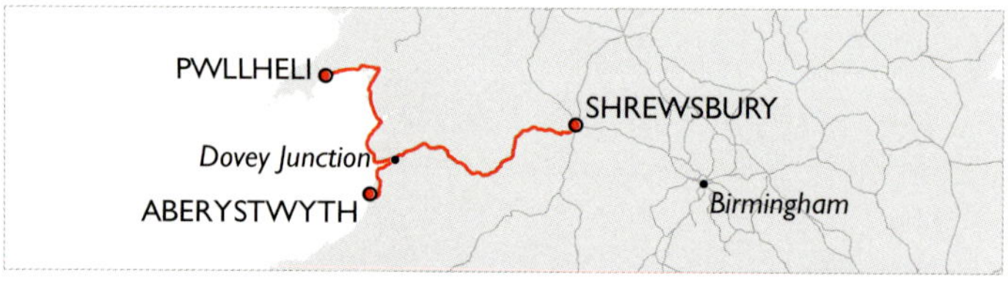

Following in the footsteps of the famous 'Cambrian Coast Express', this single-track route follows the Severn Valley before crossing the Cambrian Mountains to the sea. Bridging wide river estuaries, the coastal section north of Machynlleth must rate as one of the most scenic rail journeys in Britain.

The 118¾-mile railway route from Shrewsbury to Pwllheli was born out of early proposals to build a trunk railway to link the English Midlands with the ports and harbour towns along the West Wales coastline. While none of these ambitious schemes ever materialized the first railway to be built partly along the present-day route was a very local affair. The fourteen-mile Llanidloes & Newtown Railway (L&NR) was opened in 1859 and remained isolated from other railways until 1861, when it was joined from the east by the twenty-nine-and-a-half-mile Oswestry & Newtown Railway (O&NR). To the west, the Newtown & Machynlleth Railway (N&MR) opened from its junction with the L&NR at Moat Lane for twenty-two-and-three-quarter miles to Machynlleth. Construction along this heavily graded route through the Cambrian Mountains involved the excavation through gritstone of a 120-ft-deep cutting at the summit of the line at Talerddig. All three companies came together in 1864 to form the Cambrian Railways with its headquarters in Oswestry.

The Aberystwyth & Welch [sic] Coast Railway (A&WCR) promoted the coastal route from Machynlleth, northwards to Pwllheli and southwards to Aberystwyth. The twenty-and-a-half-mile line to Aberystwyth was completed in 1864, but the fifty-three-and-three-quarter-mile coast-hugging route from Dovey Junction to Pwllheli took much longer. Completion was held up until 1867 by the construction of the 830-yd, 121-span Barmouth Bridge across the Mawddach Estuary – by which time the A&WCR had been taken over by the Cambrian Railways. Originally built completely of timber, a steel swing bridge was incorporated into its northern section in 1901. An infestation of timber-boring insects in its timber piles during the 1980s very nearly spelt the end for the line but fortunately was successfully treated. Meanwhile, to the east, the Shrewsbury & Welshpool Railway had already opened in 1861. Jointly operated by the Great Western Railway (GWR) and the London & North Western Railway, this seventeen-mile line opened up the Cambrian route

The highly scenic coastal route to Pwllheli crosses the Afon Dyfi immediately to the west of Dovey Junction.

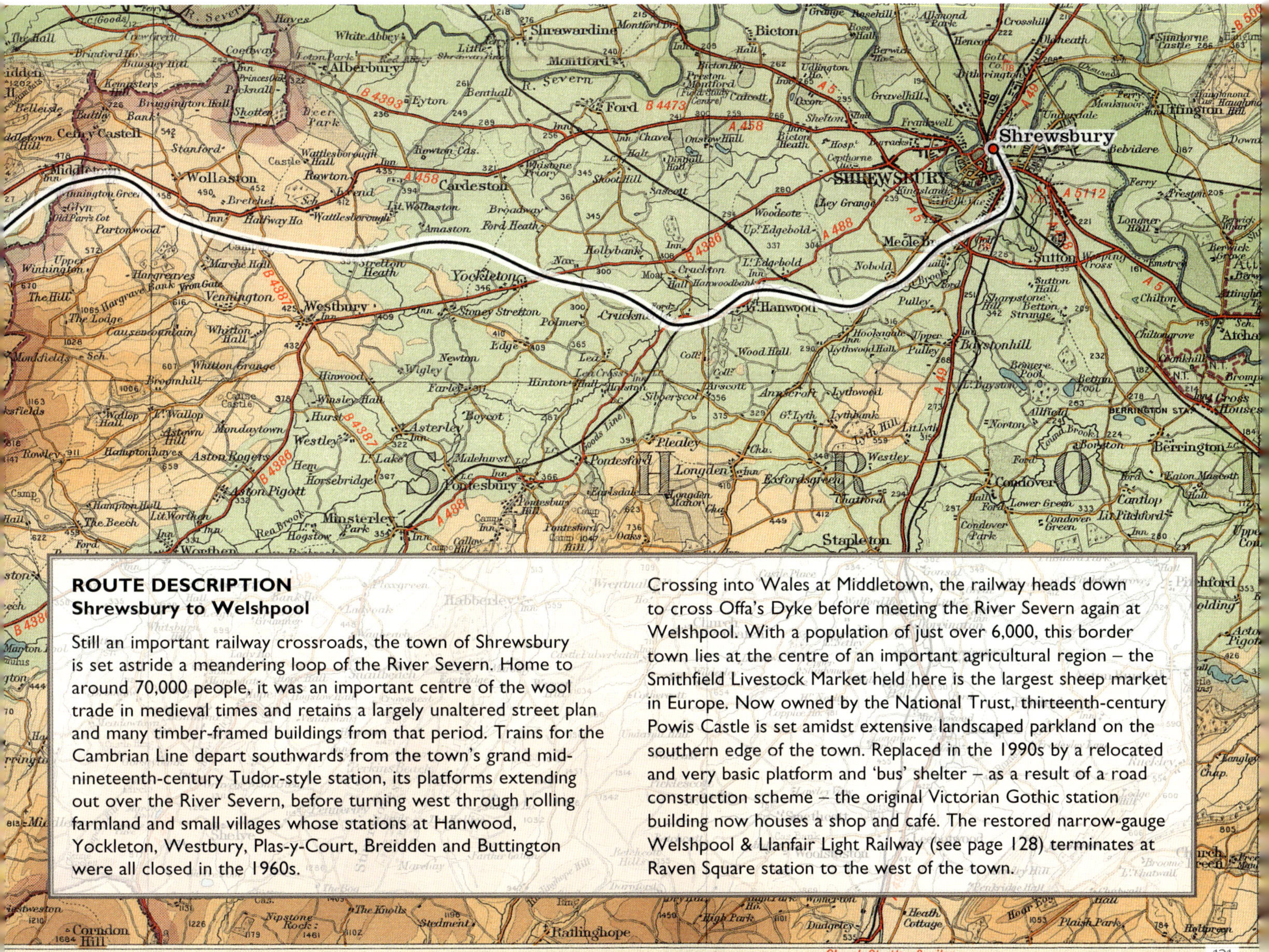

ROUTE DESCRIPTION
Shrewsbury to Welshpool

Still an important railway crossroads, the town of Shrewsbury is set astride a meandering loop of the River Severn. Home to around 70,000 people, it was an important centre of the wool trade in medieval times and retains a largely unaltered street plan and many timber-framed buildings from that period. Trains for the Cambrian Line depart southwards from the town's grand mid-nineteenth-century Tudor-style station, its platforms extending out over the River Severn, before turning west through rolling farmland and small villages whose stations at Hanwood, Yockleton, Westbury, Plas-y-Court, Breidden and Buttington were all closed in the 1960s.

Crossing into Wales at Middletown, the railway heads down to cross Offa's Dyke before meeting the River Severn again at Welshpool. With a population of just over 6,000, this border town lies at the centre of an important agricultural region – the Smithfield Livestock Market held here is the largest sheep market in Europe. Now owned by the National Trust, thirteenth-century Powis Castle is set amidst extensive landscaped parkland on the southern edge of the town. Replaced in the 1990s by a relocated and very basic platform and 'bus' shelter – as a result of a road construction scheme – the original Victorian Gothic station building now houses a shop and café. The restored narrow-gauge Welshpool & Llanfair Light Railway (see page 128) terminates at Raven Square station to the west of the town.

for through trains from London and the Midlands.

Serving only small towns and villages along its mainly single-track route, the Cambrian Railways struggled to make ends meet into the twentieth century. Already financially hamstrung, poorly equipped and with tediously slow journey times, the railway met its nemesis in 1921 when a head-on collision between two trains near Abermule killed nineteen people – while unauthorized use of a single-line tablet was to blame, the overstretched company never recovered and was taken over by the GWR in 1922.

The new broom from the GWR's London Paddington headquarters soon introduced much-needed upgrades to the route, rebuilding stations and opening up a series of halts along the coastal section north of Dovey Junction. Services were improved along with the introduction of the 'Cambrian Coast Express' in 1927 – serving holiday destinations along the West Wales coastline, this famous restaurant car express ran from Paddington to Aberystwyth and Pwllheli until the end of steam in 1967.

While the entire Cambrian line escaped the 'Beeching Axe', the more northerly Ruabon to Barmouth Junction line was sacrificed instead – many intermediate stations on the 'main line' from Shrewsbury to Aberystwyth were closed in 1965. Six years later, the Dovey Junction to Pwllheli coastal section was threatened but, despite closure notices being put up at all of the twenty-six stations and halts, a strong campaign against closure ensured that the planned end of services on 4 October 1971 never happened. A 1996 move to close many little-used halts also failed. Today, all of the stations along the line are unstaffed, with many being request stops only.

Operated by Arriva Trains Wales, today's diesel passenger trains have to cope with long single-line sections between passing loops – the longest, between Shrewsbury and Welshpool, is just over nineteen miles – and late-running trains can have a severe knock-on effect on punctuality. Recently introduced, the ERTMS (European Rail Traffic Management System) has replaced the previous Radio Electronic Token Block system and now train drivers are given instructions for movement on cab

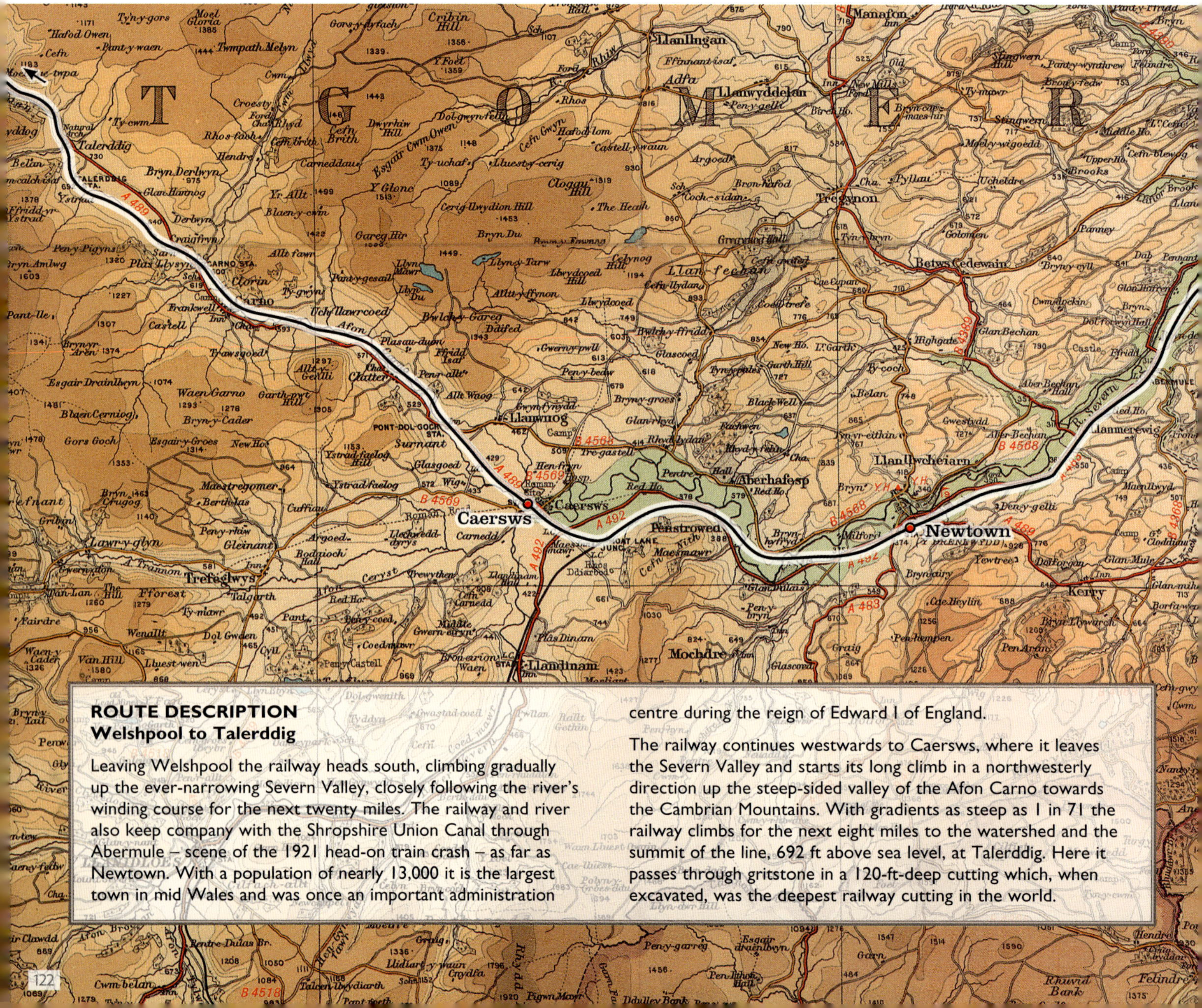

ROUTE DESCRIPTION
Welshpool to Talerddig

Leaving Welshpool the railway heads south, climbing gradually up the ever-narrowing Severn Valley, closely following the river's winding course for the next twenty miles. The railway and river also keep company with the Shropshire Union Canal through Abermule – scene of the 1921 head-on train crash – as far as Newtown. With a population of nearly 13,000 it is the largest town in mid Wales and was once an important administration centre during the reign of Edward I of England.

The railway continues westwards to Caersws, where it leaves the Severn Valley and starts its long climb in a northwesterly direction up the steep-sided valley of the Afon Carno towards the Cambrian Mountains. With gradients as steep as 1 in 71 the railway climbs for the next eight miles to the watershed and the summit of the line, 692 ft above sea level, at Talerddig. Here it passes through gritstone in a 120-ft-deep cutting which, when excavated, was the deepest railway cutting in the world.

display screens. This allows the headway between trains using the same track to be reduced without compromising safety. Once a regular fixture along the coastal route during the summer months, steam-hauled trains can no longer operate until the new and expensive ERTMS equipment is fitted to the locomotives.

Linking small market towns in the Welsh borders, faded Victorian seaside resorts along Cardigan Bay and the narrow-gauge railways of Snowdonia, this long rail journey offers the discerning traveller an unfolding vista of rivers, mountains and sea.

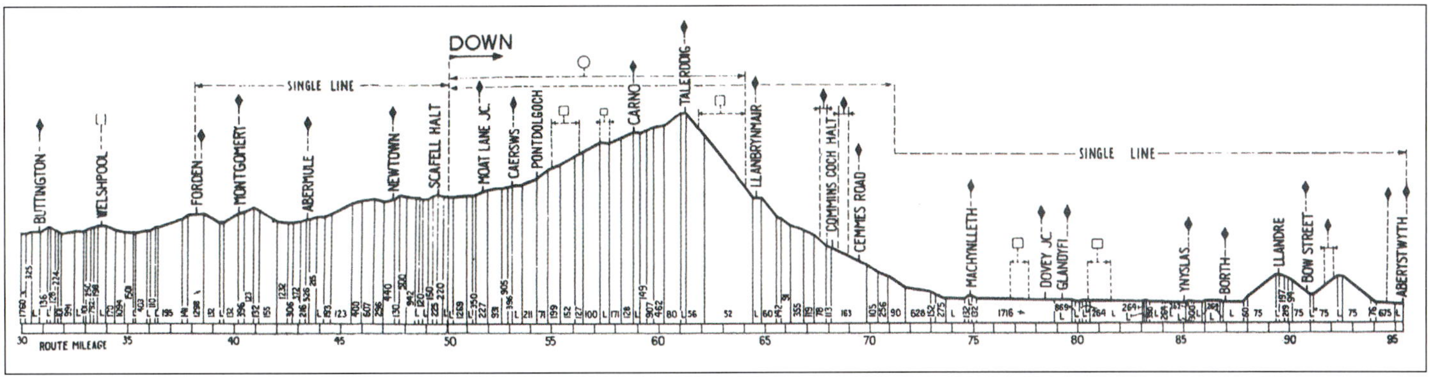

Eastbound trains on the Cambrian Line are faced with an unrelenting climb up to the 693-ft summit at Talerddig.

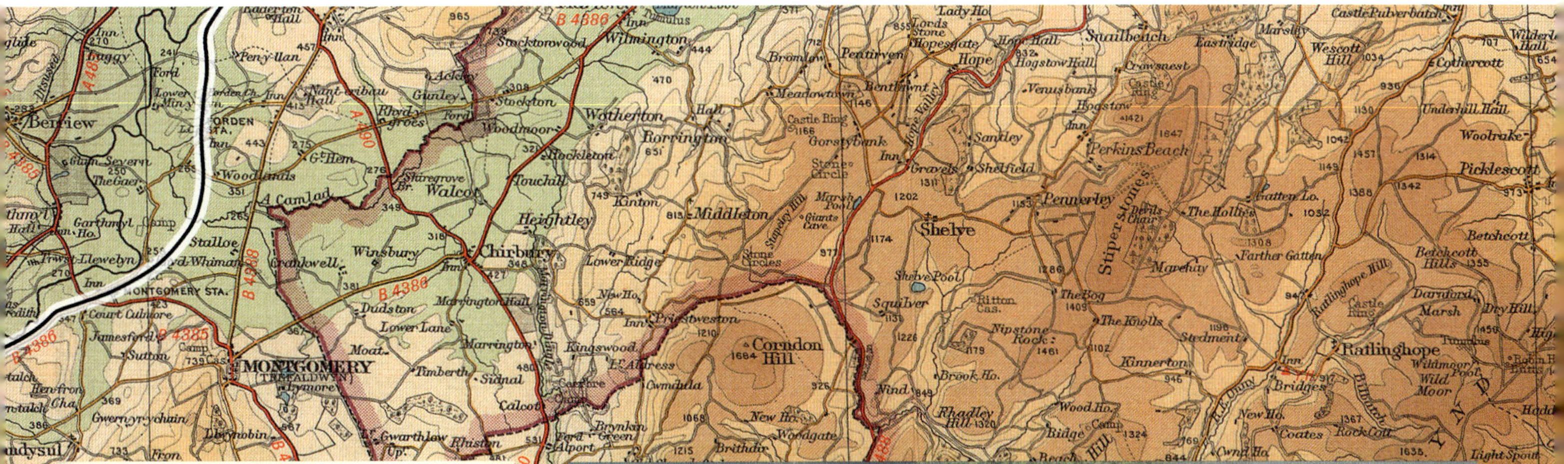

Introduced by the GWR in 1927, 'The Cambrian Coast Express' served holiday destinations along the West Wales coastline for forty years.

Ex-GWR 'Manor' Class 4-6-0 No. 7802 'Bradley Manor' climbs up to Talerddig Summit with the westbound 'Cambrian Coast Express' charter train in 2006. Sadly, this scene will not be repeated until steam locomotives are fitted with the new ERTMS equipment.

ROUTE DESCRIPTION
Dovey Junction to Fairbourne

Crossing over the Afon Dyfi from Dovey Junction, the railway sets off on its dramatic coastal route in the shadow of the Cambrian Mountains, passing through a series of four rock-cut tunnels before reaching the attractive harbour village of Aberdovey. Turning north, the railway heads behind a three-mile stretch of sand dunes and reclaimed land to arrive at the small resort town of Tywyn – here the narrow gauge Talyllyn Railway (see page 129) sets off from Wharf station on its seven-and-a-quarter-mile scenic route up the Fathew Valley to Nant Gwernol. Still heading north along the coast, the railway crosses reclaimed land and a stretch of saltmarshes at the mouth of the Afon Dysynni before passing through the little-used halt at Tonfanau, home to the Second World War army camp which was used as a transit centre for Ugandan Asian refugees in the early 1970s.

With the Cambrian Mountains ever present to the east, the railway continues along its scenic coastal route through Llwyngwril to Fairbourne. Sandwiched between cliffs, sand dunes and sandy beaches at the mouth of the Mawddach Estuary, the village was founded as a seaside resort by wealthy flour miller Arthur McDougall in the late nineteenth century. The Fairbourne Railway takes visitors through the sand dunes and along a spit at the mouth of the Mawddach, where the miniature steam-hauled trains connect with a passenger ferry to Barmouth.

ROUTE DESCRIPTION
Dovey Junction to Aberystwyth

Aberystwyth-bound trains continue to hug the marshy shoreline and skirt the unique raised peat bog of Cors Fochno – to the north the Cambrian Mountains dominate the ever-widening estuary which at low tide is exposed as a broad stretch of sand – before reaching the coast near the resort village of Borth. Here, close to the long stretch of sandy beach, the imposing station has been restored with a new museum housed in the former booking office.

Avoiding the difficult coastal terrain to the south, the railway follows an inland course from Borth, taking a switchback route through the hills at Llandre and Bow Street before descending to the broad Rheidol Valley and terminating at Aberystwyth. Despite its premier coastal location overlooking the wide sweep of Cardigan Bay, the town's isolation both by road and rail (234½ miles from Paddington) has hampered any serious development since the Victorian tourist boom of the late nineteenth century. With a population of 16,000, the town is home to a university, founded in 1872, and the National Library of Wales, founded in 1907. Here the Vale of Rheidol Light Railway (see page 129) begins its twelve-mile journey to Devil's Bridge.

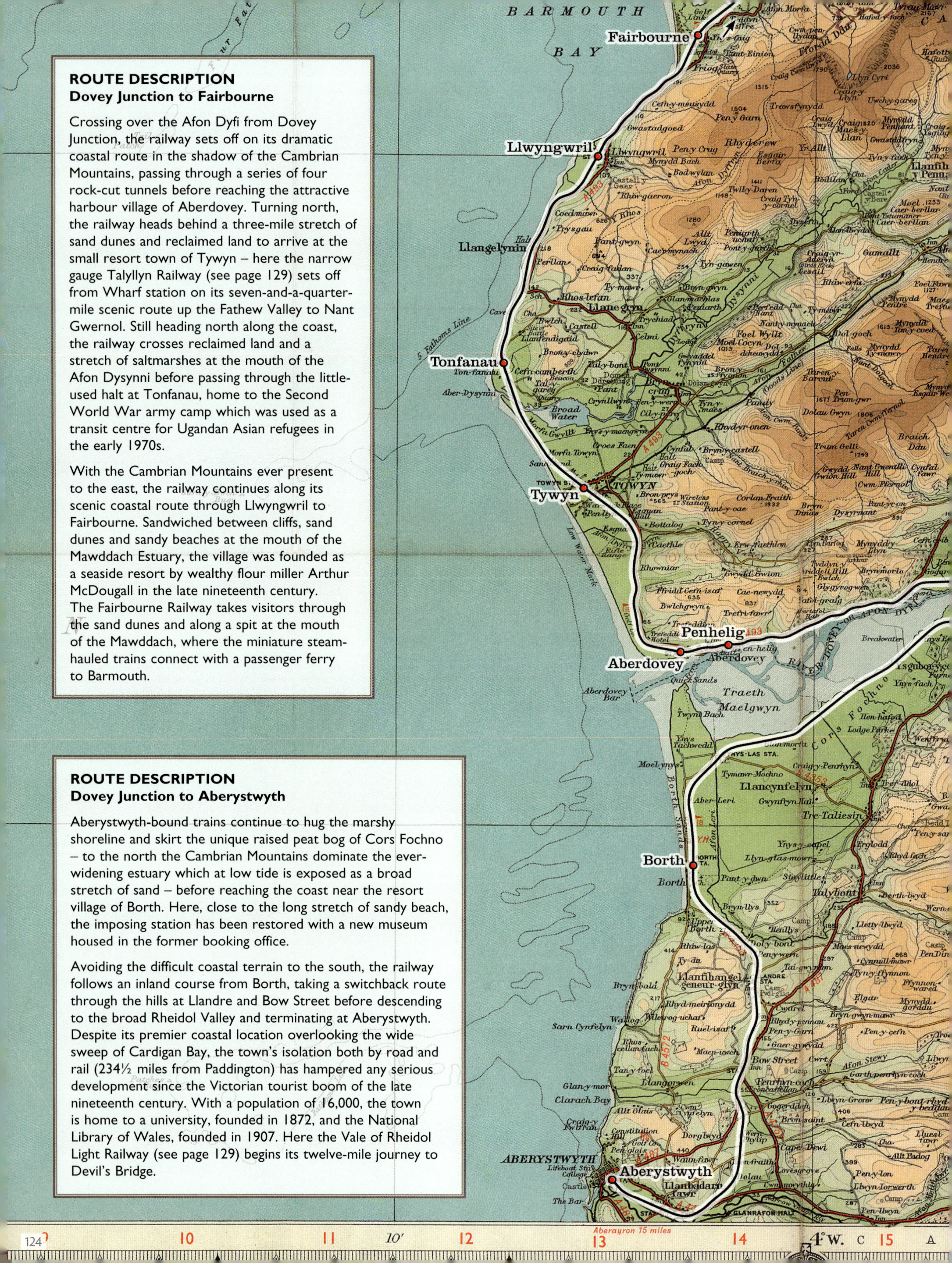

Overlooked by the forested hillside of Foel Caethle, a Class 158 diesel multiple unit leaves Tywyn station for Dovey Junction and Machynlleth. Nearby, Wharf station is the terminus of the Talyllyn Railway (see page 129).

ROUTE DESCRIPTION

Talerddig to Dovey Junction

Descending from the summit, the railway heads down the wooded and narrow valley of the Afon Iaen on gradients as steep as 1 in 52 through the closed station at Llanbrynmair – in steam days, heavy eastbound trains were often double headed up the long drag to the summit. Keeping company with the A470 road, the railway continues westwards down the wooded valley of the Afon Twymyn to Cemmaes Road where it joins the broad and lush valley of the Afon Dyfi (River Dovey) before levelling out to reach the small historic market town of Machynlleth. An important centre for copper mining during the early Bronze Age and a Roman outpost, the town was granted a royal charter by Edward I of England in 1291 to hold a weekly market which, over 700 years later, is still held in the town. Machynlleth station remains an important railway centre where eastbound trains from Aberystwyth and Pwllheli combine and westbound trains from Shrewsbury divide. It is also the main servicing depot for Cambrian Line trains and the control centre for the newly introduced ERTMS system.

Leaving Machynlleth the railway heads southwest down the broad Dyfi Valley to reach the remote outpost of Dovey Junction. Here the line divides, with Pwllheli-bound trains heading across the Dovey Estuary and Aberystwyth-bound trains skirting the southern shore of the estuary through the Dyfi National Nature Reserve. Only accessible on foot, the station is mainly used as an interchange point for passengers travelling between Aberystwyth and Pwllheli – on average around five people a day use the station.

Topped and tailed by Class 37 diesel locos 37602 and 37218, Network Rail's radio survey train heads away from Barmouth to cross the Afon Mawddach on the 121-span Barmouth Bridge.

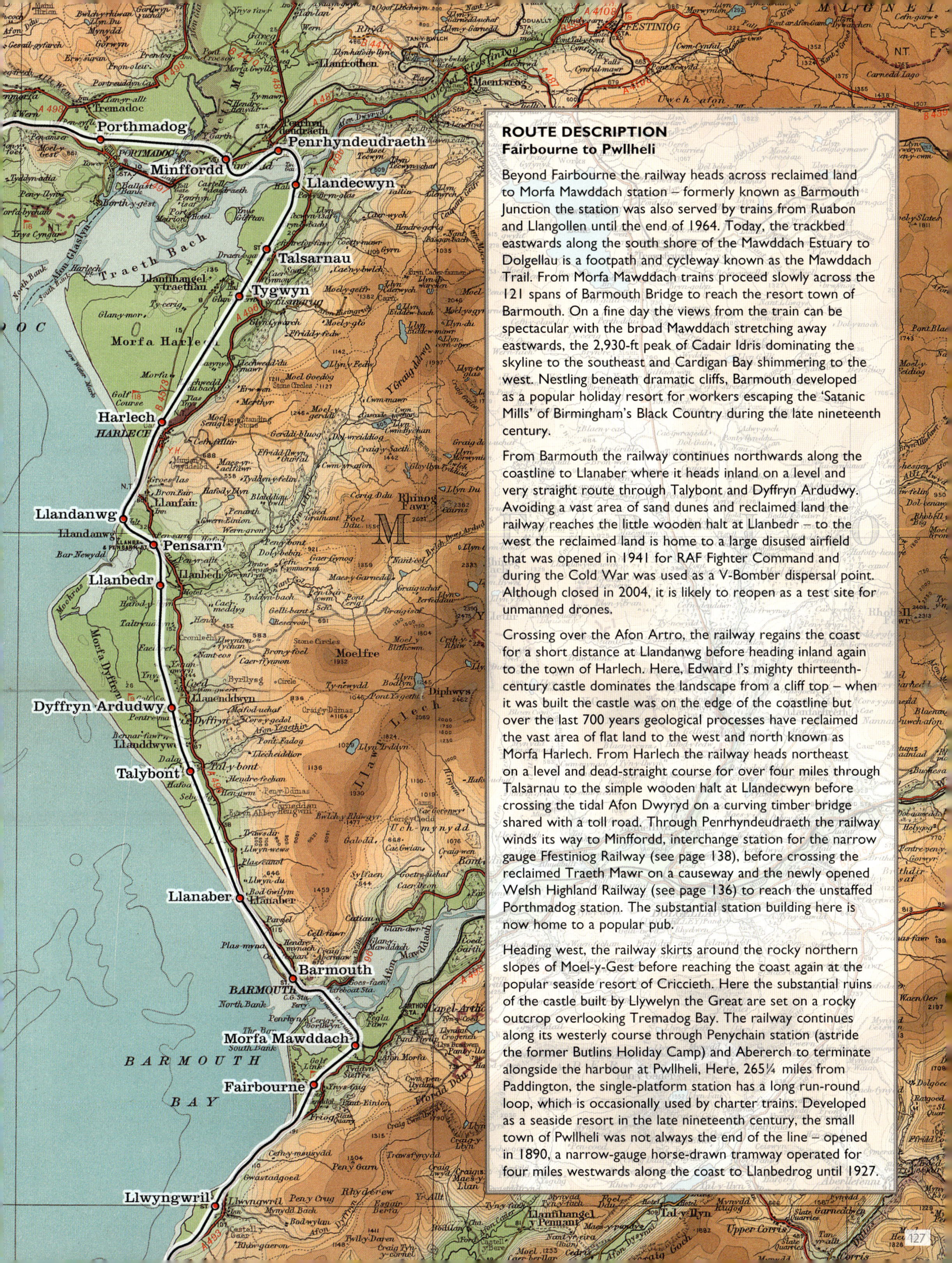

ROUTE DESCRIPTION

Fairbourne to Pwllheli

Beyond Fairbourne the railway heads across reclaimed land to Morfa Mawddach station – formerly known as Barmouth Junction the station was also served by trains from Ruabon and Llangollen until the end of 1964. Today, the trackbed eastwards along the south shore of the Mawddach Estuary to Dolgellau is a footpath and cycleway known as the Mawddach Trail. From Morfa Mawddach trains proceed slowly across the 121 spans of Barmouth Bridge to reach the resort town of Barmouth. On a fine day the views from the train can be spectacular with the broad Mawddach stretching away eastwards, the 2,930-ft peak of Cadair Idris dominating the skyline to the southeast and Cardigan Bay shimmering to the west. Nestling beneath dramatic cliffs, Barmouth developed as a popular holiday resort for workers escaping the 'Satanic Mills' of Birmingham's Black Country during the late nineteenth century.

From Barmouth the railway continues northwards along the coastline to Llanaber where it heads inland on a level and very straight route through Talybont and Dyffryn Ardudwy. Avoiding a vast area of sand dunes and reclaimed land the railway reaches the little wooden halt at Llanbedr – to the west the reclaimed land is home to a large disused airfield that was opened in 1941 for RAF Fighter Command and during the Cold War was used as a V-Bomber dispersal point. Although closed in 2004, it is likely to reopen as a test site for unmanned drones.

Crossing over the Afon Artro, the railway regains the coast for a short distance at Llandanwg before heading inland again to the town of Harlech. Here, Edward I's mighty thirteenth-century castle dominates the landscape from a cliff top – when it was built the castle was on the edge of the coastline but over the last 700 years geological processes have reclaimed the vast area of flat land to the west and north known as Morfa Harlech. From Harlech the railway heads northeast on a level and dead-straight course for over four miles through Talsarnau to the simple wooden halt at Llandecwyn before crossing the tidal Afon Dwyryd on a curving timber bridge shared with a toll road. Through Penrhyndeudraeth the railway winds its way to Minffordd, interchange station for the narrow gauge Ffestiniog Railway (see page 138), before crossing the reclaimed Traeth Mawr on a causeway and the newly opened Welsh Highland Railway (see page 136) to reach the unstaffed Porthmadog station. The substantial station building here is now home to a popular pub.

Heading west, the railway skirts around the rocky northern slopes of Moel-y-Gest before reaching the coast again at the popular seaside resort of Criccieth. Here the substantial ruins of the castle built by Llywelyn the Great are set on a rocky outcrop overlooking Tremadog Bay. The railway continues along its westerly course through Penychain station (astride the former Butlins Holiday Camp) and Abererch to terminate alongside the harbour at Pwllheli. Here, 265¼ miles from Paddington, the single-platform station has a long run-round loop, which is occasionally used by charter trains. Developed as a seaside resort in the late nineteenth century, the small town of Pwllheli was not always the end of the line – opened in 1890, a narrow-gauge horse-drawn tramway operated for four miles westwards along the coast to Llanbedrog until 1927.

NARROW-GAUGE WELSH LINES

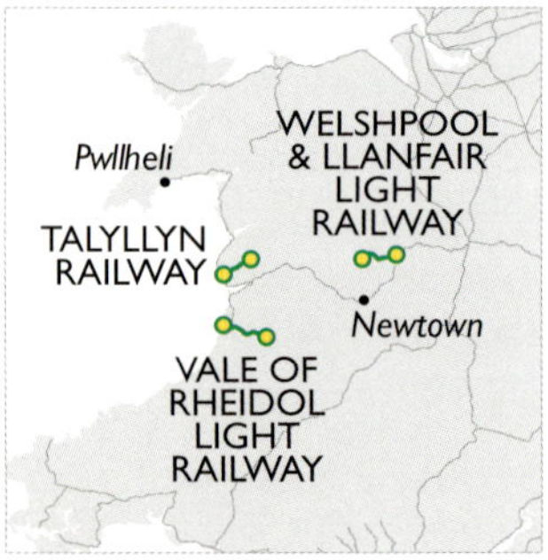

Set within unspoilt Welsh valleys, these three charming narrow-gauge heritage railways all employ historic steam locomotives and coaches to carry passengers on an unforgettable journey through stunning countryside.

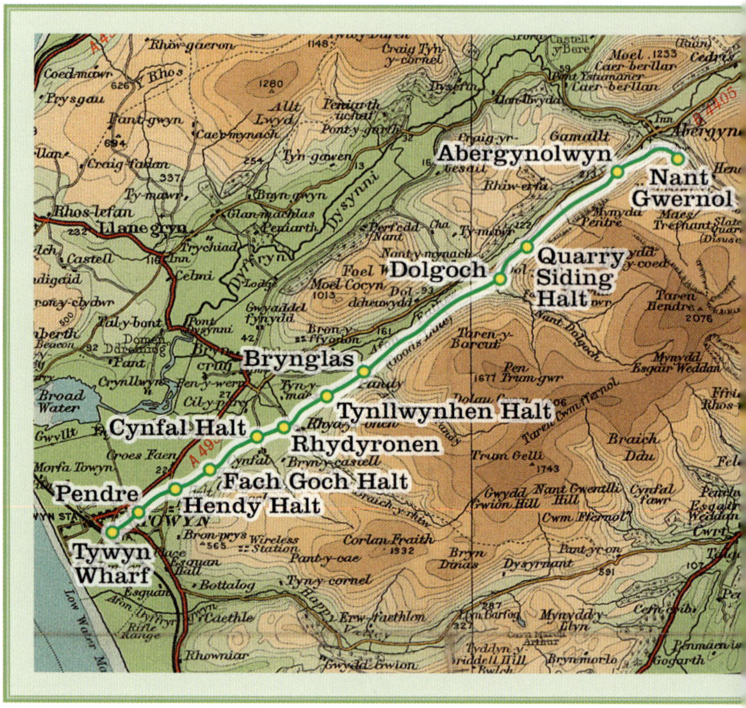

With their sharp curves and contour-hugging routes, narrow-gauge railways were the ideal solution to railway building during the nineteenth century in the mountains of North Wales. Visitors to the country today are treated to journeys on steam-hauled trains on thirteen narrow-gauge lines, of which three are featured here.

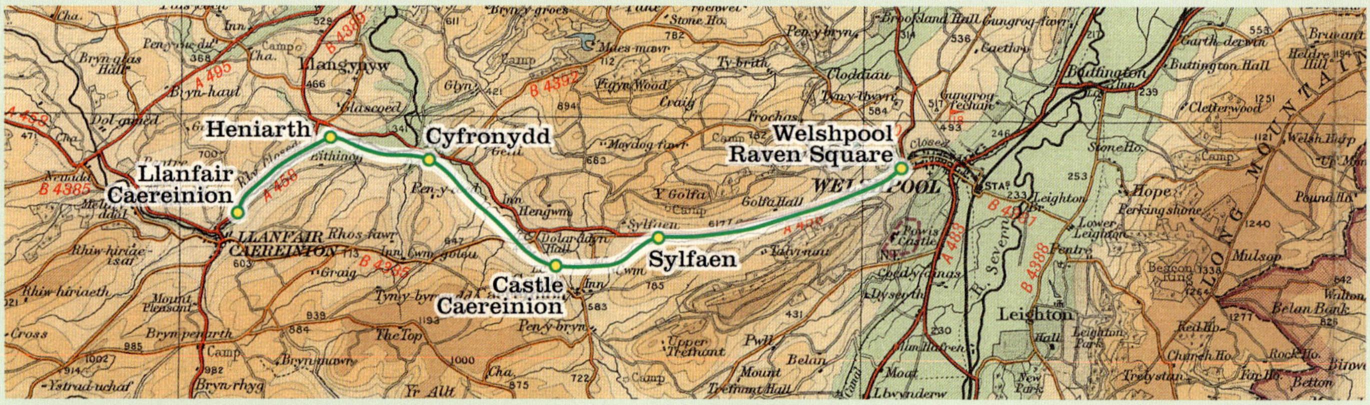

WELSHPOOL & LLANFAIR LIGHT RAILWAY
Welshpool to Llanfair Caereinion

With some of the steepest gradients in Britain, the 2-ft-6-in-gauge Welshpool & Llanfair Light Railway (W&LLR) opened along the Banwy Valley between Llanfair Caereinion and the border town of Welshpool in 1903. Built to link the farms and villages along the valley with the market at Welshpool, it was operated initially by the Cambrian Railways until the company's absorption by the Great Western Railway (GWR) in 1922. Until closure, trains ran through the back lanes of Welshpool to reach the interchange station where it met the standard-gauge Cambrian Railways' main line. Passenger trains ceased in 1931 but the unprofitable railway struggled on through nationalization in 1948 before complete closure in 1956. A preservation group started to run trains again in 1963 and services now terminate at Raven Square on the outskirts of Welshpool.

Still operated by the two original Beyer Peacock locomotives, along with some foreign interlopers and balconied coaches, steam train services along this delightful eight-and-a-half-mile line start their journey at Raven Square terminus in Welshpool, close to the grounds of Powis Castle. First climbing the notorious 1-in-24 Golfa Bank to the 600-ft summit of the line, the railway traces the contours round sharp curves, crossing country lanes on four level crossings, before reaching Cyfronydd. Here the line joins the fast-flowing River Banwy, crossing it near Heniarth and hugging the north bank to end at the little terminus at Llanfair Caereinion.

No. 822 'The Earl' crosses the River Banwy with a goods train for Llanfair Caereinion on the Welshpool & Llanfair Light Railway.

TALYLLYN RAILWAY
Tywyn to Nant Gwernol

The 2-ft-3-in-gauge Talyllyn Railway was opened in 1866 to connect slate quarries at Bryn Eglwys and Nant Gwernol with the standard-gauge railway at the coastal town of Tywyn. Steam operated from the beginning, the seven-and-a-half-mile privately owned railway was saved from closure by a group of preservationists after its owner, Sir Henry Haydn Jones, died in 1950. Reopened in 1951, the Talyllyn became the first preserved railway in the world. Today, passengers are carried up the valley of the Afon Fathew in restored Victorian coaches hauled by diminutive steam locomotives including 1865-built No. 1 'Talyllyn'.

Leaving Tywyn Wharf station and its Narrow Gauge Railway Museum, the railway climbs up through the town, passing the railway's workshops at Pendre before heading off up the tranquil Afon Fathew Valley. Clinging to contours on the southern side of the lush valley, the railway passes small farms nestling in the shadows of the Cambrian Mountains. Threading through oak woodland it reaches Dolgoch where locomotives take water from a mountain stream – passengers can alight here to enjoy a walk to the nearby beauty spot of Dolgoch Falls. Still climbing steadily up the side of the valley the railway reaches Abergynolwyn station, formerly the passenger terminus until the line was extended in 1976 along the route of a mineral line to the Nant Gwernol ravine. While the station here has no road access, walkers can explore tracks and footpaths in the surrounding forest and up the incline of the old slate workings at Bryn Eglwys.

Hauled by No. 1 'Talyllyn', a train of vintage coaches heads up the valley of the Afon Fathew to Nant Gwernol.

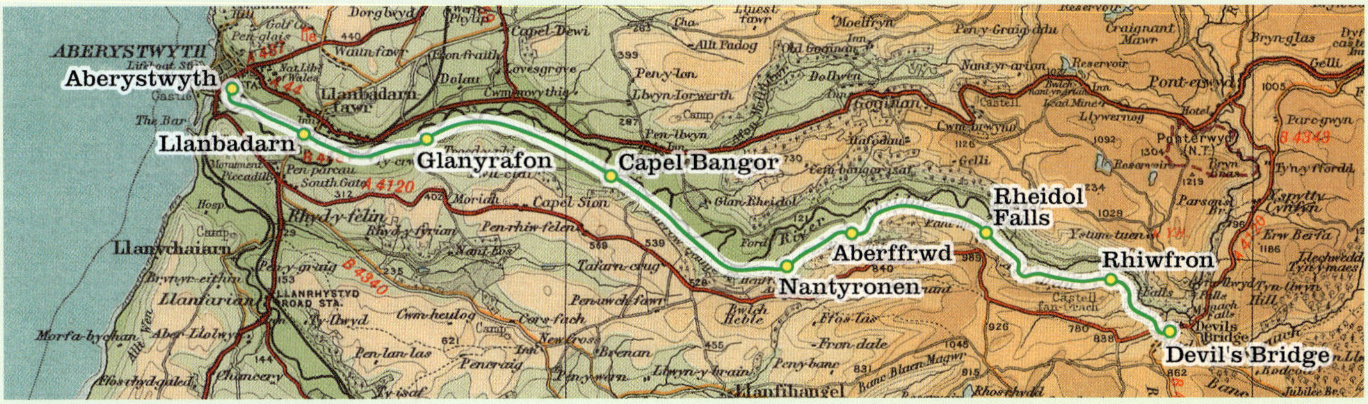

VALE OF RHEIDOL LIGHT RAILWAY
Aberystwyth to Devil's Bridge

The 1-ft-11¾-in gauge Vale of Rheidol Light Railway was opened between the seaside resort of Aberystwyth and Devil's Bridge in 1902. Climbing high into the Rheidol Valley the railway was built to serve lead mines and carry tourists to the beauty spot of Devil's Bridge. The railway was taken over by the Cambrian Railways in 1913 before becoming part of the Great Western Railway in 1922. What little goods traffic there was ceased as early as 1920 and the withdrawal of winter passenger services followed in 1931, leaving a summer-only operation. The railway was nationalized in 1948 and in 1968 became the only steam-hauled railway on British Railways. It was privatized in 1989 and services are currently operated by three 2-6-2 tank locomotives that were built at Swindon in the 1920s and 1930s.

Departing from Aberystwyth's terminus, the railway initially runs beside the standard-gauge line before crossing the Afon Rheidol and heading up the southern slopes of the valley. Following the contours on sharp curves, the railway climbs continuously, hugging the wooded slopes along a ledge through Nantyronen, Aberffrwd and Rheidol Falls with the meandering Rheidol far below. The final dramatic approach on a 1-in-50 gradient to Devil's Bridge passes through a narrow rock cutting. From here, deep in the Cambrian Mountains, eleven and three quarter miles from Aberystwyth and 639 ft above sea level, there are walks to nearby beauty spots such as Jacob's Ladder, the Devil's Punchbowl and Mynach Falls.

Painted in Cambrian Railways colours, 2-6-2T No. 9 'Prince of Wales' approaches Devil's Bridge with a Vale of Rheidol train from Aberystwyth in pre-privatization days.

ROUTE DESCRIPTION

Llandudno Junction to Betws-y-Coed

With railway connections from London, Crewe, Holyhead and Llandudno, Conwy Valley Line trains leave Llandudno Junction station in an easterly direction before striking south away from the main line along the east shore of the tidal Conwy Estuary, its mudflats and sandbanks teeming with wading birds at low tide. On the west bank sits the historic town of Conwy with its thirteenth-century castle built by King Edward I of England. Further to the west lie the 823 square miles of the Snowdonia National Park, its mountains reaching a height of 3,560 ft at Snowdon. Served by a small station and once the site of a Roman river crossing and a medieval ferry, the tiny village of Tal-y-Cafn (meaning 'place opposite the ferry boat') is now the site of the only road bridge across the river between Conwy and Llanrwst. On the opposite bank, one mile south of the village, is the site of the first-century Roman fort of Canovium, where its 500 soldiers were strategically placed to guard this once-important river crossing into the mineral-rich mountains.

Now flanked by woodland on either bank, the railway heads up the floodplain of the Vale of Conwy to Dolgarrog where, on the opposite bank, a redundant aluminium smelter was once powered by hydroelectricity generated from water held in reservoirs high up in the mountains. This was the site of a tragic dam burst in 1925 that killed sixteen people in the village below. Three miles beyond Dolgarrog the railway reaches the small town of Llanrwst (population 3,000), once an important centre for the wool trade and for harp manufacturing and the site of a fifteenth-century fortified manor house and a seventeenth-century narrow stone bridge across the Conwy.

One mile south of Llanrwst the railway makes its only crossing of the River Conwy, entering the Snowdonia National Park before hugging the heavily forested west bank to the village of Betws-y-Coed. Located at an important junction of river valleys and roads, the village was once a major staging post for the horse-drawn Irish Mail until the coming of the railways. The village was soon transformed into a popular resort for tourists, anglers and artists – today Victorian hotels, boarding houses and gift shops line its main street while the Conwy Valley Railway Museum can be found next to the railway station.

LLANDUDNO JUNCTION TO BLAENAU FFESTINIOG

CONWY VALLEY LINE

This twenty-seven-and-a-quarter-mile single-track railway into the heart of Snowdonia is a fortunate survivor of the 'Beeching Axe'. Running along the unspoilt Conwy and Lledr valleys it ends its journey dramatically through one of the longest railway tunnels in Britain.

First conceived as a narrow-gauge line, the single-track standard-gauge Conway & Llanrwst Railway (CLR) was opened in 1863. It covered the eleven-and-a-quarter-mile distance to the town of Llanrwst from Llandudno Junction, along the east bank of the River Conwy on the Chester to Holyhead main line. The London & North Western Railway (LNWR) operated the line from the start, taking over the CLR in 1867. A three-and-three-quarter-mile extension up the narrowing valley to the village of Betws-y-Coed was opened in 1868. Here, set astride Telford's London to Holyhead Irish Mail coach route, the arrival of the railway brought with it intrepid Victorian tourists eager to discover the scenic delights of the mountains of Snowdonia.

Meanwhile, across the mountains, nine miles to the southwest as the crow flies, the slate-quarrying industry around the town of Blaenau Ffestiniog had been served by the narrow-gauge Ffestiniog Railway (FR) since 1836. Initially operated by gravity and horses, the railway introduced steam locomotives in 1865. However, the LNWR saw an opportunity to tap into this lucrative trade and first proposed a narrow-gauge line to link up with the FR at Blaenau – whichever route this took from Betws-y-Coed would have entailed tunnelling through the mountains. In the end, sense prevailed and a twelve-and-a-quarter-mile standard gauge extension of the Conwy Valley Line from Betws-y-Coed to Blaenau was opened via the Lledr Valley in 1879. A triumph of Victorian engineering, the line not only involved the building of four short tunnels, a viaduct and seventeen bridges but also a 3,858-yd single-bore tunnel under the 1,719-ft-high mountain of Moel Dyrnogydd. A permanent station in the town of Blaenau replaced the temporary station at the southern end of the tunnel in 1881.

A two-car diesel multiple unit heads along the east bank of the Conwy Estuary near Glan Conwy with a Llandudno Junction to Blaenau Ffestiniog train.

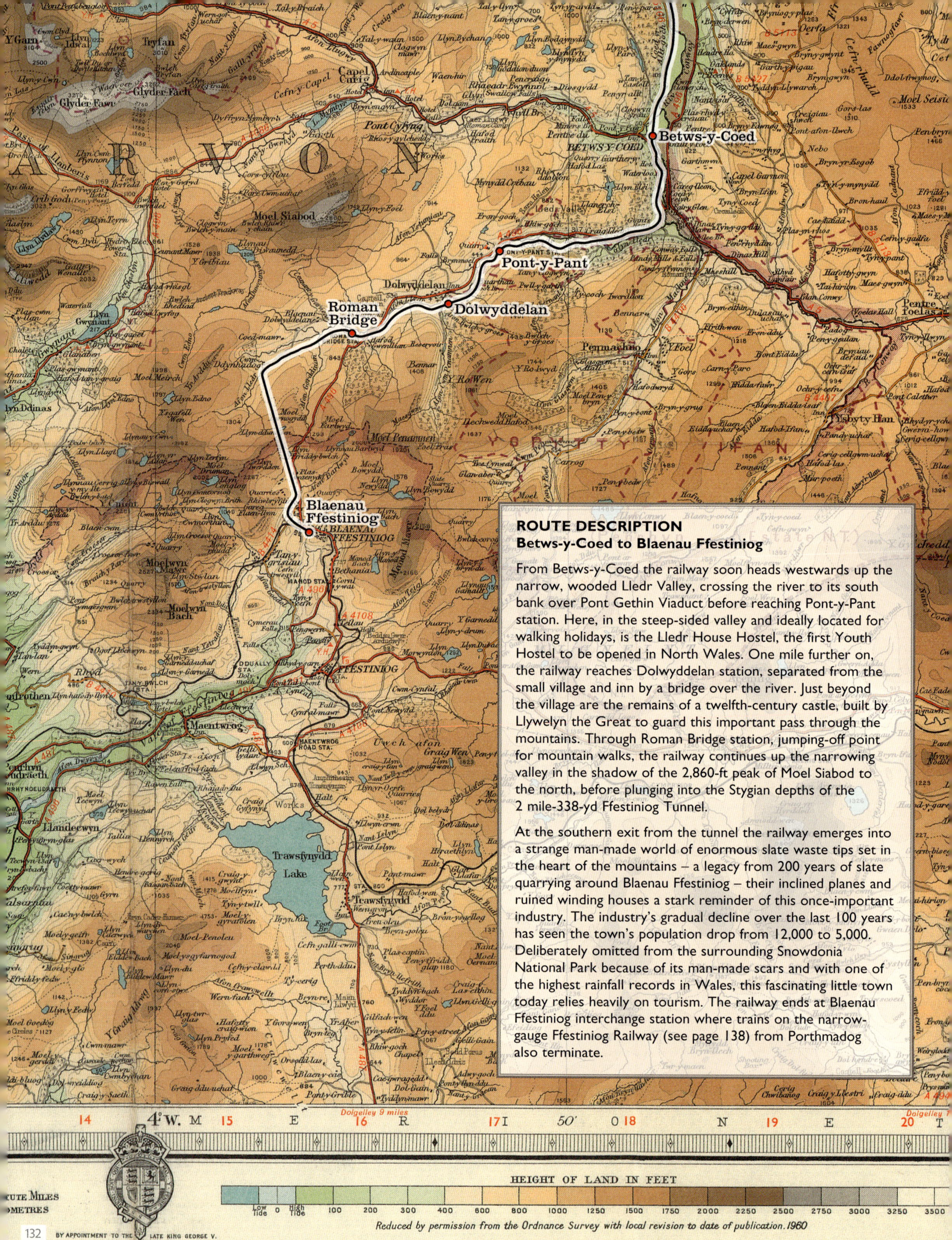

Reduced by permission from the Ordnance Survey with local revision to date of publication. 1960

ROUTE DESCRIPTION

Betws-y-Coed to Blaenau Ffestiniog

From Betws-y-Coed the railway soon heads westwards up the narrow, wooded Lledr Valley, crossing the river to its south bank over Pont Gethin Viaduct before reaching Pont-y-Pant station. Here, in the steep-sided valley and ideally located for walking holidays, is the Lledr House Hostel, the first Youth Hostel to be opened in North Wales. One mile further on, the railway reaches Dolwyddelan station, separated from the small village and inn by a bridge over the river. Just beyond the village are the remains of a twelfth-century castle, built by Llywelyn the Great to guard this important pass through the mountains. Through Roman Bridge station, jumping-off point for mountain walks, the railway continues up the narrowing valley in the shadow of the 2,860-ft peak of Moel Siabod to the north, before plunging into the Stygian depths of the 2 mile-338-yd Ffestiniog Tunnel.

At the southern exit from the tunnel the railway emerges into a strange man-made world of enormous slate waste tips set in the heart of the mountains – a legacy from 200 years of slate quarrying around Blaenau Ffestiniog – their inclined planes and ruined winding houses a stark reminder of this once-important industry. The industry's gradual decline over the last 100 years has seen the town's population drop from 12,000 to 5,000. Deliberately omitted from the surrounding Snowdonia National Park because of its man-made scars and with one of the highest rainfall records in Wales, this fascinating little town today relies heavily on tourism. The railway ends at Blaenau Ffestiniog interchange station where trains on the narrow-gauge Ffestiniog Railway (see page 138) from Porthmadog also terminate.

By 1882, with the arrival of the Great Western Railway's (GWR) line from Bala, Blaenau was well served by railways and this continued until slate traffic had seriously declined in the 1930s. The FR withdrew passenger services to Porthmadog in 1939 and had completely closed by 1946. While the ex-GWR line to Bala lost its passenger service in 1960 and closed completely in 1961, the opening of a nuclear power station at Trawsfynydd in 1964, eight and a half miles south of Blaenau, saw this section reopened. This was probably the saving grace for the Conwy Valley Line, already listed for closure in the 1963 'Beeching Report', as it was now the only route for nuclear waste trains to Sellafield. Fortunately reprieved, this scenic railway still serves the small towns and villages along its route and is especially popular with walkers during the summer months.

As part of the national rail network, train services between Llandudno Junction and Blaenau Ffestiniog are operated by Arriva Trains Wales. With one passing loop at North Llanrwst, the single-track line occasionally sees charter trains in connection with excursions to the Ffestiniog and Welsh Highland railways.

Following river valleys and burrowing under mountains the Conwy Valley Line is perhaps the most scenic standard-gauge railway in Wales.

Headed by Stanier Class '5' 4-6-0 No. 45231, a charter train emerges from the long, single-bore Ffestiniog Tunnel into a man-made landscape of old slate workings.

Overlooked by mountains of old slate workings, Class 67 diesel No. 67002 is seen at the head of a charter train at Blaenau Ffestiniog station while on the right a Ffestiniog Railway Double-Fairlie locomotive waits to leave with a train for Porthmadog.

CAERNARFON TO BLAENAU FFESTINIOG

WELSH HIGHLAND AND FFESTINIOG RAILWAYS

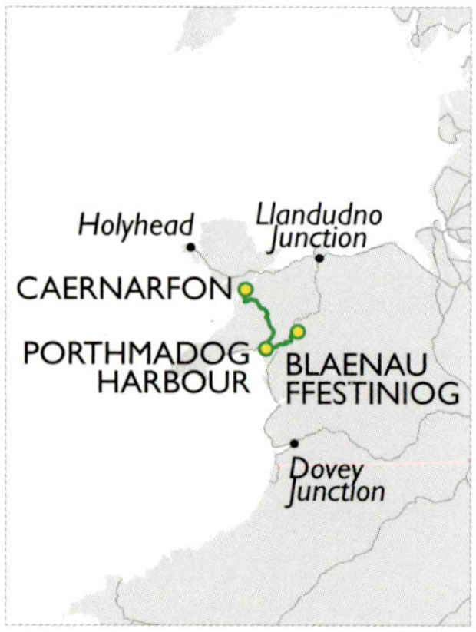

Steam hauled throughout, this thirty-eight-and-a-half-mile narrow-gauge route through the mountains of Snowdonia owes much of its existence to the once-mighty slate industry. Today, it provides tourists with an unforgettable journey through stunning scenery on one of Britain's most spectacular railways.

This is the story of three separate railways that today form the route of one of Britain's most scenic railway journeys. Two of these railways were built to a narrow gauge, which allowed them to hug the contours of the difficult terrain of this mountainous part of North Wales. With their sharp curves and cheaper construction costs they were the forerunners of other narrow-gauge railways built around the British Empire. Chronologically, the 1-ft-11½-in-gauge Ffestiniog Railway (FR) was the first on the scene when it opened as a horse-drawn tramway between slate quarries at Blaenau Ffestiniog and the harbour town of Porthmadog in 1836. Enormous tonnages of slate were transported down the thirteen-and-a-half-mile railway by gravity to Porthmadog for onward shipment around the coast of Britain – the loaded trains were controlled by brakesmen who rode precariously on the wagons while the empties were hauled back to Blaenau by horse.

Steam replaced horse power in 1865 and four years later the FR introduced the first of its powerful double-ended locomotives. Designed by Robert Fairlie, they were in effect two steam locomotives attached back to back with the cab in the centre and supported on two swivelling powered bogies. The first, 'Little Wonder', was such a success that Fairlie went on to stage a series of demonstrations on the railway in 1870. Witnessed by railway engineers from around the world, the trials led to many export orders for his unique locomotives. The FR eventually owned six, and three are still in service on the line today. Passenger services were introduced at the same time as steam power, and the FR went on to order the first iron-framed (as opposed to wooden) bogie coaches in Britain.

Slate traffic peaked in 1897, with 139,000 tons being carried, but by the time passenger services were withdrawn in 1939, this had dropped to 30,000 tons. Independent to the end, the FR struggled through the Second World War only to close completely in 1946. However, spurred on by the success of the world's first railway preservation scheme further down the coast at Tywyn (see page 129), a group of early preservationists set about reopening the railway. The first trains ran across the Cob (a man-made embankment) from Porthmadog Harbour station to Boston Lodge in 1955 and since then the railway has re-opened in stages, finally reaching Blaenau Ffestiniog in 1982. Here a new station was built alongside the terminus of the standard-gauge Conwy Valley Line (see page 131). Meanwhile, a reservoir for a new hydroelectric scheme near Tanygrisiau had already flooded part of the line, necessitating the building of a spiral to gain height at Dduallt and construction of a new tunnel on a higher route along the valley.

The second railway to be built was the standard-gauge Carnarvonshire Railway (CR) which opened between Caernarfon and Afon Wen in 1867. The northern section between

Overlooked by the 3,560-ft peak of Snowdon, a Welsh Highland Railway train heads southwards up the Gwyfrai Valley near Snowdon Ranger station.

Caernarfon and Penygroes was built along the course of the narrow-gauge horse-drawn Nantlle Railway, which had opened in 1828. The Bangor & Carnarvon Railway (B&CR) had already reached Caernarfon from the north by 1852, while the Cambrian Railways coastal route from Machynlleth to Pwllheli already served Afon Wen. Both the CR and the B&CR had been taken over by the London & North Western Railway (LNWR) by 1870. Listed for closure in the 'Beeching Report', the railway between Caernarfon and Afon Wen saw its last train at the end of 1964. Today, the twelve and a half miles of tracked between Caernarfon and Bryncir is a footpath and cycleway known as Lon Eifion Cycleway – as far as Dinas it keeps company with the recently opened Welsh Highland Railway (see page 136).

The third railway to form part of this modern day route started life as the 1-ft-11½-in-gauge North Wales Narrow Gauge Railways Company (NWNGR). Originally seen as an ambitious narrow-gauge system it ended up as a much smaller operation with a four-and-a-quarter-mile line from Dinas Junction, on the LNWR's Caernarfon to Afon Wen standard gauge line, to slate quarries at Bryngwyn opening in 1877. A branch from Tryfan Junction to Rhyd Ddu, on the lower slopes of Snowdon, was opened in 1881. An extension southwards to Beddgelert never materialized and the company, facing mounting losses due to a recession in the slate industry, went into receivership. Meanwhile a completely separate company, the Portmadoc, Beddgelert & South Snowdon Railway (PB&SSR), had plans to reach Beddgelert from the south using part of the route of the horse drawn Croesor Tramway. This was all to no avail and by 1916 the NWNGR had suspended passenger services and work had stopped on building the PB&SSR.

Both railways were saved from extinction in 1921 – with the financial support of the Government and local government they were reformed as the Welsh Highland Railway (WHR). Construction work was started anew and by 1923 trains were able to run for the first time between Dinas Junction and Porthmadog. Operated on a shoestring by the indomitable

Colonel H F Stephens, the WHR lurched from one financial crisis to another, only being saved from closure when the neighbouring Ffestiniog Railway leased it in 1934. Despite some improvements and the injection of FR motive power and carriages, services were painfully slow, with passengers having to change trains at Beddgelert. The end came in 1936 when passenger services ceased and by the following year the railway had closed completely.

Six decades later, after years of legal wrangling with a competing preservation group, the Ffestiniog Railway once again came to the rescue. With substantial funding provided by the Millennium Commission, the Welsh Assembly Government and regional authorities, the FR purchased the old narrow-gauge trackbed and a new Welsh Highland Railway rose like a phoenix from the ashes. The northern section, from Caernarfon to Dinas Junction, follows the trackbed of the former standard-gauge line, which closed in 1964.

One of the most ambitious railway preservation schemes in the world, the reborn WHR was opened in stages between 1997 and 2011, when for the first time in seventy-five years narrow-gauge trains, this time hauled by restored South African Railways Garratt articulated steam locomotives (regauged from 2 ft to 1 ft 11½ in), once again ran through the streets of Porthmadog.

Steam trains on the thirteen-and-a-half-mile Ffestiniog Railway between Porthmadog and Blaenau Ffestiniog operate daily between the end of March and early November. Steam trains on the twenty-five-mile Welsh Highland Railway between Caernarfon and Porthmadog operate daily (except certain Mondays) between the end of March and early November.

Without any doubt this is one of the most remarkable scenic rail journeys in Britain, if not Europe. Unfortunately, even during the peak summer months, it is not possible to make the complete journey from Caernarfon to Blaenau and return in one day – hopefully sensible timetable connections at Porthmadog will eventually allow this to be a 'railway day to remember'.

South of Beddgelert the Welsh Highland Railway enters the spectacular Aberglaslyn Pass on its route to Porthmadog.

WELSH HIGHLAND RAILWAY

Caernarfon to Porthmadog

This epic thirty-eight-and-a-half-mile journey on two narrow-gauge steam railways through the heart of Snowdonia starts in the historic town of Caernarfon. Set on the south shore of the Menai Strait and fringed by Snowdonia to the east, the town has a fascinating history dating back to the Celts and the Romans – the latter subjugating the former in their fort of Segontium. Dominating the town's population of 9,600, its huge castle was built on the waterfront in the thirteenth century by Edward I of England, again with the purpose of subjugating the Welsh. Today, along with the medieval city walls, the castle forms part of a World Heritage Site.

Welsh Highland Railway trains leave Caernarfon in a southerly direction along the trackbed of the former standard-gauge line, keeping company with the Lon Eifion Cycleway for the next three miles through Bontnewydd to Dinas. At Dinas, where the WHR has its workshops, engine and carriage sheds, the railway turns eastwards along the old route of the North Wales Narrow Gauge Railway to join the Gwyfrai Valley, following it deep into the heart of the Snowdonia National Park. After passing through the former slate-mining village of Waunfawr (where the former station master's house is now a hotel) and Betws Garmon the railway skirts the north shore of Llyn Cwellyn to reach Snowdon Ranger station. From here the well-worn but steep path to the 3,560-ft summit of Snowdon stretches away eastwards, while across the lake to the west the Beddgelert Forest envelops the lower slopes of Mynydd Mawr (2,290 ft).

Following the contours up the side of the valley, the WHR winds into Rhyd Ddu station, a passing loop and jumping-off point for an alternative footpath up to the summit of Snowdon. Leaving Rhyd Ddu the railway continues to climb the valley, closely paralleling the A4085 road on a long straight to the summit of the line at Pitt's Head. From here it heads downhill once more along the narrow, wooded valley of the Afon Colwyn to the picturesque village of Beddgelert. Located at the confluence of the Colwyn and Glaslyn rivers and surrounded by mountains, the village was once an important centre for copper mining but today depends heavily on tourism.

Leaving Beddgelert's curving station platform and passing loop, the railway continues southwards to cross the fast-flowing Afon Glaslyn into the narrow Aberglaslyn Pass. Here, on a ledge set on the east bank of the river, the railway passes through a series of three rock-cut tunnels, forming the most spectacular part of its journey. Fed by mountain streams, the Glaslyn can become a roaring torrent during the winter and, until the early nineteenth century, was navigable by small boats from the sea at Porthmadog up to this point.

Breaking out of the Aberglaslyn Pass and leaving the Snowdonia National Park, the railway finally levels out to cross a wide area of flat pastureland reclaimed from the sea by the construction of a long stone embankment across the Traeth Mawr Estuary in 1811 – known as the Cob, this embankment also carries a road and the Ffestiniog Railway (see page 138). After crossing the standard-gauge Cambrian Coast Line on a level diamond crossing (the only mixed-gauge railway crossing in Britain) the WHR nears the end of its journey running behind the back streets of Porthmadog before crossing the main road on the level to terminate at Harbour Station.

CAERNARVON
Caernarfon
Bontnewydd
Dinas
Tryfan Junction
Waunfawr
Plas-y-Nant
Snowdon Ranger
Rhyd Ddu
Meillionen
Beddgelert
Nantmor
Pont Croesor
Penrhyn
Minffordd
Porthmadog Harbour
Boston Lodge Halt
LLANBERIS
CAERNA
PORTMADOC
CRICCIETH
PWLLHELI
TREMADOC BAY
Newborough Warren
Carnedd goch
Moel Hebog
Traeth Bach
Morfa Harlech

FFESTINIOG RAILWAY
Porthmadog to Blaenau Ffestiniog

Named after the landowner and MP William Madocks who completed the Cob embankment in 1811, the coastal town of Porthmadog sprang to prominence in the mid-nineteenth century as a major exporter of slate, carried down to the harbour from inland quarries by several narrow-gauge tramways and the Ffestiniog Railway. The slate-exporting market went into terminal decline in the 1920s and the town's current population of around 4,000 depends heavily on tourism for its livelihoods.

Leaving the restored Porthmadog Harbour station, Ffestiniog Railway trains for Blaenau Ffestiniog immediately head eastwards for nearly one mile across the windy top of the Cob – to the north, across the reclaimed land of Traeth Mawr, loom the Snowdonia mountains while to the south, exposed at low tide, lies a vast expanse of sand that stretches across the estuary of the Afon Glaslyn and its tributary, the Afon Dwyryd. At the end of the Cob is Boston Lodge, site of the railway's historic nineteenth-century workshops, engine and carriage sheds, where the little railway starts its climb inland before crossing the standard-gauge Cambrian Coast Line at Minffordd. Jumping-off point for Clough Williams-Ellis's Italianate village of Portmeirion, the station unusually acts as an interchange for travellers to and from the standard-gauge line below.

On leaving Minffordd the railway continues to ascend the side of the Vale of Ffestiniog and by Penrhyn station has already climbed 160 ft above sea level. With its streets of slate-roofed terraced houses, the village below was once the site of a large munitions factory and is still home to an active granite quarry.

Still climbing up the valley, the railway heads into woodland high above the meandering Afon Dwyryd through Plas Halt (for walkers to the gardens at Plas Tan-y-Bwlch), before looping up through the contours and crossing a cast-iron bridge to halt at Tan-y-Bwlch station. A passing loop and the principle intermediate station on the line, this tranquil spot, 430 ft above sea level, is the jumping-off point for woodland walks.

From Tan-y-Bwlch to Dduallt station the railway is set on a ledge cut in the mountainside affording passengers fine views of the valley far below. After passing Campbell's Platform, a private halt serving Plas y Dduallt, the railway soon dives under the spiral 360-degree loop to climb up to Dduallt station, 540 ft above sea level. This remote spot is the start of a two-and-a-half-mile deviation through the new Moelwyn Tunnel completed in 1978 to take the railway away from a hydroelectric reservoir that had flooded the original trackbed – sections of this can still be seen today below the present route. Following the west bank of Tanygrisiau Reservoir past the pumped storage power station, the railway reaches Tanygrisiau station and its passing place, 669 ft above sea level.

The last fairly level mile to Blaenau Ffestiniog is characterized by the enormous man-made mountains of slate spoil and the remains of inclined planes that dominate the landscape. The *raison d'être* for the Ffestiniog Railway, the output of the slate quarries and mines at Blaenau reached a peak in the late nineteenth century but they now lie silent apart from two, Gloddfa Ganol and Llechwedd, which are tourist attractions. Ffestiniog Railway trains terminate at Blaenau's interchange station, which it shares with the standard-gauge Conwy Valley Line (see page 131) from Llandudno Junction.

A Ffestiniog Railway Double-Fairlie loco prepares to haul a long train of slate wagons with their human cargo across the Cob causeway near Porthmadog Harbour station.

Ffestiniog Railway's new Double-Fairlie loco 'David Lloyd George' hauls a train of vintage carriages up the spiral deviation near Dduallt.

In driving rain in the shadow of Snowdon, the first and last locomotives built by Beyer Peacock to the Garratt patent, head a freight train along the horseshoe curve near Rhyd Dhu on the Welsh Highland Railway.

LLANBERIS TO SNOWDON SUMMIT

SNOWDON MOUNTAIN RAILWAY

Still employing veteran Swiss steam engines, Britain's only 'rack-and-pinion' mountain railway has been carrying tourists to the summit of Snowdon, the highest mountain in England and Wales, for over 115 years.

Invented by Swiss locomotive engineer Roman Abt, the Abt rack system allows steam trains to ascend mountain railways on gradients (up to 1 in 4) that are too steep for adhesion locomotives. In a rack-and-pinion system locomotives are fitted with cogs placed between the driving wheels that mesh with a double rack laid within the running rails. First used on the Harzbahn in Germany in 1885, the Abt system was chosen for Britain's first and only tourist steam mountain railway, which opened from Llanberis to the summit of Snowdon in 1896.

Employing 0-4-2 tank locomotives with sloping boilers, this 2-ft-7½-in-gauge railway climbs for four and a half miles on gradients as steep as 1 in 5.5 to the 3,560-ft summit of the mountain. From here, on clear days, tourists are treated to a magnificent view over Snowdonia and the Irish Sea. Apart from a tragic accident on the railway's opening day when a locomotive became derailed and a passenger died, it has been operating without incident for 115 years. Diesel locomotives took over some of the duties of the ageing steam locomotives in the 1990s while a new summit station was opened in 2009.

Subject to weather conditions, trains on the Snowdon Mountain Railway depart daily from Llanberis from late March to the end of October. From March to early May trains will normally terminate at either Clogwyn or Rocky Valley. Trains may be operated by steam or diesel traction.

By any standards, the hour-long railway journey to the summit of Snowdon is ,weather permitting, certainly the most breathtaking in Britain, if not in Europe.

High above the Llanberis Pass, two ascending steam trains halt at Clogwyn station before resuming their journey up to the summit of Snowdon.

With breathtaking views for the passengers, a diesel locomotive pushes a carriage up the exposed ridge on the final leg of its journey from Llanberis to Summit station.

ROUTE DESCRIPTION

Trains for Snowdon depart from the former slate-mining village of Llanberis, located on the south shore of Llyn Padarn on the northern boundary of the Snowdonia National Park. Home to the National Slate Museum, the Padarn Country Park and the Llanberis Lake Railway, this mainly Welsh-speaking village is overlooked by disused slate quarries that scar the southwest slopes of 3,028-ft Elidyr Fawr – deep below the mountain is the modern Dinorwig pumped-storage hydroelectric power station.

Attracting over six million visitors each year, Snowdonia National Park was established in 1951 and its 838 square miles include the highest mountains in Wales, many of them, including Snowdon, over 3,000 ft above sea level. Formed over 450 million years ago by volcanic activity, the mountains, valleys and lakes have all been shaped by the action of glaciation. Today they have one of the wettest climates in the UK and their unique environment is home to many rare Alpine plant species including the 'Snowdon Lily'.

Ascending from Llanberis terminus, the Snowdon Mountain Railway follows the well-worn Llanberis Footpath to the summit, climbing steadily through Hebron station and Halfway station, where steam locos take on more water, to Rocky Valley and Clogwyn. Here the railway teeters on the edge of an exposed black volcanic ridge. Far below to the east is the Llanberis Pass while, to the west, are the glaciated valleys, or cwms, and their lakes that characterize this impressive mountain range. Clogwyn's spectacular cliffs are popular with experienced rock climbers – Edmund Hillary trained here before his ascent of Everest in 1953 – the many routes that are used to reach the summit can be dangerous or even fatal for inexperienced walkers during the winter.

From Clogwyn, the final approach to the new Summit station is along the steepest and most exposed parts of the railway. Described as 'probably the busiest mountain in Britain', Snowdon's 3,560-ft summit can offer some of the most awe-inspiring views in Britain – England, Scotland, the Isle of Man and the Wicklow Mountains in Ireland can all be seen on a very clear day.

SWANSEA TO PEMBROKE DOCK

WEST WALES LINE

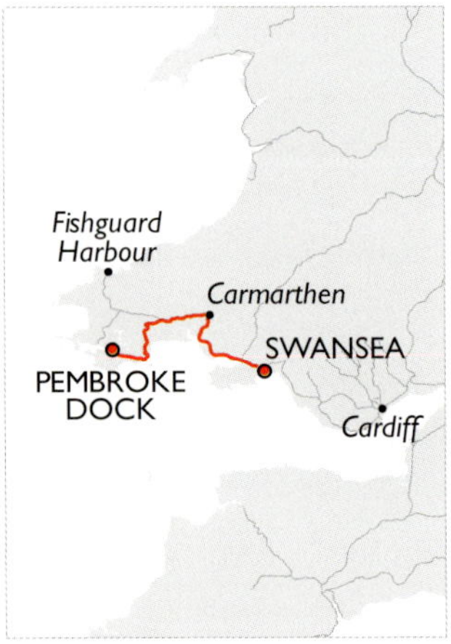

One of Brunel's 'grand designs', this once-major trunk route linking London with Ireland took over sixty years to complete. Surviving the Beeching years fairly intact, the seventy-mile scenic railway journey to Pembroke Dock traces coastline and river estuaries, linking the towns, villages and seaside resorts of Pembrokeshire.

Backed by the Great Western Railway (GWR), the broad-gauge South Wales Railway (SWR) was seen as the western half of a major trunk route linking London to Fishguard, where sea connections could be made to Ireland. Branches from Whitland to Pembroke and from Clarbeston Road to Neyland were also included in this scheme. The GWR had already reached Gloucester via Swindon in 1845 and the 172½-mile extension, financially backed by the GWR, was granted Royal Assent in the same year. In modern parlance the railway was definitely a 'grand design' – surveyed and engineered by Isambard Kingdom Brunel, it followed a very level route down the west shore of the Severn Estuary to Chepstow, Newport and Cardiff. The only challenging gradient to the west of Swansea is the 1-in-52 climb up to Cockett Tunnel between Landore and Gowerton.

The railway opened in stages: Chepstow to Swansea in 1850; Swansea to Carmarthen and Gloucester to Chepstow in 1852; to Haverfordwest in 1854; and to Neyland (for Milford Haven) in 1856.

However, Brunel's superbly engineered but truncated route had one serious drawback – it was built to his unique broad gauge of 7 ft ¼ in, while the majority of the other railways that served the South Wales coalfield were built to the standard gauge of 4 ft 8½ in. The SWR was soon paying dearly for its different gauge, losing out on lucrative coal traffic that was being sent by standard-gauge lines direct to the ports of South Wales for onward transhipment by boat. Missing out on both the Irish and coal traffic the SWR was soon in trouble and was amalgamated with the GWR in 1863. Sense eventually prevailed at the GWR's headquarters at Paddington and the whole route was converted to standard gauge in 1872, twenty years before the end of the broad gauge in southwest England.

Meanwhile another company, the Pembroke & Tenby Railway, had been authorized in 1859 to construct a line between Pembroke Dock and the expanding seaside resort of Tenby. Although this opened in 1864, the railway had no physical connection with the rest of the railway network and remained isolated until 1866 when it was extended northwards from Tenby to Whitland to meet the GWR's line from Carmarthen.

Ex-GWR 'King' Class 4-6-0 No. 6024 heads a long charter train southwards along the east shore of the Tywi Estuary at Ferryside.

Once the extension opened in 1906 to Fishguard Harbour, with its steamer service to Rosslare, the railway west of Swansea assumed far greater importance. Six years later the Swansea District Lines opened, thus allowing Ocean Liner Specials, Irish boat trains and heavy cattle trains to and from Fishguard to avoid the choke point of Swansea and the notorious Cockett Bank gradient. In 1953, during the British Railways' era, 'The Pembroke Coast Express' was introduced. Running non-stop at

more than a mile-a-minute between Paddington and Newport, this steam-hauled restaurant car express served the popular holiday destinations along the south Pembrokeshire coast, calling at all stations between Whitland and Pembroke Dock until it was withdrawn in 1963.

It is interesting to note that while none of southwest Pembrokeshire's railways were listed for closure in the 'Beeching Report', the four-and-a-half-mile section of the original SWR main line from Johnston to Neyland succumbed in 1964.

In modern times the railway west of Swansea has seen some rationalization: from Cockett West to Duffryn West via Loughor Viaduct, the line has been singled (but is now being redoubled) as has the entire branch from Whitland to Pembroke Dock. The only passing loops on the branch are at Tenby and Pembroke Dock stations. Diesel train services between Swansea and Pembroke Dock are operated by Arriva Trains Wales. Freight services, in particular tinplate from the Tata works at Trostre and oil trains from Robeston-Elf near Milford Haven, take the Swansea District Lines to Briton Ferry.

First sweeping along the South Wales coastline, Brunel's visionary route to Ireland via Fishguard continues westwards across the Pembrokeshire Hills, with the branch line to historic Pembroke Dock still serving the seaside resorts and golden sands overlooking the wide sweep of Carmarthen Bay.

Introduced in 1953, the 'Pembroke Coast Express' was the premier train from Paddington to the Pembrokeshire resorts until it was withdrawn ten years later.

Heading for Carmarthen, two Class 153 diesel units catch the early morning sunshine as they cross the Gwendraeth Fach near Kidwelly.

ROUTE DESCRIPTION
Swansea to Carmarthen

Trains for Carmarthen and Pembroke Dock depart from Swansea's railway terminus on the site of the original station opened by the South Wales Railway in 1850. With a population of 170,000, Wales's second-biggest city was an important centre for the copper and tin industries until the late nineteenth century – opened to serve the nearby metals industries, Swansea Docks are now but a shadow of their former importance, with only one of the five docks still in operation for cargo handling.

Initially heading north, the railway for Carmarthen turns west at Landore before tackling the 1-in-52 gradient up to 789-yd Cockett Tunnel. Soon leaving behind the suburbs of Swansea, the line descends through Gowerton, first skirting marshland before crossing the tidal reaches of the River Loughor, a noted salmon and trout river until industrial pollution took its toll in the nineteenth century. Its estuary to the southwest is a vast expanse of sand at low tide and still supports a thriving cockle industry.

After crossing the Loughor, the railway approaches the steel town of Llanelli, which although still an important producer of tin, is probably better known today for its rugby team. Heading westwards on Brunel's level route along the north shore of the Loughor Estuary the railway passes through the small town of Burry Port, which in the eighteenth century was an important seaport for the shipping out of coal and limestone carried down from the valleys on horse-drawn waggonways and a canal. Today, the vast area of sand dunes to the west known as Pembrey Burrows is a popular country park.

Leaving Burry Port the railway turns northwards following an inland route for five miles to Kidwelly – when surveying the South Wales Railway, Brunel took this route to avoid the shifting sands along the coastline. After crossing the River Gwendraeth at Kidwelly the railway rejoins its scenic coastal route, closely following the east bank of the Tywi Estuary to the seaside village of Ferryside, where the vast area of sand exposed at low tide still supports a protected cockle industry. From here the railway continues to hug the east bank of the Tywi for the next seven miles until its arrival at Carmarthen's small terminus station. Located at an important river crossing, Carmarthen is one of the oldest towns in Wales and was once an important strategic outpost for both the Romans and Normans. Formerly an important railway crossroads, the town lost its railway links eastwards to Llandeilo in 1963 and northwards to Aberystwyth in 1965.

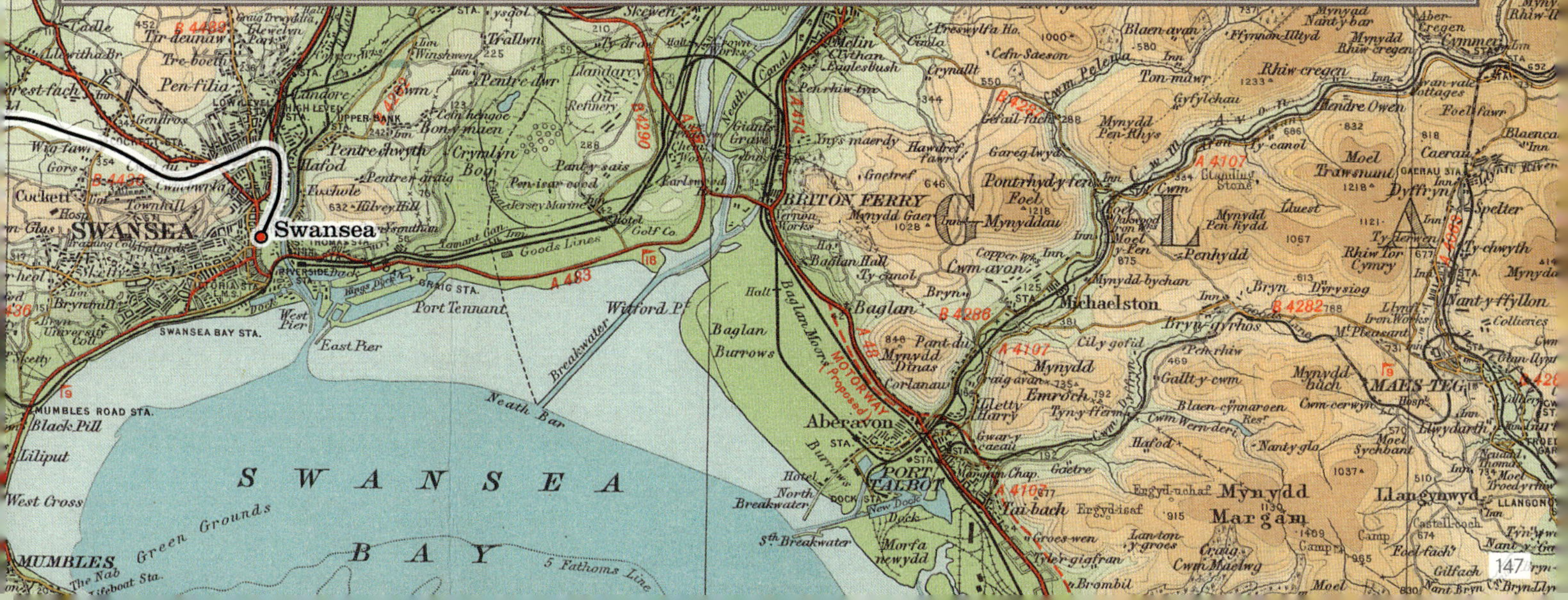

A two-car Class 158 diesel multiple unit speeds westwards into the lush valley of the Afon Cywyn near Sarnau, between Carmarthen and St Clears.

ROUTE DESCRIPTION

Carmarthen to Pembroke Dock

Reversing direction at Carmarthen, trains for Pembroke Dock head westwards for thirteen and a quarter miles along and across small river valleys before following the meandering River Taf to the small town of Whitland. Set in an important dairy farming region, the junction station here was once a busy railway centre, daily despatching milk trains to London until this traffic was lost to road transport in the 1970s.

Leaving the main line to the west of Whitland, the branch line to Pembroke Dock heads off up Lampeter Vale before climbing through hills, skirting the Norman hill town of Narberth and heading south to Kilgetty and Saundersfoot. Set in the Pembrokeshire Coast National Park, Saundersfoot is today a popular seaside resort, its miles of golden sands overlooking the wide sweep of Carmarthen Bay. Back in the nineteenth century the harbour shipped out coal and iron ore, carried there from mines by a narrow-gauge horse-drawn tramway.

Leaving Saundersfoot station (over one mile inland from the coastal resort it serves), the railway descends southwards through the hills to reach the coast at Tenby. With its natural sheltered harbour the town has been an important seaport since the Normans built a stone-walled castle here in the mid-twelfth century. Town walls were added in the thirteenth century and royal financial patronage brought it much prosperity as an important trading port. Captured by Royalists in the English Civil War, the town soon fell into decline after a plague epidemic but later revived as a seaside resort in the early nineteenth century.

The railway's arrival soon led to Tenby becoming one of the most popular resorts in Wales – today, its 5,000 population depends largely on tourism. Blessed with miles of sandy beaches, an equable climate and a fascinating mix of thirteenth- and nineteenth-century architecture the town is also a good jumping-off point for the 186-mile Pembrokeshire Coast Path.

Heading westwards from Tenby, the railway first hugs the sandy shoreline before striking inland to Manorbier station, located a mile north of the village itself. Here the spectacular eleventh-century Norman castle, which dominates the rugged coastline, is a popular venue for wedding ceremonies. Continuing westwards through Lamphey – the site of a ruined thirteenth-century Bishops Palace – the railway arrives at Pembroke station. Dominated by the massive ramparts of a Norman castle set high on a rocky outcrop overlooking the mouth of the River Pembroke, the town was the birthplace of King Henry VII and also an important naval shipbuilding centre until the early nineteenth century.

Heading north on an embankment and through Pembroke Tunnel, the final leg of this fascinating railway journey ends at Pembroke Dock. Once a small fishing village on the south shore of Milford Haven, the town grew up around a new naval dockyard that was established here in 1814 – two Martello towers stand testimony today to the town's importance during the years of conflict with France. Shipbuilding ended in 1922 but it became an RAF base for Sunderland flying boats during the Second World War. Today, only the grid-like pattern of its streets gives a clue to the town's once illustrious military past.

The dramatic Victorian structure which has come to epitomise the character, and survival, of the Settle & Carlisle line – Ribblehead Viaduct, graced by a steam-hauled train heading north towards Ais Gill summit.

NORTHERN ENGLAND

MANCHESTER TO BUXTON

This intriguing branch line climbs from Manchester and Stockport up the valley of the River Goyt to England's second-highest railway summit before reaching the distinguished spa town of Buxton, surrounded by the Peak District National Park.

By the time the Railway Age dawned, Buxton had had a long history as a spa town. Located at over 1,000 ft above sea level, on the western slopes of the Pennines, it was initially developed by the Romans around its geothermal spring and became an increasingly popular tourist attraction with the rise of interest in the therapeutic qualities of mineral waters during the eighteenth and nineteenth centuries.

With a population of just 1,800 in 1861, Buxton was, however, still a small settlement compared to the towns and cities that were in the vanguard of the railway revolution in the 1830s and 1840s, and it was not until 1863 that the first two local railways arrived.

During the Railway Mania of the 1840s a number of schemes were promoted to take railways to Buxton – notably the Manchester, Sheffield and Lincolnshire Railway (MS&L), which planned to extend its Peak Forest Tramway and convert it to a railway, and the Manchester, Buxton, Matlock and Midlands Junction Railway (MBM&MJ), with its grand plan for a trans-Pennine line from Ambergate (on the Midland Railway's Derby-Sheffield main line), through Matlock and Bakewell to Buxton and onwards via Whaley Bridge to Stockport, where it would join the Manchester & Birmingham Railway (M&B). An MBM&MJ Act was passed and building began, only to be brought to a halt by the financial crisis following the Railway

A Manchester to Buxton service passes Combs Reservoir on the approach to Chapel-en-le-Frith in March 2012.

Mania. Work commenced on the MS&L extension, but funds ran out once the line had reached only as far as New Mills, to the north of Buxton.

Three companies were thereafter hoping to reach the spa town – the giant London & North Western Railway (LNWR), formed in 1844 and incorporating the M&B; the equally large Midland, as a backer of the MBM&MJ (which was nevertheless jointly owned by the LNWR) in a bid to reach Manchester, deep in LNWR territory; and the more modest MS&L. Despite the so-called Euston Square Confederacy between the MS&L, LNWR and Midland, which stated that none of them would promote a line to Buxton without the agreement of the others, this eventually fell apart in a battle to reach the spa town.

The potentially strategic MBM&MJ route was blocked and the Midland instead reached Buxton in 1863 by means of a branch line from Miller's Dale, on its main line from Derby through Matlock to New Mills (with running powers thence over the MS&L to Manchester). Just two weeks later, the LNWR opened its steeply graded branch line from Stockport on the Manchester-Crewe main line, passing through New Mills (on the south side of the valley of the River Goyt, opposite the Midland to the north), Whaley Bridge and Chapel-en-le-Frith. Ironically, the two companies' stations at Buxton were sited side by side and were of similar external appearance.

Unsurprisingly, the arrival of these railways stimulated the growth of the town – its Pavilion and gardens were completed in 1875 and Buxton enjoyed increasing popularity as a residential location for businessmen working in Manchester. The town's population rose to over 6,000 by 1881 and the LNWR's presence in the town was then strengthened by the opening of the line from Ashbourne to the south in 1899, but – despite the early aspirations of the MBM&MJ – Buxton was fated never to find itself on a strategic through route.

Passenger services on all three routes converging on Buxton survived until the withdrawal of trains from Ashbourne in 1954, although ramblers' excursions continued until 1963. The 'Beeching Report' of the same year proposed withdrawal of all passenger services to Buxton (over the former LNWR and Midland routes), plus the latter's associated Manchester-Sheffield ('Hope Valley') main line – but only withdrawal of stopping trains on the nearby former Midland main line from Derby via Matlock and Bakewell to Manchester. In practice, both the LNWR route to Buxton and the Hope Valley line (see pages 156–61) were reprieved by the Government, but Buxton's Midland Railway station closed in 1967, as did the through line from Matlock to Chinley in 1968 – a 'post-Beeching' closure echoed by the loss of passenger services over the main Manchester-Woodhead-Sheffield line in 1970.

Northern operates today's diesel rail services from Manchester and Stockport to Buxton hourly throughout the day – an important commuter link as well as a sustainable means of access to the nearby Peak District National Park, whose territory encloses the town to the east, south and west.

The start of the journey to Buxton – one of Manchester Piccadilly station's four dramatic, and sympathetically refurbished, arched roofs.

BR Standard Class 4 No. 76079 and ex-LMS 'Black Five' No. 45407 begin the climb from Buxton to Bibbington Summit with the 'Derbyshire Dalesman' rail tour in October 2007.

ROUTE DESCRIPTION

Buxton trains share the four-track electrified main line from Manchester's Piccadilly station to Stockport with a wide range of local and inter-city services, including the 125-mph tilting Pendolino trains to London. The twenty-seven red brick spans of the imposing 597-yd Stockport Viaduct – the largest viaduct in the world when built in 1840 – take the railway across the River Goyt to Stockport station. Half a mile to the south, at Edgeley Junction No. 1, local trains swing east to the commuter station of Hazel Grove, where a chord line opened in 1986 allowing Manchester-Sheffield trains to be routed via Stockport (population 136,000) rather than the previous routing through Reddish and New Mills.

Beyond Hazel Grove, the double-track branch line is diesel only, with no less than seven intermediate stops over the next sixteen and a half miles to Buxton, offering many options for exploration of the countryside on foot and bike. The railway climbs steadily through wooded countryside, crossed by the Macclesfield Canal aqueduct just beyond Middlewood station, with the summits of the Peak District hills increasingly evident to the east.

New Mills (population 10,000) is perhaps unique among small British towns in that no less than three different passenger railways pass through it – along with the River Goyt and the Peak Forest Canal! Between Whaley Bridge and Dove Holes, the gradient is as steep as 1 in 58 – one reason why freight trains serving limestone works south and east of Buxton were diverted away to the more easily graded former Midland route in the 1960s. The passenger line passes through tunnels and steep cuttings, crossing over the Midland line (now a freight-only railway) at two separate points, firstly on a bridge and then above the 1-mile 1,224-yd Dove Holes Tunnel.

Chapel-en-le-Frith was the scene of the line's greatest tragedy in 1957 when the brakes on the locomotive hauling a goods train failed on the northbound descent from Dove Holes, crashing into the rear of a stationary goods train at Chapel-en-le-Frith station. The driver of the runaway stayed in his cab to the end, and he and the guard of the other goods train were both killed – as recorded in The Ballad of John Axon and commemorated on a memorial plaque at Chapel station. A voluntary group, the Friends of Chapel Station, now tend the station gardens and publicize the railway locally.

Beyond Dove Holes is the least known of Britain's highest standard-gauge passenger railway summits – the 1,140-ft Bibbington, which is higher than the famous Beattock and Shap on the West Coast Main Line and exceeded only by three summits in Scotland and Ais Gill on the Settle & Carlisle line. The railway then drops down to Buxton's station – designed by Joseph Paxton, best known for his design of the Crystal Palace in London – conveniently located in the heart of the town.

Buxton (population 21,000) is overlooked by Grinlow Hill (1,441 ft) and Corbar Hill (1,433 ft), and is the highest market town in England. The town – which lies within Derbyshire – is an ideal base for walking, cycling or travelling the Peak District by bus. It also has many classic attractions in its own right, including Buxton Opera House, The Crescent (modelled on Bath's Royal Crescent) and the Devonshire Dome, as well as its Natural Baths and Pump Room.

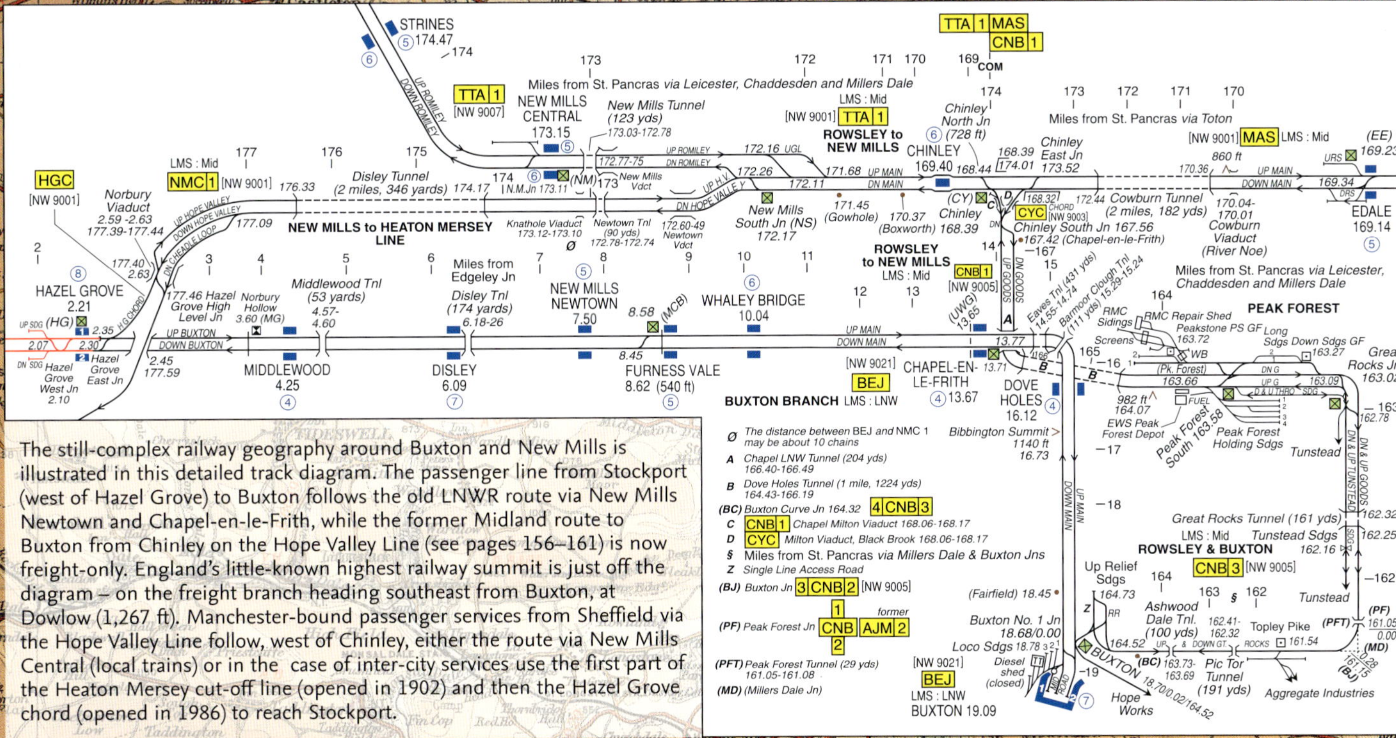

The still-complex railway geography around Buxton and New Mills is illustrated in this detailed track diagram. The passenger line from Stockport (west of Hazel Grove) to Buxton follows the old LNWR route via New Mills Newtown and Chapel-en-le-Frith, while the former Midland route to Buxton from Chinley on the Hope Valley Line (see pages 156–161) is now freight-only. England's little-known highest railway summit is just off the diagram – on the freight branch heading southeast from Buxton, at Dowlow (1,267 ft). Manchester-bound passenger services from Sheffield via the Hope Valley Line follow, west of Chinley, either the route via New Mills Central (local trains) or in the case of inter-city services use the first part of the Heaton Mersey cut-off line (opened in 1902) and then the Hazel Grove chord (opened in 1986) to reach Stockport.

MANCHESTER TO SHEFFIELD

HOPE VALLEY LINE

This late trans-Pennine arrival on the Victorian railway scene survived a 1960s closure threat to become the only main line between Manchester and Sheffield, and is now an attractive gateway to the Peak District National Park.

The first railway from Manchester to Sheffield – via Woodhead – was opened as early as 1845 by the Manchester, Sheffield & Lincolnshire Railway, but it would not be until 1894 that the giant Midland Railway would open its route through the Hope Valley to challenge its northern competitor (soon to be renamed as the Great Central).

The twenty-one-mile line from Dore (on the Midland's Chesterfield-Sheffield main line which had opened in 1870) to Chinley (on its Derby-Matlock-Manchester main line which opened in 1867) involved six years of construction through some arduous mountainous country. Two major tunnels were required – Totley, which at 3 miles 950 yds was the longest in the world at the time, and Cowburn at 2 miles 182 yds – and the route needed a climb of up to 1 in 100 westbound and 1 in 87 eastbound to reach the summit at Edale.

In 1902 the journey north of Chinley to Manchester was speeded up by construction of a nine-mile cut-off line linking New Mills South Junction with Heaton Mersey. Over just four miles in the New Mills area, where it drops from one side of the Goyt Valley to the other, this chord line required no less than two viaducts and two tunnels, including the 2 miles 346 yds of the Disley Tunnel.

A major new traffic generator for the line was established with the 1929 opening of the Hope cement works, close to easily quarried deposits of limestone and thereafter served by a one-and-a-half-mile branch line from Earle's Sidings, west of Hope station. This long predated the creation of the Peak District National Park in 1951, but the railway has always been an important means of minimizing the impact of the works' output on local roads and the amenity of the surrounding area.

By the early 1960s the Hope Valley line was very much overshadowed by its northern 'rival' – the Woodhead line – which had been electrified in 1954 and carried all the main inter-city services between Manchester and Sheffield, as well as a steady procession of heavy coal trains from Yorkshire mines to Lancashire power stations. Hope Valley diesel services were infrequent and often required a change of train at Chinley – and all passenger services were proposed for withdrawal in the 1963 'Beeching Report'. In an ironic twist of fate, however, the line was reprieved, and in 1970 the Woodhead line's passenger services succumbed to 'post-Beeching' closure, leaving the Hope Valley as the main route between Manchester and Sheffield. Trains were diverted from Manchester Central station (which closed, but survives today as an exhibition centre) to Piccadilly and the Hope Valley line's fortunes gradually revived thereafter.

In 1986 a short chord line was opened at Hazel Grove, linking the 1902 cut-off line and the former LNWR Buxton branch (see page 155), thereby allowing Manchester-Sheffield express services to run via the major traffic generator of Stockport. The

Taking the pressure off the roads, 60 094 hauls a heavy limestone train eastwards between Edale and Hope in January 2001, overlooked by the distinctive mountains rising up to Edale Moor.

Hope Valley & High Peak Community Rail Partnership was one of the first of such partnerships in England, established in 1996, and now comprises Derbyshire County Council, the Peak Park Planning Board, train operator Northern, Cheshire East Council and Transport for Greater Manchester. The Partnership's achievements have included securing funding for station improvements, organizing guided walks from Peak District stations and running their renowned 'folk trains' to rural pubs convenient for the railway.

Today's passenger trains are run by three different operators – Manchester-Sheffield expresses by First TransPennine and East Midlands Trains, and local trains by Northern. There are two fast trains per hour, but due to intensive freight operations – other than at weekends – the line only has sufficient capacity for a local train every two hours.

The eye-catching evening exterior of Sheffield (formerly Sheffield Midland) station following refurbishment and enhancement carried out in recent years as part of the city's urban renewal.

Overlooked by Mam Tor, an East Midlands Trains diesel unit approaches Edale station with a special service from Derby to Liverpool in August 2008.

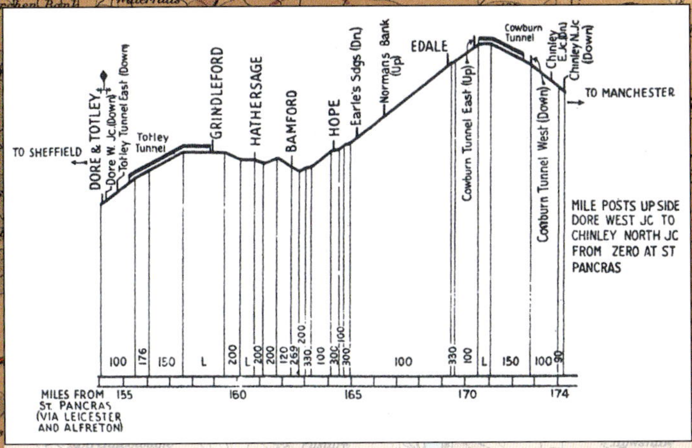

Steep gradients and heavy freight traffic on the Hope Valley Line currently limit the scope for additional Monday-Friday local passenger trains serving the Peak District National Park.

ROUTE DESCRIPTION
Manchester to Hope

The journey through the Hope Valley by express or local train begins in Manchester's fourteen-platform Piccadilly station. Built in 1842, the station was named London Road in 1847 by the Manchester, Sheffield and Lincolnshire Railway and was substantially enlarged in the 1860s and 1880s by the London & North Western Railway, incorporating the four arched roofs of its imposing 'train shed', now a Grade II-listed building. Renamed 'Piccadilly' in 1960 – after the nearby square, and in preparation for the electrification of the line to London – further expansion and refurbishment in the 1980s and 1990s saw platforms lengthened and the entire glass roof replaced. It is now the fourth-busiest station in Britain outside London, used by more than twenty-eight million travellers annually.

Express trains run via Stockport (see Buxton line feature on pages 152–155), but ramblers heading for the Peak District will normally catch a local service following the traditional route via Marple. These trains swing eastwards away from the six-track London route less than a mile south of Piccadilly, then leave the main line to Leeds to run south through the Manchester suburbs of Reddish and Romiley. At Marple, ten miles from Piccadilly, the train is sharing the valley with the River Goyt and the Peak Forest Canal, and the foothills of the Pennines emerge ahead.

New Mills Central retains its station building dating from 1864, featuring a recent mural in the waiting room depicting the Kinder Scout mass trespass of 1932 which ultimately led to the creation of the Peak District National Park. This is one of the few original station buildings on the line to have survived, the majority disappearing during the 1960s when the line's fortunes reached their lowest ebb.

New Mills is a typical former mill town of narrow streets and stone-built cottages, now popular with Manchester commuters. The rivers Goyt and Sett join at a spectacular sandstone gorge called The Torrs, and nearby is a community-owned – but controversial – modern hydro-power scheme, utilizing the same natural resource which allowed early industrialists to develop cotton mills in the eighteenth century.

The railway continues its long climb through, in places, dense woodland – with progressively more spectacular views opening out across the Goyt Valley towards the Buxton branch line opposite – and is joined at New Mills South Junction by the express route from Manchester and Stockport. At Chinley the freight route to Buxton and surrounding aggregate quarries diverges, as did main-line trains to Derby via Bakewell and Matlock until their 1968 withdrawal.

The next five miles of the Hope Valley line is over a gruelling – mostly 1 in 90 – gradient, through the lengthy Cowburn Tunnel to reach the line's 860-ft summit immediately beyond the eastern tunnel mouth. The long down gradient extends through Edale station – one of the most popular stations on the line, sited as it is near to the attractive village of the same name, as well as the start of the Pennine Way long-distance footpath. Numerous local walks take the rambler through beautiful countryside of stone-wall and tree-fringed pasture, rising towards open moorland and the dramatic limestone ridges of Kinder Scout (2,087 ft), Lose Hill (1,562 ft) and Mam Tor (1,696 ft).

ROUTE DESCRIPTION
Hope to Sheffield

Down the Vale of Edale, accompanied by the River Noe and just a minor through road, the railway is then joined by the branch line from the Hope cement works, some half of whose 750,000 tonnes of annual production is moved out by rail. Together, the quarries in the Buxton area and the Hope works generate more than twenty long and heavy trainloads daily on the Hope Valley line – some of more than 1,500 tonnes each. These stretch the limits of the route's capacity, not least as the line has traditional signalling, controlled by seven conventional signal boxes over the nineteen miles from New Mills Central to the junctions at Dore.

Just before Bamford station, after eight miles of descent, the railway begins to climb again, through woodland and away from the valley of the River Derwent to plunge into the lengthy Totley Tunnel, whence the line drops down to Dore and Totley station – on the one short stretch of single track on the route – before joining the 'Midland main line' for the final descent into Sheffield.

The city of Sheffield (population 555,000) underwent massive growth in the Industrial Revolution, as a steel-making centre fed by intensive coal mining to the north, south and east of the city. Production of steel cutlery earned a worldwide reputation – and at various times from 1947 to the present day a named train, 'The Master Cutler', has linked Sheffield and London. While coal mining has ended, steel is still made in Sheffield but the city's economy has moved in new directions, aided by the development of a modern tram network.

Sheffield sits at the confluence of five rivers, surrounded by hills, in what has been described as a natural amphitheatre. More than half of the city is green space, one-third lies within the Peak District National Park and it is claimed to have the highest ratio of trees to people of any city in Europe!

Dominated by the 1,562-ft mass of Lose Hill, a Manchester-Sheffield local nears its stop at Hope in September 2010.

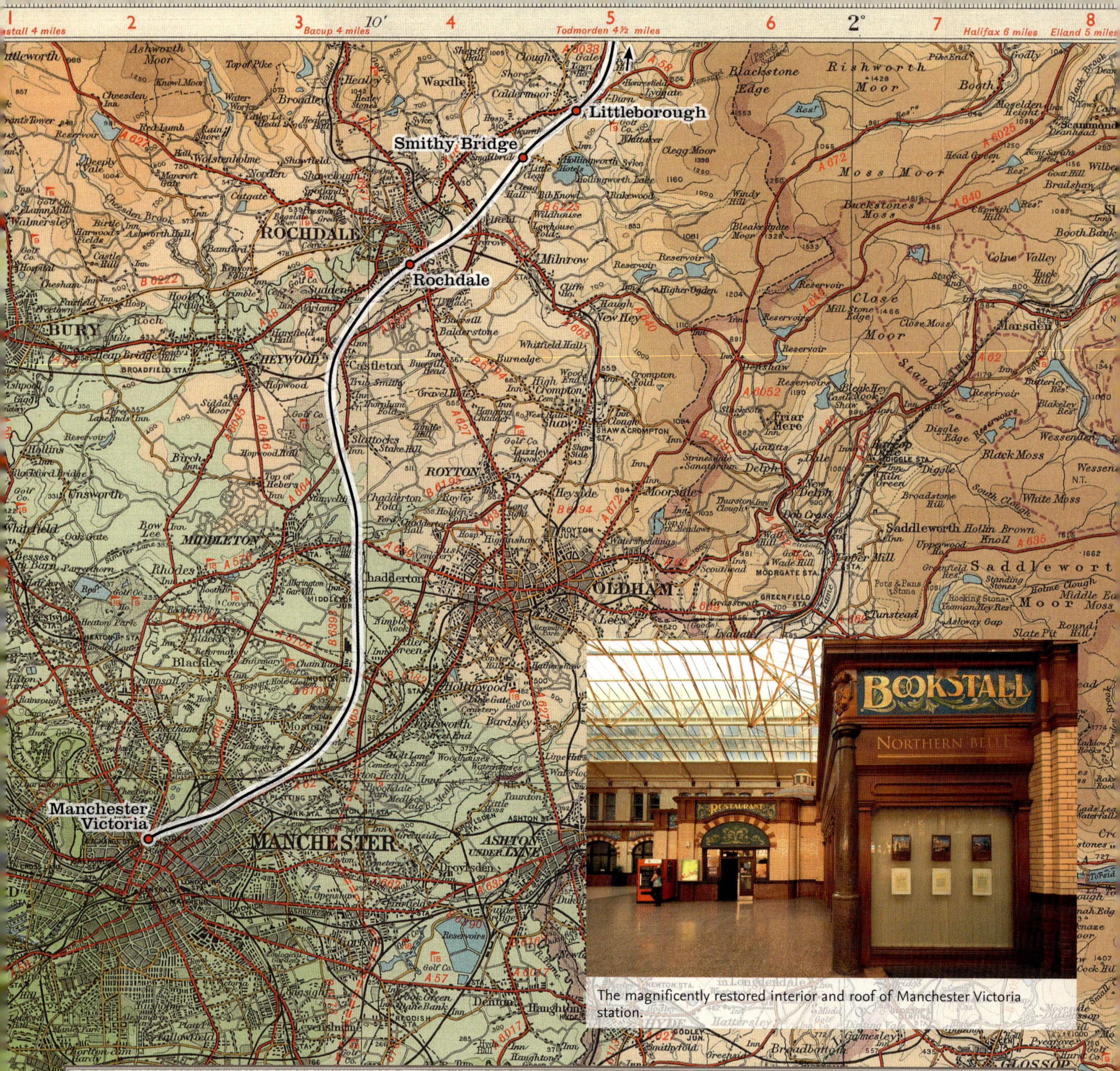

The magnificently restored interior and roof of Manchester Victoria station.

ROUTE DESCRIPTION

Manchester to Littleborough

Manchester Victoria station sees fewer inter-city trains today than prior to 1988 when the 'Windsor Link' was opened between former LYR and LNWR lines west of Victoria, allowing trains from Preston and northeast Lancashire to reach Manchester's main station at Piccadilly. It is still however a grand Victorian station, now sharing some of its space with the Manchester Metrolink line from Bury, converted from BR tracks to light-rail operation in 1992.

Heading away from Victoria on the initially quadruple-track route towards Bradford and Leeds, main line trains quickly encounter a short but very severe gradient of 1 in 39 – and other than a two-mile level stretch just north of Rochdale, virtually all of the seventeen miles to the line's summit is on a continuous gradient.

It is not until beyond Rochdale, heading towards Littleborough, that the railway leaves behind the Greater Manchester conurbation (population 2.2 million, of which the city represents 499,000) and encounters the fields and moorlands of the eastern edge of traditional Lancashire. Most of the line between Manchester and Leeds – both in its rural and urban sections – is under the control of traditional manual signal boxes and a number of the original station buildings survive, including Todmorden, Sowerby Bridge (with its Jubilee Refreshment Rooms, which have become an established stopping-off point on the 'West Yorkshire Rail Ale Trail') and the gem that is Hebden Bridge, with its LYR period signage, canopies, replica gas lamps, working toilets, booking office and station buffet.

MANCHESTER TO BRADFORD

CALDERVALE LINE

This cross-country railway passes through the south Pennines, linking the former powerhouses of the Industrial Revolution – Lancashire and Yorkshire – and is a classic illustration of Victorian railway endeavour, with a succession of viaducts and tunnels carving through difficult and distinctive terrain.

The bleak fells and moorlands of the limestone- and millstone-grit South Pennines have represented a challenge to human movement across northern England since earliest times. Tracks became roads as industry grew and trade in commodities such as cotton, wool, coal and salt developed, but the first real strides in transport were brought about by the opening of the Rochdale Canal in 1804 from Manchester to Sowerby Bridge (where it connected with the Calder and Hebble Navigation), creating the principal artery between Lancashire and Yorkshire. The canal followed the easiest route through the uplands by way of Littleborough and Todmorden, but still required ninety-two locks – and its era of ascendancy was to be relatively short lived, brought to an end by the opening of the parallel railway in 1841.

The prospects for a line linking Manchester and Leeds were being discussed as early as 1825, five years before the opening of the pioneering Liverpool & Manchester Railway. Both towns were expanding at an enormous rate, driven by the rapid development of the Industrial Revolution. While population statistic sources vary in their definition, Manchester's population certainly increased from some 80,000 in 1801 to 150,000 or more in 1831, with that of Leeds increasing from under 100,000 to nearly 200,000 just thirty years later. Both were centres of textile manufacturing, Manchester specializing in cotton and Leeds in woollens, although Manchester's pre-eminence in cotton would eventually be overtaken by its rise as a financial centre dominating the surrounding region.

There was much opposition to the proposed railway from the Rochdale Canal Company and other interests, and it was not until 1836 that the Manchester & Leeds Railway (MLR) received the royal assent. Work began in 1837 under control of the illustrious engineer George Stephenson – and although the most straightforward route across the Pennines was chosen, it did involve difficult terrain and working conditions. It was not until 1841 that the fifty-one-mile route opened throughout from Manchester (Oldham Road) to the North Midland Railway at Normanton, over which it had 'running powers' to Leeds. Many tunnels (including Summit Tunnel, which was briefly the longest in the world) and viaducts were needed to take the

A Selby to Manchester Victoria train passes Eastwood, high in the Calder Valley between Hebden Bridge and Todmorden, in July 2010.

ROUTE DESCRIPTION
Littleborough to Halifax

The final climb from Littleborough to the summit is not severe, at 1 in 330, but much of it is through the 2,885-yd length of the mighty Summit Tunnel. From the tunnel mouth the railway drops at 1 in 182 to the town of Todmorden, where the railway, the road, the River Calder and the Rochdale Canal all share the steep-sided valley. This is classic West Yorkshire valley territory, with large and small settlements stretching from valley floor up the hillsides to fields and rolling tops above. That railways should have carved this and a number of other similar paths through such difficult terrain is succinctly explained by David Joy in *A Regional History of the Railways of Great Britain Volume 8*:

> *The West Riding had all the prerequisites of industrial boom: water transport; good quality stone to build factories and back-to-back houses; coal and fast-flowing streams to provide power; and iron to construct machinery. It is not surprising that together with Lancashire, it led Britain and the world into the Industrial Revolution.*

The descent of the railway along the Calder Valley is almost continuous to beyond Mirfield on the direct Leeds line, but immediately after Milner Row Junction, Bradford trains swing north and have to climb gradients of up to 1 in 91– and there are no less than three viaducts and seven tunnels over the eleven miles between the junction and Bradford Interchange station. The original station at Halifax (population 82,000) was opened in 1855 but was completely rebuilt in the 1880s, opening in full in 1886. Part of this grand and Grade II-listed station building now houses the nursery associated with the nearby Eureka! children's museum.

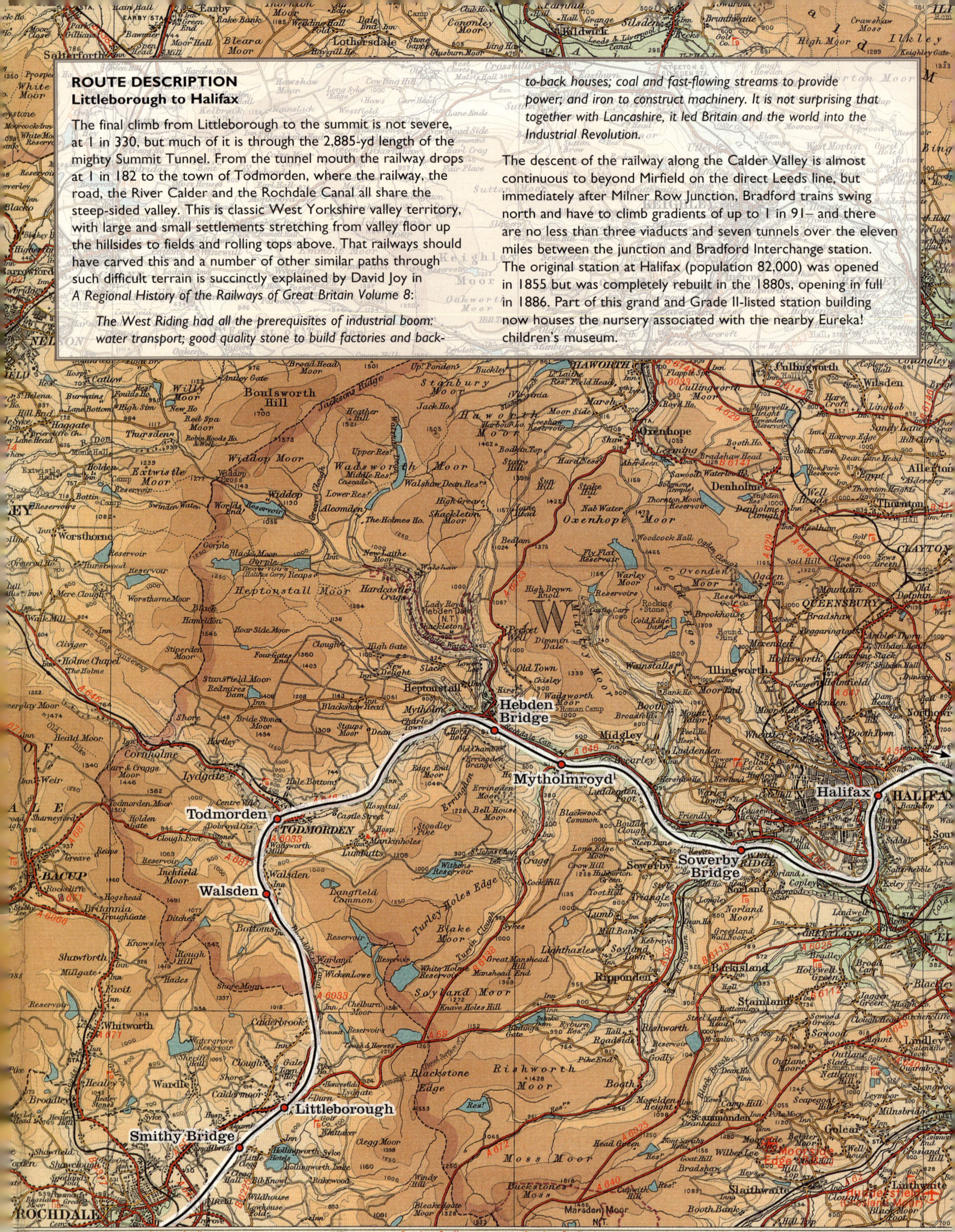

railway across the 537-ft summit, but this has bequeathed a dramatic and attractive route for today's travellers.

The gap between the MLR and the Liverpool & Manchester Railway was filled in 1844 with the opening of a more centrally located station at Manchester Victoria, which was enlarged to broadly its current scale in 1909. Various branch lines and connecting routes were built after the opening of the MLR, and all of these along with the MLR became incorporated as the Lancashire and Yorkshire Railway (LYR) in 1847. The MLR had opened a branch to Halifax off their main line in 1844 and in 1850 this was extended onwards to Bradford by the LYR.

The LYR's route from Manchester to Leeds would prove to be the longest of the three that connected the two great cities, so the company concentrated on business west from Bradford, whence no other line ran across the Pennines. Tourist potential was soon identified and two special viewing coaches were provided for First Class passengers. The *Railway Magazine* wrote of the "Alpine grandeur" of the route. Another distinctive feature of the railway was its pioneering use of pasteboard tickets and iron dating presses, as developed by Thomas Edmonson – a system that did not finally disappear from Britain's national network until 1990.

The early promise of the railway – from a passenger perspective – had deteriorated by the mid- to late-nineteenth century, the *Oxford Companion to British Railway History* (1997) noting of the LYR that:

> *From MLR days the company adopted a cavalier attitude towards its passengers; trains were slow, dirty and unpunctual. Significantly, the dividend was highest when services were worst: 8 3/8 per cent in 1872.*

In 1922 the LYR amalgamated with the London & North Western Railway, which in turn was 'grouped' into the London Midland & Scottish Railway the following year. With the growth of motor transport and tram competition towards each end of the line after the First World War, the lesser branch lines and cross-country routes radiating from the Manchester-Leeds line began disappearing as early as the 1920s.

By the time of the 1963 'Beeching Report' most of the small intermediate stations had already disappeared and the infamous report posed a threat only to some stopping services on the eastern side of the Pennines towards Bradford and Leeds. The connecting Copy Pit route from Hebden Bridge to Burnley had lost its local passenger services in 1961, but the by-now secondary main line along the Calder Valley continued as an important inter-urban link serving a string of otherwise relatively isolated towns and villages on either side of the summit. Passenger services on the original main line to Leeds between Sowerby Bridge, Brighouse and Mirfield were withdrawn in 1970 and the 'Calder Valley' trains from Manchester to Leeds were re-routed via Halifax and Bradford Exchange (later Bradford Interchange), reversing direction there to reach Leeds via Bramley rather than Normanton.

This rail corridor shared in the nationwide revival of rail passenger services from the 1970s onward – through passenger trains were reinstated by British Rail (BR) on the Copy Pit line in 1984 and local passenger services were reintroduced between Sowerby Bridge, Brighouse and Mirfield in 2000. Today, frequent and regular train services are provided by Northern, linking Leeds and Manchester via Mirfield and via Bradford and Halifax, while their Blackpool, Preston and Burnley to York trains traverse the section of the line east of Hall Royd Junction to Leeds.

BR Standard Class 4 No. 76079 and LMS Black Five No. 45407 haul a charter train across Gauxholme Viaduct, west of Todmorden, in March 2009.

ROUTE DESCRIPTION

Halifax to Bradford

The railway then passes through the heart of the West Yorkshire conurbation, a virtually continuous built-up area stretching from Sowerby Bridge in the west to well east of Leeds. There are however glimpses of surviving slices of open countryside as the railway carves its way through the steeply undulating terrain, so characteristic of the region, before finally dropping down at 1 in 50 to Bradford Interchange. One of two main-line termini in the city (the other is Forster Square), Interchange is sited to the south of the original Victorian station at Exchange, which was demolished in the 1970s. Trains continue from here to Leeds, reversing out of the four-platform terminus to reach the 'capital' of West Yorkshire by way of New Pudsey and Bramley.

Bradford (population 294,000) came to prominence during the Industrial Revolution, notably as a centre of textile manufacturing. Although the boom years of woollen manufacture are long gone, the city has been bequeathed fine Victorian architecture in municipal buildings such as the grand City Hall, and in more recent times has diversified into tourism through attractions including the National Media Museum, formerly the National Museum of Photography, Film and Television.

A York-Blackpool North service emerges from the 1,105-yd Beacon Hill Tunnel on the approach to Halifax in November 2007.

HUDDERSFIELD TO BARNSLEY

PENISTONE LINE

This twenty-one-mile single-track route through the foothills of the eastern Pennines has survived several threats of closure, coming to prominence with its early 'community rail partnership' in which local people promote the line – including its pioneering 'jazz' and 'real ale' trains.

When the Huddersfield & Sheffield Junction Railway (HSJR) was opened in 1850, there were no large settlements en route to boost revenue. Furthermore, costly engineering works had been required to cut across the grain of the country, over watersheds and the tributary valleys of the rivers Colne, Dearne and Don – with no less than six tunnels and four viaducts over the first thirteen and a half miles from Huddersfield to Penistone.

At Penistone the line – which eventually became part of the Lancashire & Yorkshire Railway empire – joined the Manchester-Woodhead-Sheffield trans-Pennine main line of the Manchester, Sheffield and Lincolnshire Railway (MS&LR), which had opened in 1845. In 1900, the Penistone line reached its pinnacle as a double-track main line, with the introduction of through trains between Bradford Exchange and Marylebone over the new London extension of the Great Central Railway (whose name was changed from MS&LR in 1897), which lasted until 1960. The Penistone-Huddersfield section faced closure in the 1963 'Beeching Report', but was reprieved, as was its short branch line to Clayton West. In practice it was the former Great Central main line over Woodhead, along with its eastwards extension from Penistone to Barnsley, which were lost in 'post-Beeching' closures in 1970.

The remaining route was progressively singled and the Clayton West branch closed in 1983, with a further closure threat finally averted when South Yorkshire Passenger Transport Executive agreed to continue funding the Penistone end of the route that year. As part of this deal, the trains which had continued to use the old Great Central main line down the Don Valley to Sheffield since 1970 were diverted via Barnsley, over the route which had been disused for some years but was now refurbished to passenger standards.

Thereafter the line flourished, not least through its pioneering 'community rail partnership' initially led by ex-railwayman, activist and author Paul Salveson. The current Penistone Line Partnership is a voluntary organization which brings together train operator Northern Rail, Network Rail, local authorities and users to raise awareness of the railway in the local community, promote and market it more effectively, develop ideas and arrange funding for local improvements. The hourly Sheffield-Huddersfield train service, calling at all ten stations between Barnsley and Huddersfield, is operated by Northern.

In October 2009, 'Black Fives' Nos. 45407 and 45231 head an excursion towards Huddersfield across the four iron spans of Paddock Viaduct some 70 ft above the River Colne.

A diesel unit heads away from Berry Brow near Huddersfield en route to Barnsley and Sheffield in June 2010; the station was opened in 1989.

ROUTE DESCRIPTION

Few railway passengers in Britain can have as grand a start to their train journey as those approaching Huddersfield station on foot. The station square is fully pedestrianized, allowing the best possible appreciation of the neo-classical station frontage designed by architect James Pigott Pritchett and completed in 1850. This Grade I-listed building was described by the poet, writer and broadcaster John Betjeman as "the finest façade of any such building in the country", while architectural historian Sir Nikolaus Pevsner wrote of "one of the best early railway stations in England".

Immediately beyond the southern end of the platforms at Huddersfield (population 146,000), the railway plunges into the 696-yd Huddersfield Tunnels, the single track for Barnsley, emerging to swing southwards away from the principal Leeds-Manchester main line. The Paddock Viaduct takes the railway high above the River Colne, then shortly after comes the Lockwood Viaduct – one of the largest in Britain with its thirty-two arches towering over the River Holme, downstream from nearby Holmfirth which was the setting for the classic television series *Last of the Summer Wine*.

Huddersfield's superb station frontage.

From Huddersfield the railway climbs almost continuously to Penistone, over gradients as steep as 1 in 96 and through Thurstonland Tunnel, the line's longest at 1,631 yds. Just beyond the tunnel begins the one-and-three-quarter-mile crossing loop which requires both Stocksmoor and Shepley to have twin-platform stations. From the latter, a short and pleasant walk through rolling wooded countryside leads to the Shelley terminus of the Kirklees Light Railway, a 15-in-gauge steam heritage line, which follows the impressive earthworks of the old branch line to Clayton West.

A single track takes the Barnsley and Sheffield trains on to Penistone station which, despite its crossing loop and twin platforms, is a shadow of its former four-platform status prior to 1970, when electrified main line passenger trains and vast coal trains plied the Woodhead route from Sheffield to Manchester. Here the current line swings eastwards through local stations opened in 1989 at Silkstone Common and Dodworth to reach Barnsley, whence the trains continue to Sheffield or main line connections can be made for Wakefield and Leeds.

Barnsley (population 218,000) was formerly a major centre for coal mining – the mines have all gone, but its major (and rail-connected) glass works remains a key local manufacturing employer.

LEEMING BAR TO REDMIRE

WENSLEYDALE RAILWAY

Currently operating over seventeen miles, this is one of the longest heritage railways in Britain – linking the flat fertile lands of the Vale of York, by way of small market towns and villages, with Redmire, high in scenic Wensleydale.

The twenty-two-mile single-track Wensleydale Railway was originally part of a through route from Northallerton on the East Coast Main Line to Garsdale on the Settle & Carlisle Railway. As early as 1845, railway surveyors had descended on the lonely dale, staking out a potential alignment following the grain of local geography to create an east-west route, linking the trunk lines pushing north on either side of the Pennines. However, with the ebb and flow of funding available for rail construction – and the knowledge that this was unlikely to be a highly profitable exercise – it was not until more than thirty years later that the through route was completed.

ROUTE DESCRIPTION

At present the timetabled Wensleydale service begins at Leeming Bar station, close to the A1 road. A passenger chord line is now being planned, permitting direct interchange with East Coast Main Line trains at Northallerton station as well as the operation of tourist charter trains from throughout the national network.

The first section of the route runs west across the rich arable fields of the Vale of York to the historic market town of Bedale. Beyond here cropland gives way to rolling pasture, and the railway reaches tiny well-preserved Finghall station, the delightful rural setting which was used for filming of the television series *All Creatures Great and Small* about the life of Yorkshire-based vet James Herriot.

Through Leyburn – the main market town for lower Wensleydale – and beyond, the railway passes through increasingly wooded country as it gains yet more height along the northern slopes of the valley. The line's 626-ft summit is reached shortly before the terminus at Redmire village, with spectacular views of the dale southwards towards Penhill (1,792 ft) and the Great Whernside massif beyond. Nearby is a wide variety of rural walks and bike rides and the notable attraction of Castle Bolton, one of Britain's best preserved medieval castles, dating back to 1399.

Military trains still run to the sidings by Redmire station and initial plans are being drawn up to extend the Wensleydale Railway some three miles westwards, towards Aysgarth Falls, as the first phase of a long-term plan to reach Hawes and Garsdale, connecting once again with the Settle-Carlisle railway.

K1 No. 62005 heads away from Redmire with a rail tour returning to Middlesbrough in May 2007.

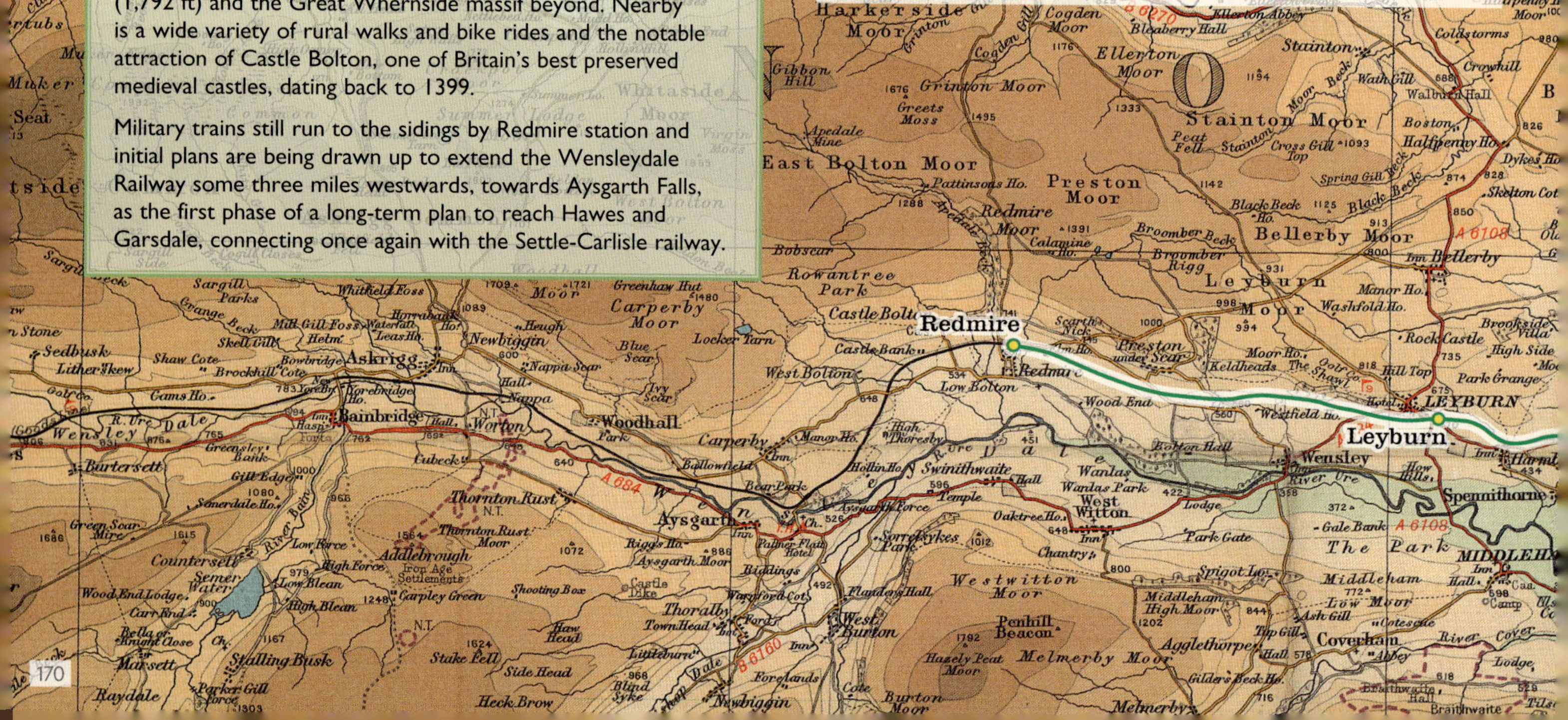

The first section from Northallerton to Leeming Bar opened in 1848, but it took until 1878 for the North Eastern Railway (NER) to reach Hawes, thirty-four miles up Wensleydale. In that same year the Midland Railway opened a short (five-mile) branch from Garsdale, on its trunk route from Settle to Carlisle, to an interchange with the NER at Hawes.

As well as passengers, the railway developed a steady business conveying milk and stone from its hinterland. Despite its rural nature, the line survived the 1923 'Grouping' – when east of Hawes became part of the London & North Eastern Railway and west fell to the London Midland & Scottish Railway – and lasted until the withdrawal of passenger services, first from Northallerton to Hawes in 1954, then from Hawes to Garsdale in 1959; freight then ended west of the large Redmire quarry in 1964.

Rail services to the quarry ceased in 1992, but the prospect of complete closure was averted when the Ministry of Defence funded line refurbishment and installation of loading facilities at Redmire for conveyance of armoured vehicles to and from Catterick Garrison. The Wensleydale Railway Association (WRA) had earlier been formed with the aim of developing heritage passenger services, and in 2000 WRA created a separate operating company, the Wensleydale Railway plc (WRC), which signed a ninety-nine-year lease with Railtrack in 2003.

Seasonal passenger services resumed between Leeming Bar and Leyburn in 2003, followed by re-opening to Redmire in 2004. The regular timetable is operated by first-generation diesel multiple unit trains, with fine views of the line ahead or behind from seats in the end sections of the train. There are bus links from Northallerton station on the East Coast Main Line to Leeming Bar, as well as a vintage bus service from Redmire to Aysgarth Falls and Garsdale.

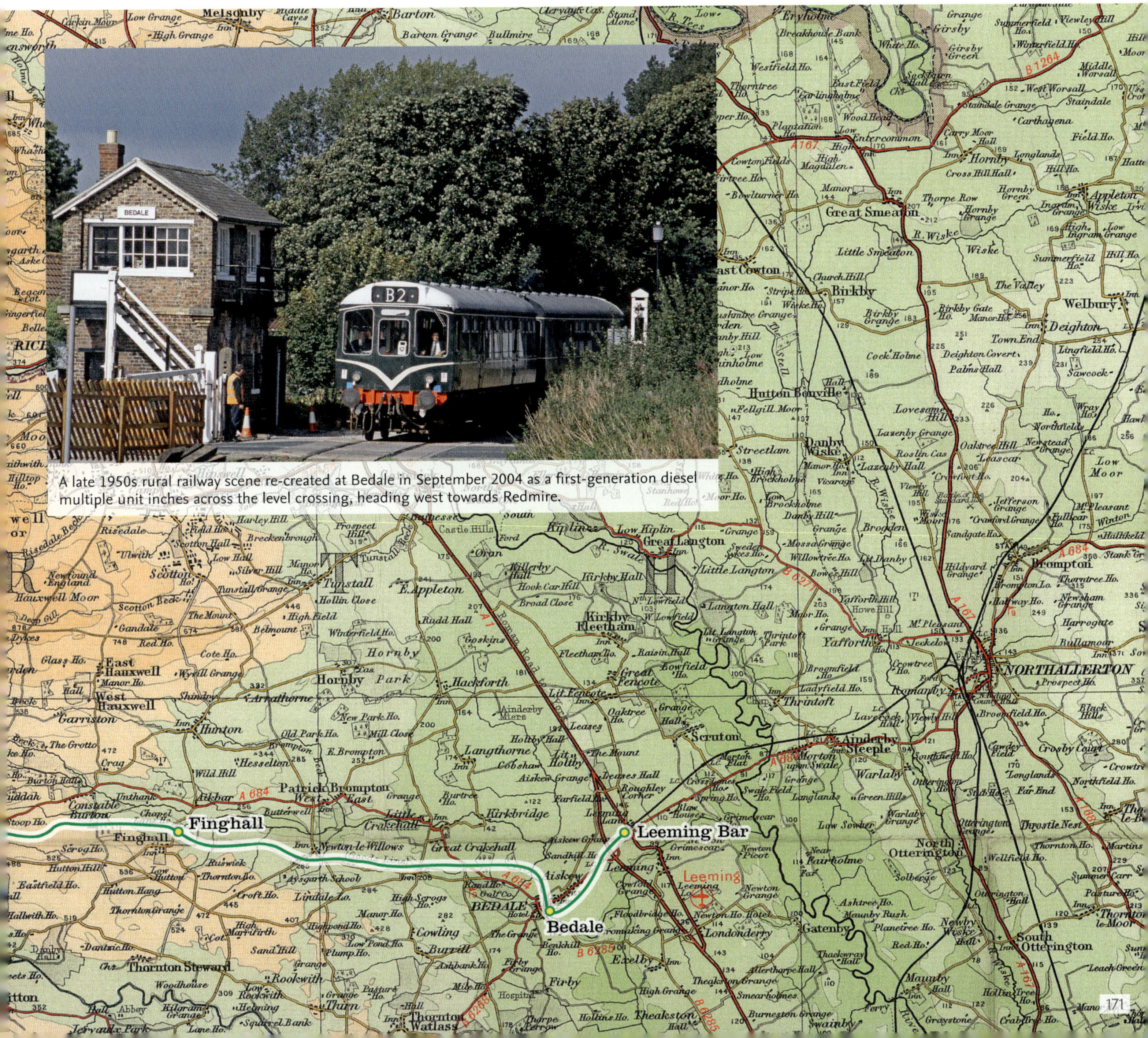

A late 1950s rural railway scene re-created at Bedale in September 2004 as a first-generation diesel multiple unit inches across the level crossing, heading west towards Redmire.

MIDDLESBROUGH TO WHITBY

ESK VALLEY LINE

This thirty-five-mile railway through the North York Moors provides a lifeline for villages along the valley of the River Esk, and now sees steam-hauled services on its eastern section into the attractive coastal town of Whitby.

Whitby was for a century served by four different rail routes, and the Esk Valley line itself has its origins in four separate railways opened at different stages between 1836 and 1864.

The Whitby-Pickering line, south via Grosmont across the North York Moors, was completed by engineer George Stephenson – in isolation from the emerging national rail network – in 1835, and had the distinction of being operated by horse haulage for the first decade of its life. In 1854 this line became part of the North Eastern Railway (NER) empire and was then joined in 1865 at the Grosmont junction by the NER's west-east branch line from Picton on the cross-country route from Northallerton to Eaglescliffe on Tees-side. The surviving seventeen-and-three-quarter-mile section from Grosmont to Battersby forms just over half of today's Middlesbrough-Whitby rail route.

The Middlesbrough-Nunthorpe section was opened in 1854 as part of a line that continued east – with a branch to the town of Guisborough – to join the coastal Whitby, Redcar & Middlesbrough Union Railway. The final link in today's railway chain came in the form of the 1864 opening of the Grosmont-Battersby link.

The North York Moors were an important – but relatively short-lived – Victorian source of ironstone, and the Esk Valley line soon found itself conveying raw materials and finished products from local iron works, as well as stone from quarries served by the railway. The Rosedale Railway, extending eleven miles from Battersby into the high moorlands to tap iron traffic for Tees-side mills, incorporated the remarkable Ingleby Incline – a three-quarter-mile section at gradients of 1 in 5 and 1 in 6, built on the 'self-acting' principle, with a brakeman at the 1,200-ft summit controlling the speed of descending loaded wagons linked by cable to ascending empty wagons. This most unusual railway – whose annual traffic peaked at 561,000 tons in 1873 – survived until 1929, and its formation today makes for a delightful hill path for walkers.

All Whitby's railways began to feel serious competition from the car and the lorry in the 1950s. The through route to Middlesbrough via the coast and Nunthorpe lost its services in 1958, with the remaining stub west from Guisborough to Nunthorpe closing six years later. The 'Beeching Report' brought very bad news for Whitby – all three of its remaining railways were proposed for closure. On 8 March 1965, both the coastal route to Scarborough via Robin Hood's Bay and the Pickering line were closed, leaving the reprieved Esk Valley line as Whitby's sole link with the national rail network. Fortunately the line between Grosmont and Pickering was soon taken over by the volunteer-led North Yorkshire Moors Railway (NYMR) as a successful steam and diesel heritage operation (see pages 176–179).

The Esk Valley line saw much rationalization in the subsequent three decades, with all but three intermediate crossing loops removed and the double-track section from Grosmont to Whitby reduced to a single line. However, the line has benefitted from a community rail partnership – the Esk Valley Railway Development Company set up in 2003 – and was one of the first lines to be designated a pilot 'Community Railway' by the Government in 2005. Although Northern operate only four trains in each direction daily (with a fifth on summer Fridays and Sundays), traffic is growing – not least since the NYMR secured Network Rail authority to extend some of its steam-hauled trains from Pickering to Grosmont onwards to Whitby.

A Whitby to Pickering train headed by 31 128 'Charybdis' skirts the estuary of the River Esk at Whitby in August 2010.

A Class 156 unit pauses at well-maintained Lealholm station deep in the Esk Valley on a Middlesbrough to Whitby service in April 2006.

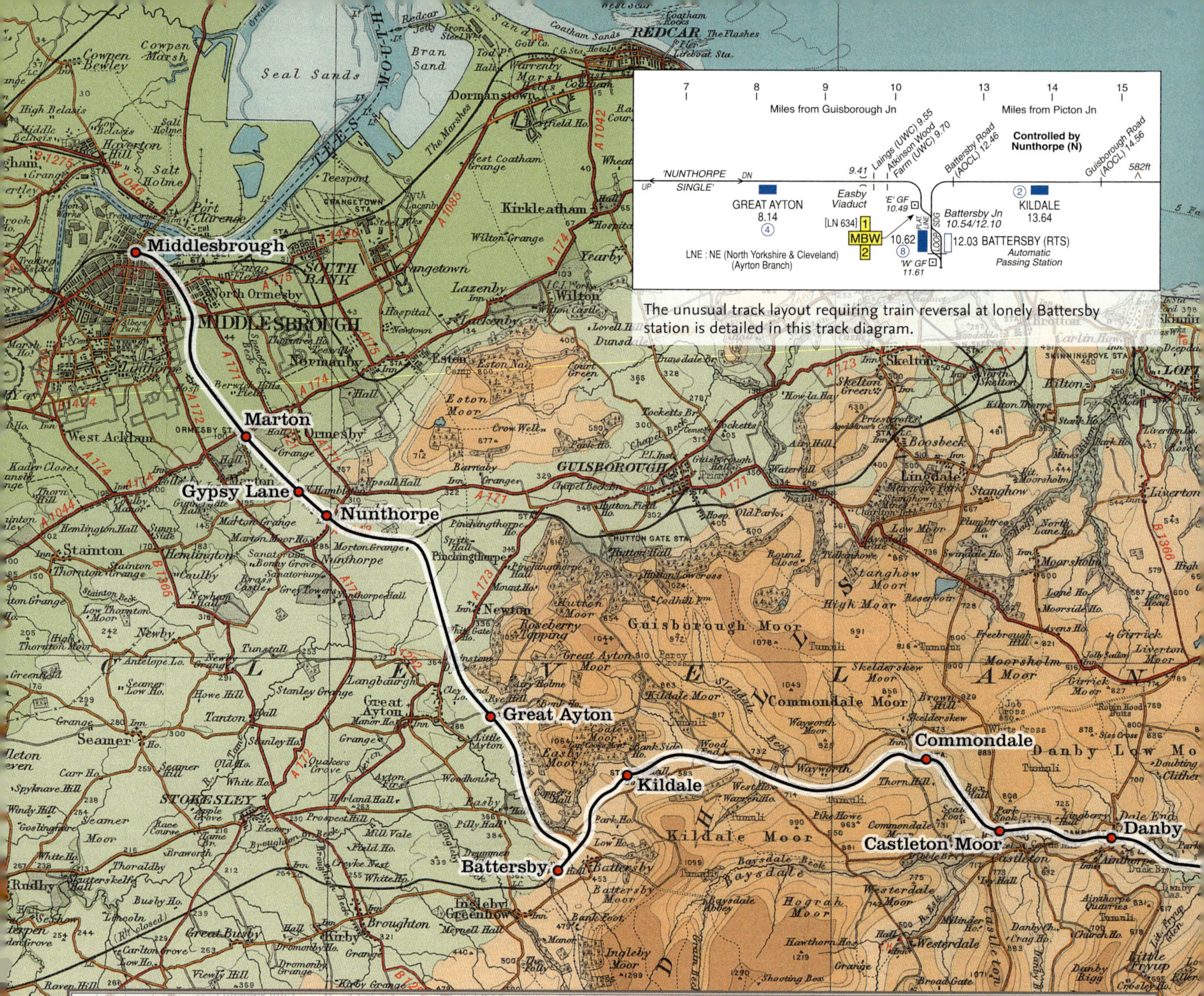

The unusual track layout requiring train reversal at lonely Battersby station is detailed in this track diagram.

ROUTE DESCRIPTION

The industrial town of Middlesbrough (population 142,000) has an imposing station of Gothic design, dating back to 1877. It is also one of twelve stations in eastern England where there survives an original glazed tile map. Dating from the early twentieth century, it shows the entire North Eastern Railway network, stretching from Berwick and Carlisle in the north to southern outposts at Swinton, near Doncaster, and Withernsea on the North Sea coast.

Within sight of the iconic 'transporter bridge' over the River Tees, Whitby trains quickly swing southeast, away from the double-track line to Redcar (famed for its steel works) and Saltburn, on to the single-track branch railway. Marton is Captain Cook's birthplace, and beyond Nunthorpe, the line leaves behind Middlesbrough's suburbs to become a classic country branch railway – with the North York Moors coming into view ahead.

At the former Nunthorpe Junction, the railway curves sharply south, close to the steep escarpment of the Cleveland Hills around Roseberry Topping (1,050 ft) then climbs through the foothills towards Battersby, originally the junction for the line west to Picton and the mineral line to Rosedale. This unusual rural terminus – where the train has to reverse to proceed to Whitby – sits at the end of a minor cul-de-sac road. Other than Battersby Farm, the railway station *is* Battersby!

With Kildale Moor to the south, the line climbs to the watershed between rivers draining south to the Leven and east to the Esk. From Great Ayton to within a few miles of Whitby, this railway corridor serving attractive scattered villages has no parallel 'A' or 'B' roads – and the difficulty of running replacement buses (in particular for schoolchildren) along narrow unclassified roads has been one of the keys to the line's survival. At Castleton Moor the railway joins the valley of the River Esk, then crosses and re-crosses the river on a number of occasions past woods and farm land, with the moors rising to the north and south. Many of the line's original station buildings have survived, including Glaisdale, which is the one place between Battersby and Whitby where trains can cross. The railway has an unusual signalling system – officially known as 'No Signalman Token with Remote Token Stations' – in which train drivers access a signalling cabin on the station platform in order to electrically exchange, with the controlling signaller at Nunthorpe signal box, the physical 'token' required for authority to proceed safely on to each single-line section.

K4 No. 61994 'The Great Marquess' heads a westbound steam-hauled charter away from Glaisdale station and crossing loop in September 2007.

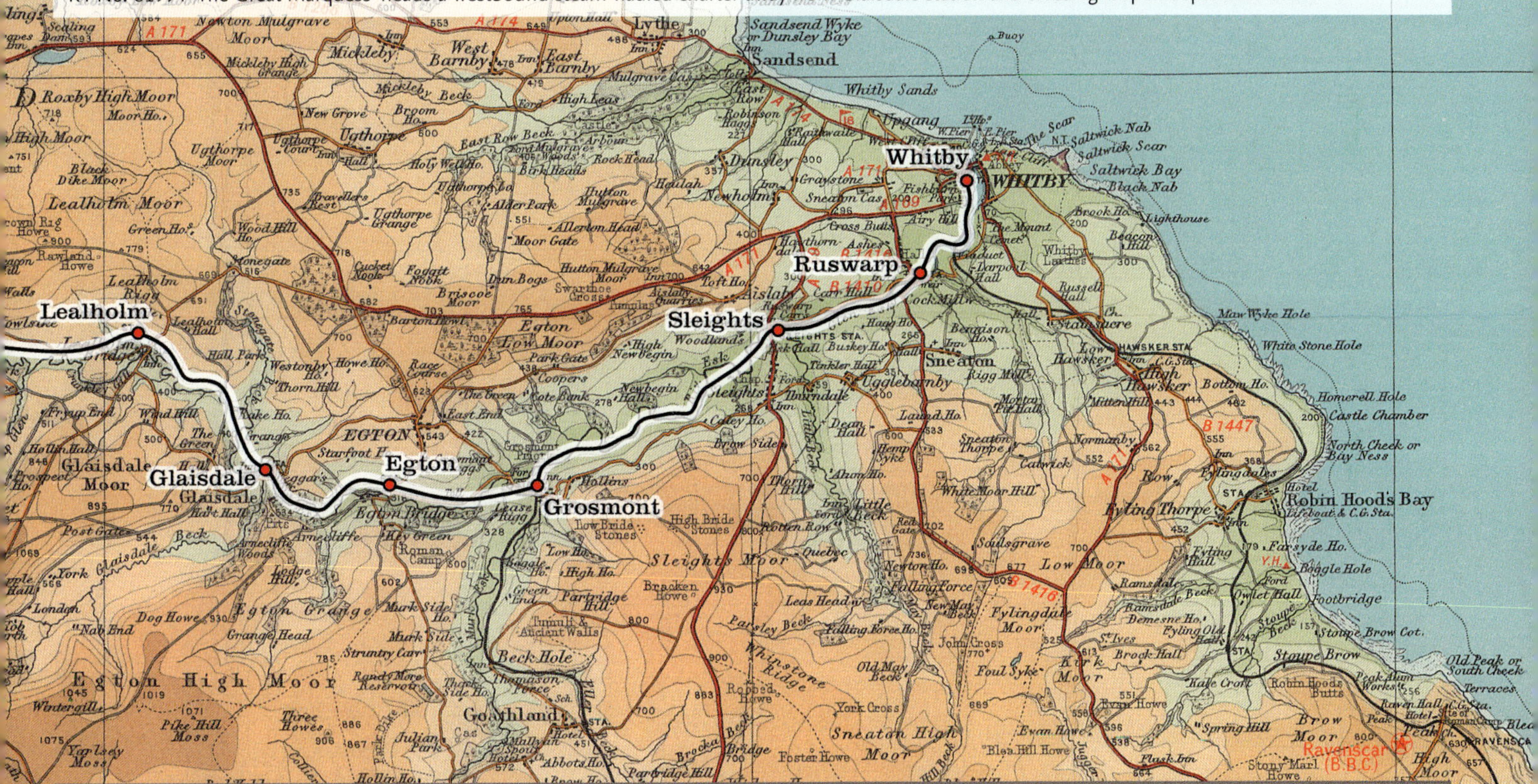

Along the by-now heavily wooded valley of the Esk – and parallel to the Esk Valley Walk – the railway serves the village of Egton Bridge, famed for its annual Gooseberry Show, and crosses high above the river to reach the junction station of Grosmont. There is just a single platform for Esk Valley trains but the NYMR operation (see pages 176–179) at this delightful rural location includes no less than four platforms plus the original station building and extensive sidings serving a locomotive shed and workshop.

The railway, now also used by steam-hauled trains from Pickering, drops down along the steep-sided gorge formed by Esk Dale, still crossing and re-crossing the river as it widens out towards the harbour and basin at Whitby. The terminus station – conveniently located at the heart of the town – has been rationalized down to a single platform, but its attractive buildings survive, as does the historic NER glazed tile map. Whitby (population 14,000) has a long fishing tradition, and is characterized by red pantiled roofs and narrow streets climbing the steep slopes which make for a memorable urban setting. Coastal bus routes run north and south, together with the two railways making Whitby – also known for its associations with Captain Cook, Bram Stoker and whaling – an ideal holiday base from which to explore this beautiful corner of Yorkshire.

PICKERING TO GROSMONT

NORTH YORKSHIRE MOORS RAILWAY

Britain's busiest heritage line takes railway travellers on a dramatic route through the heart of the North York Moors National Park – with selected steam-hauled trains now continuing over Network Rail tracks to the coastal town of Whitby.

Designed by George Stephenson, the Whitby & Pickering Railway was planned to open up trade through the then-busy sea port of Whitby. Its construction included one of the first railway tunnels in the world (at Grosmont) and a rope-worked incline at Beck Hole, between Grosmont and Goathland. Worked by horses in its early years after opening for business in 1836, the railway was taken over by George Hudson's York & North Midland Railway in 1845, re-engineered to allow steam locomotives to be introduced and extended south from Pickering to Malton to connect with the burgeoning national rail network.

In 1854 the York & North Midland Railway became part of the North Eastern Railway, which in 1865 opened a new alignment by passing the Beck Hole incline – whose line of route is now an attractive walking trail. A deeply rural railway linking Gilling (on the cross-country route from Pilmoor on the East Coast Main Line to Malton) and Pickering was opened in 1875 and a branch east from Pickering to Seamer on the York-Malton-Scarborough line was finally completed in 1882.

A busy snow-clad scene at Goathland in December 2005, as S15 No. 825 departs from Goathland with a Pullman dining train to Pickering.

B1 No. 61264 hauls a North Yorkshire Moors Railway service north through the distinctive landscape of Newtondale in October 2007.

Both these marginal routes were early casualties of road competition and had lost their passenger services by 1953 and 1950 respectively. Much worse was to come for Pickering, however, as the 1963 'Beeching Report' proposed withdrawal of all passenger trains through the town. Despite a campaign for reprieve, the line closed between Grosmont and Malton in 1965, but a group of rail enthusiasts saw the potential for a 'preserved line' running through such stunning scenery – and in 1967 the North Yorkshire Moors Railway Preservation Society was formed. Heritage train services, initially operated overwhelmingly by volunteers, began in 1973 between Pickering and Grosmont. Regrettably the section from Pickering to Malton was not protected against development and has been breached by a new road and several houses – but there have long been aspirations to find a way of reconnecting the two towns by rail and allowing a more direct national rail connection to both the NYMR and Whitby itself.

The NYMR is now a major tourist attraction, carrying a third of a million passengers annually – more than any other heritage line in Britain. Its eighteen-mile length is second only to the West Somerset Railway (see page 38–39) and it today employs a substantial team of full-time and part-time staff as well as enjoying ongoing volunteer support. The NYMR's locomotive fleet is very large, comprising some twenty-five steam locomotives (the oldest dating from 1909), sixteen diesel locos, assorted diesel multiple units and scores of passenger coaches and goods wagons. Its seasonal timetable sees up to nine steam- or diesel-hauled trains daily in each direction – with three services continuing over Network Rail tracks from Grosmont to Whitby – although trains do not run on all Sundays.

Q6 No.63395 works a Whitby to Pickering train past Esk Cottages in September 2009.

ROUTE DESCRIPTION

Pickering (population 7,000) is an ancient market town lying just outside the southern boundary of the North York Moors National Park. The station – close to the centre of the town – has been restored to its late 1930s' condition and includes an overall roof in the original style of the 1846 structure, which was removed by British Railways in the 1950s.

Less than a mile north of the station the railway reaches the National Park, on the long climb up the increasingly narrow defile of Newtondale, surrounded by mixed woodland for much of the way. The line is single track, with two intermediate crossing loops, and is controlled by traditional lineside signal boxes operating an electro-mechanical signalling system with semaphore signals.

Lonely Levisham station lies at the end of an unclassified hill road – a suitably peaceful location for the NYMR's Artist in Residence and a 'camping coach' for holidaymakers. Beyond here, the railway twists and turns along the edge of coniferous plantations, with the distinctive shape of a Cold War survivor – the characteristic 'golf balls' or geodesic domes of RAF Fylingdales' Ballistic Missile Early Warning System on nearby Snod Hill – soon coming incongruously into view. Newton Dale Halt has no road access, but is an ideal jumping-off point for walkers exploring the surrounding moors. The line is crossed nearby by the route of the Lyke Wake Walk – famous for its challenge of walking forty miles across the North York Moors from Scarth Wood Moor, near Osmotherley, to Ravenscar on the North Sea coast in less than twenty hours.

Finally out in open country, the railway runs through the highly distinctive road-less trench formed by the sandstone cliff of Northdale Scar to the northwest and Fen Moor to the southeast, before reaching the 532-ft summit on Goathland Moor and dropping down to the village of Goathland. The station here has been beautifully restored – and featured as 'Aidensfield' in the TV series *Heartbeat*, set in the early to mid 1960s, then later as 'Hogsmeade' in the first *Harry Potter* film. Nearby are Mallyan Spout waterfall, the remains of a Roman road and numerous hill walks.

Beyond Goathland, the railway follows the narrowing valleys of the Eller Beck and Murk Esk over a downhill gradient as steep as 1 in 49 to reach the NYMR terminus and headquarters at Grosmont station, where there is interchange with Network Rail's Esk Valley line (see pages 172–175). This location – despite being just a small village – is a hive of railway activity, with four platforms and a traditional tea room. The line's original tunnel, built by George Stephenson for horse haulage of trains, is now used as the walking route from the station to the NYMR's locomotive shed and workshops. At Grosmont both the NYMR and the Esk Valley line are crossed by the famous 'Coast to Coast' walk from St Bees in Cumbria to Robin Hood's Bay on the North Sea coast.

Grosmont
Goathland
Newton Dale Halt
Levisham
Pickering

NEWCASTLE TO BERWICK-UPON-TWEED

EAST COAST MAIN LINE

The final English section of the East Coast Main Line through rural Northumberland affords delightful coastal views – and an unforgettable urban panorama on the final approaches to the historic town of Berwick.

There were many schemes for railways between the Tyne and the Tweed, the coastal strip east of the Cheviots being a natural contender for a line forming part of a route from London to Scotland. During the 'Railway Mania' of the early 1840s there were two principal contenders – the Newcastle & Berwick Railway (backed by the railway baron George Hudson, with the technical support of the railway engineer George Stephenson) and the Northumberland Railway supported by Lord Howick. According to Ken Hoole in *A Regional History of the Railways of Great Britain Volume 4* (1965):

> *His lordship's antagonism to the Newcastle & Berwick was aroused because its line, as surveyed by Stephenson, passed close to his residence [Howick Hall, four miles north of Alnmouth]. Hudson did eventually agree to alter the route, but by this time Lord Howick was far too committed to the opposing line and the two schemes were fought out in Parliament.*

The Northumberland Railway planned to use the 'atmospheric' method of working (as pioneered by Brunel between Exeter and Dawlish, described on page 24), but this did not find favour at Westminster. The Northumberland Railway bill was withdrawn and the Newcastle & Berwick Act received the Royal Assent in 1845. The railway (authorized from a junction on the south side of the River Tyne to join the North British Railway line from Edinburgh "near the ruins of Berwick Castle") opened from Heaton – north of the Tyne – as far as Tweedmouth on the south bank of the Tweed in 1847. Through passengers were conveyed by horse bus between Tweedmouth and Berwick stations for the next three years.

Queen Victoria formally opened a permanent bridge over the Tyne on 28 September 1849 and Her Majesty returned to the North East on 29 August 1850 to open the imposing Newcastle Central station and, later that day, the majestic Royal Border Bridge from Tweedmouth to Berwick. The era of through Anglo-Scottish rail travel by the East Coast had begun.

Four years later the York, Newcastle & Berwick Railway (which had already absorbed the Newcastle & Berwick Railway) became part of the amalgamation that formed the North Eastern Railway (NER). The main line was soon extended from York to join the Great Northern Railway at Doncaster, thereby completing the East Coast Main Line (ECML) from London King's Cross to Edinburgh Waverley. The three companies which collectively

ABOVE: Newcastle Central station's imposing roof soars high above the platforms serving the East Coast Main Line and connecting routes.

RIGHT: The original tile map of the North Eastern Railway network still adorns Morpeth station – the story of the tile maps is told in *The Times Mapping the Railways* (2011) by David Spaven and Julian Holland.

ROUTE DESCRIPTION

Newcastle to Morpeth

Leaving the imposing Central station in Newcastle (population 292,000), the railway runs high above much of the tightly packed city centre, offering glimpses of the banks of the Tyne to the south. En route to Heaton, where trains are maintained across a wide array of sidings, the distinctive Tyne & Wear Metro trains can be seen – operating on one of a number of dedicated routes, some of which follow the original conventional rail routes to the coast.

Through the landscape of suburban and post-war industrial development around Cramlington 'new town', the railway then leaves behind the Newcastle conurbation, with trains picking up speed through attractive farming country before slowing to just 50 mph for the sharply curved alignment through Morpeth, the most severe curve (seventeen chains or 1,122-ft radius) of any main railway line in Britain. The slower speed – forced on railway engineers by the geographical position of the town – allows the traveller to appreciate Morpeth's imposing station, built in 1847 and perhaps even catch a glimpse of the original North Eastern Railway 'tile map' (see opposite) on the Newcastle-bound platform. Morpeth is still a railway junction, with the freight-only 'Blyth & Tyne' line curving away eastwards to former coal-mining towns such as Ashington, Blyth and Lynemouth. The 'deep mines' have all closed, but the ECML passes two large rail-connected open-cast sites between Morpeth and Alnmouth.

A southbound Voyager service passes Spittal at the southeastern edge of Berwick-upon-Tweed in August 2007, with Meadow Haven lighthouse prominent against the North Sea.

linked Edinburgh and London soon recognized the important role of through journeys and established special rolling stock – 'East Coast Joint Stock' – in 1860. Ever since, the ECML has been one of Britain's premier rail routes, linking the two capitals, always with speed and often with style.

In 1906, operations at Newcastle Central station were simplified with the opening of the King Edward VII Bridge across the Tyne, eliminating a reversal of London-Edinburgh trains and reducing congestion at this key location. In 1923 the three companies were grouped into the London & North Eastern Railway (LNER), and the 1930s saw a brief period of increasing competition between the East Coast and West Coast routes, with the LNER and the London Midland & Scottish Railway (LMS) vying for business with ever-faster and more luxurious trains.

Speed and style had to be abandoned during the Second World War, but by the late 1950s operations on the Newcastle-Berwick line were still characterized on the one hand by the famous Pacific locomotives – designed by Sir Nigel Gresley in the 1930s and Arthur Peppercorn in the 1940s – hauling express trains, and on the other hand by the more mundane but financially lucrative business of intensive coal traffic workings from branch lines serving the Northumberland coalfield. The major changes of the modern era began in 1961 with the introduction of the distinctive 'Deltic' diesel-electric locomotives on London-Edinburgh services. At 3,300 horse power these were the most powerful diesels in Britain and enabled the through-journey time to be reduced from seven hours to six.

During the 1970s, improvements to the track allowed 100-mph running in stretches north of Newcastle, a prelude to the progressive introduction from 1978 of British Rail's new High Speed Trains (InterCity 125s), which knocked a further hour off London-Edinburgh journey times. The subsequent electrification of the ECML was completed in 1991, and the fastest train – 'The Flying Scotsman' – now completes the journey in exactly four hours, running non-stop between Edinburgh and Newcastle.

Today, Anglo-Scottish trains operated by the East Coast company operate at least hourly throughout the day, as do Cross Country trains linking Edinburgh and Glasgow with Birmingham and southwest England. Regular local trains between Newcastle and Morpeth are operated by Northern and a skeleton local service survives to stations beyond Morpeth as far as Chathill. The inter-city trains typically serve Berwick, Alnmouth (for Alnwick) and Newcastle, with occasional calls at Morpeth.

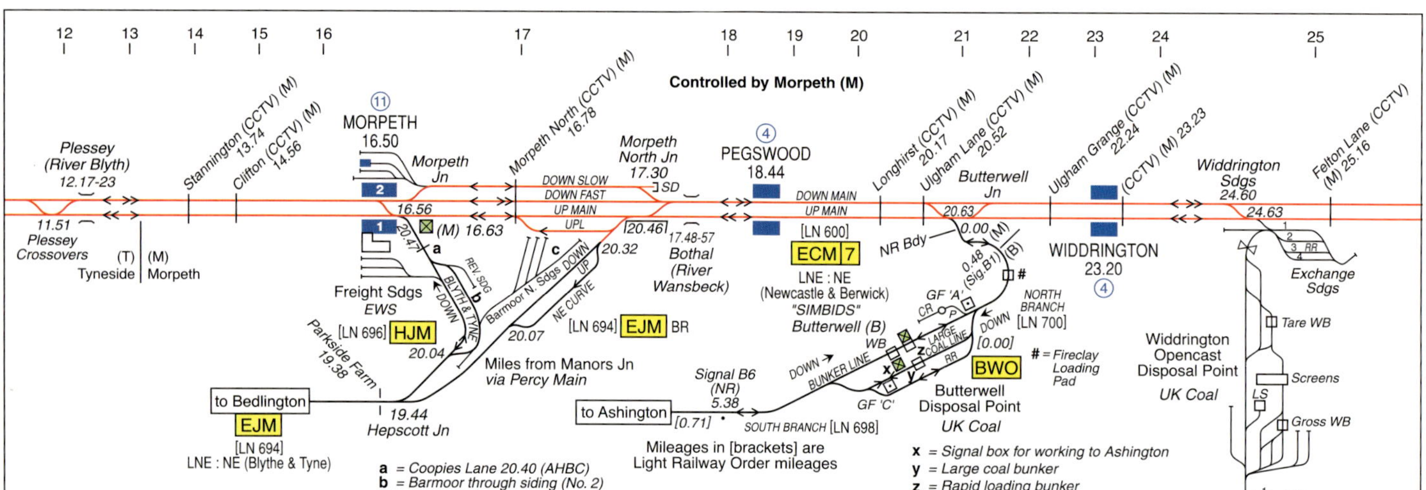

This diagram shows Morpeth's ongoing role as junction for the 'Blyth & Tyne' railway, and the continuing importance of open-cast coal to the East Coast Main Line.

ROUTE DESCRIPTION

Morpeth to Alnmouth

The countryside is characterized by rolling farmland and woodland, with a classic coastal view opening out towards the small town and one-time seaport of Alnmouth and the North Sea beyond, shortly before the railway reaches the town's station. The short branch line to the market town of Alnwick closed in 1968, but its graceful station building has since become the home of one of the largest second-hand bookshops in Britain.

The first phase of construction of a 'heritage railway' operation east of Alnwick began in 2012.

The arrival from the south by train at Berwick offers one of the most charming and memorable urban panoramas in Britain – here the seventeenth century Berwick Bridge is seen looking north from Tweedmouth.

A4 No. 60007 'Sir Nigel Gresley' eases the 'Great Britain' rail tour across the Grade I-listed Royal Border Bridge into Berwick station in April 2011.

A delightful floral welcome awaits passengers at Chathill on its two daily train services from Newcastle.

ROUTE DESCRIPTION
Alnmouth to Berwick-upon-Tweed

Through the tiny station at Chathill (with its original buildings dating from 1847), the railway passes within sight of the prominent mass of Bamburgh Castle. Fine views of the Cheviots then appear to the west, and the distinctive shape of Holy Island and Lindisfarne Castle can be seen to the east as the railway runs past the village of Belford – where there are plans to re-open its station which closed in 1968. Sea and sand dunes dominate the view on the approaches to Berwick-upon-Tweed, where the train slows for a sweeping curve leading to the Royal Border Bridge. The unspoilt roofscape of the town leads down to the Tweed estuary and the port of Tweedmouth, offering glimpses of parts of Berwick's outstanding Elizabethan walls and providing one of *the* outstanding urban views from a train anywhere in Europe.

The magnificent twenty-eight-arch structure across the Tweed is 2,160 ft long and 120 ft high – and on clear days affords a glimpse of the distinctive Eildon Hills some thirty miles to the west. Berwick (population 12,000) has an unusual and attractive island platform station, from which buses serve a wide range of destinations in Northumberland and the Scottish Borders. The town (which has changed hands between England and Scotland many times, and is now just two and a half miles inside England) is thus an ideal base for a holiday on foot and by public transport. Berwick itself – England's northernmost town – is fascinating, and is said to have more listed buildings for its size than any other in the country.

NEWCASTLE TO CARLISLE

TYNE VALLEY LINE

This double-track cross-country railway follows the Rivers Tyne and South Tyne for much of its sixty-mile length across England's narrowest point, and is noted for its striking original station buildings and signal boxes – and easy access to Hadrian's Wall.

Incorporated in May 1829, five months before the famous Rainhill Trials for the Liverpool & Manchester Railway, the Newcastle & Carlisle Railway (NCR) was one of the very first in Britain. The Act authorizing the line had one serious flaw, in that objections from a local landowner resulted in the use of locomotives being forbidden – horses were to haul trains instead – but this was soon overcome by an amendment to the original Act in 1835, the year the first section of the railway opened from Hexham to Blaydon. The line was opened throughout from Carlisle to the south bank of the Tyne at Redheugh (Gateshead)

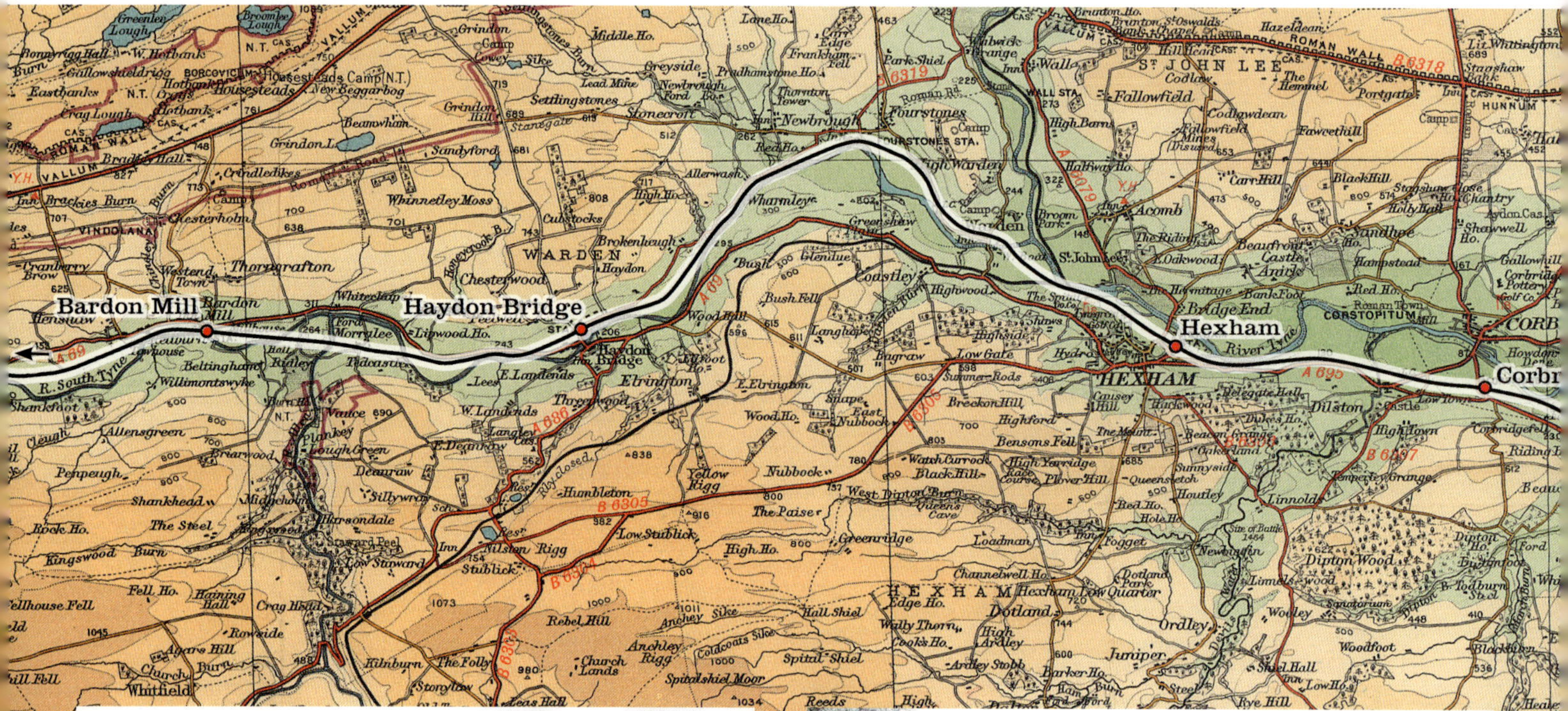

A traditional railway scene captured at Prudhoe station, with semaphore signal, lattice metal footbridge and electro-mechanical signal box.

A highly unusual gantry signal box frames the eastbound platform at Wylam station.

in 1838, then reaching a temporary terminus on the north bank in 1839 – but it was not until the 1851 opening of the west end of Newcastle Central station that the line secured a permanent terminus.

Following the broad valley of the Tyne for much of its length, the railway did not require particularly heavy engineering works over the eastern half of its route and gradients are predominantly gentle. Only at the western end, where the line climbs over the watershed between the Tyne and the Eden, do trains face a significant gradient – notably four miles of 1 in 107 between Wetheral and the former junction station of Brampton. The eastern section of the railway included a highly symbolic stretch of alignment, as noted by Ken Hoole in *A Regional History of the Railways of Great Britain Volume 4* (1965):

Part of the line was laid along the site of the Wylam wagonway, alongside which stood the house in which George Stephenson was born and spent his childhood. There is no doubt that the horse-drawn wagons trundling past the door kindled his interest in railways – an interest which was destined to play its part in changing the face of the world.

The NCR was absorbed into the North Eastern Railway in 1862 and two years later trains from Newcastle were diverted away from the original London Road terminus in Carlisle to the Citadel station, which had been opened by the London & North Western and Caledonian Railways in 1847.

Over the next 120 years Newcastle-Carlisle trains followed the north bank of the Tyne for the first four miles, before crossing the river at Scotswood to rejoin the current route along the

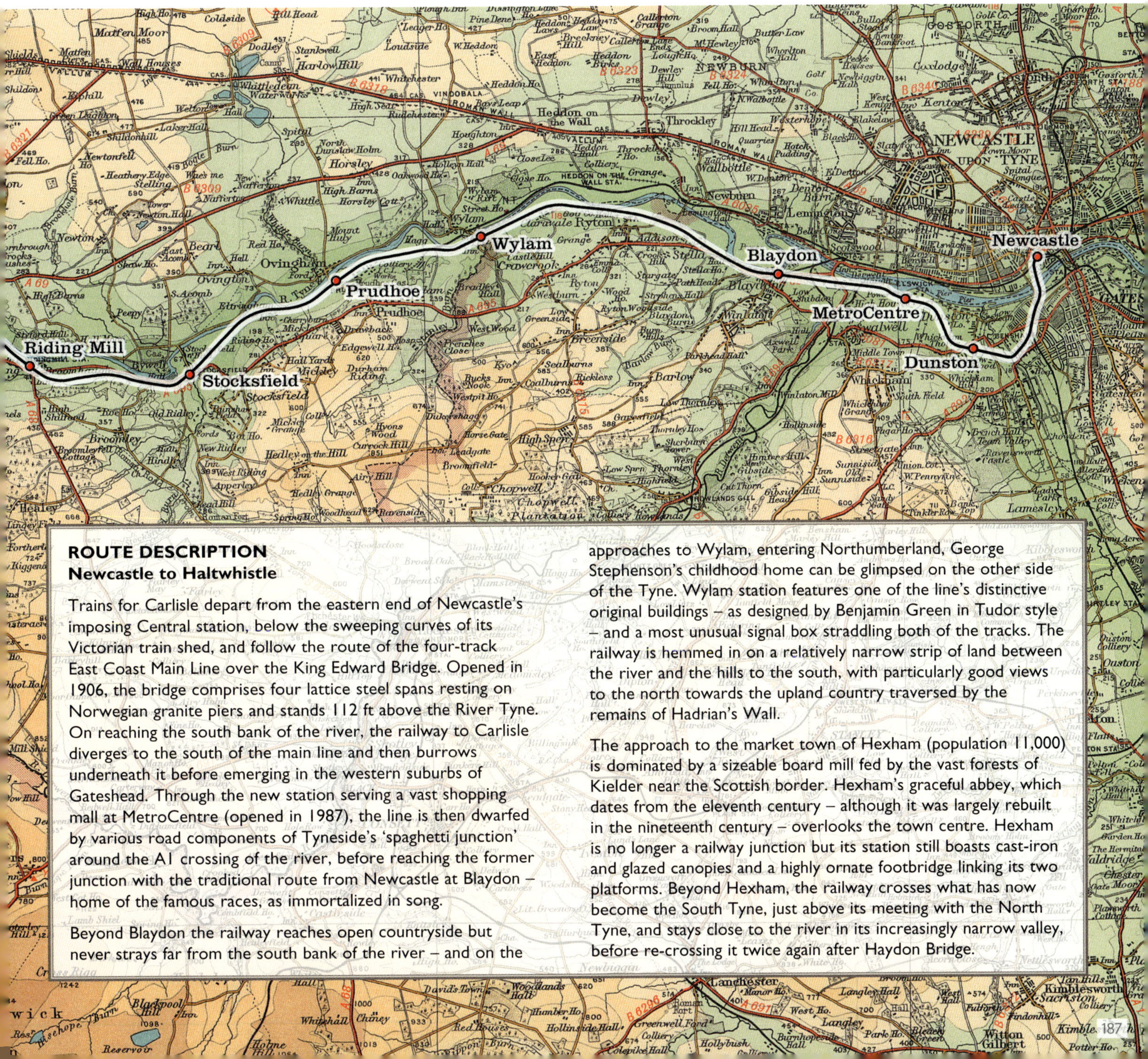

ROUTE DESCRIPTION
Newcastle to Haltwhistle

Trains for Carlisle depart from the eastern end of Newcastle's imposing Central station, below the sweeping curves of its Victorian train shed, and follow the route of the four-track East Coast Main Line over the King Edward Bridge. Opened in 1906, the bridge comprises four lattice steel spans resting on Norwegian granite piers and stands 112 ft above the River Tyne. On reaching the south bank of the river, the railway to Carlisle diverges to the south of the main line and then burrows underneath it before emerging in the western suburbs of Gateshead. Through the new station serving a vast shopping mall at MetroCentre (opened in 1987), the line is then dwarfed by various road components of Tyneside's 'spaghetti junction' around the A1 crossing of the river, before reaching the former junction with the traditional route from Newcastle at Blaydon – home of the famous races, as immortalized in song.

Beyond Blaydon the railway reaches open countryside but never strays far from the south bank of the river – and on the approaches to Wylam, entering Northumberland, George Stephenson's childhood home can be glimpsed on the other side of the Tyne. Wylam station features one of the line's distinctive original buildings – as designed by Benjamin Green in Tudor style – and a most unusual signal box straddling both of the tracks. The railway is hemmed in on a relatively narrow strip of land between the river and the hills to the south, with particularly good views to the north towards the upland country traversed by the remains of Hadrian's Wall.

The approach to the market town of Hexham (population 11,000) is dominated by a sizeable board mill fed by the vast forests of Kielder near the Scottish border. Hexham's graceful abbey, which dates from the eleventh century – although it was largely rebuilt in the nineteenth century – overlooks the town centre. Hexham is no longer a railway junction but its station still boasts cast-iron and glazed canopies and a highly ornate footbridge linking its two platforms. Beyond Hexham, the railway crosses what has now become the South Tyne, just above its meeting with the North Tyne, and stays close to the river in its increasingly narrow valley, before re-crossing it twice again after Haydon Bridge.

south bank at Blaydon. As well as the original line from Redheugh to Blaydon (which survived as a freight railway until its reincarnation as a passenger route in 1982), this main line had no less than five branch lines:

- the North Wylam loop, a six-and-a-half-mile double-track line following the course of an old coal-carrying wooden waggonway, which opened for colliery and passenger traffic in 1876 and survived until passenger closure under the 'Beeching Axe' in 1968 and final closure to freight in 1992

- the Border Counties Railway from Hexham via the underpopulated North Tyne valley and Kielder to Riccarton Junction on the Carlisle-Edinburgh 'Waverley Route', which opened in 1862 as an envisaged strategic route to allow the North British Railway to reach Newcastle from Edinburgh with minimum use of the tracks of its North Eastern Railway rival, but closed to passengers as early as 1956 and to freight north of Bellingham in 1958

- a branch line from west of Hexham to Allendale – opened in 1868 primarily to carry mineral traffic until closed completely in 1950

- the seventeen-mile branch line from Haltwhistle to the market town of Alston, high in the Pennines, which opened in 1852 and survived (due to poor road alternatives) until 1976, when it became the very last of the 'Beeching' closures (although the Alston end of the line has now been reincarnated as a narrow-gauge heritage line, the South Tynedale Railway)

- a short branch line from Brampton Junction to the small town of Brampton, which began as an eighteenth-century wooden waggonway and featured horse-hauled passenger trains for a time – but closed completely as early as 1923.

In 1982 – in order to avoid the upkeep costs of the Scotswood Bridge over the river – Newcastle-Carlisle trains were permanently diverted away from the route north of the Tyne to Blaydon. The formerly freight-only line south of the Tyne was upgraded to passenger standards and new intermediate stations were subsequently opened at Dunston and MetroCentre. Northern and ScotRail operate today's train service, with departures half hourly from Newcastle (population 292,000) to Hexham, and hourly to Carlisle (population 100,000), with a number of services extended through to Glasgow via Dumfries and Kilmarnock.

The Tyne Valley line is also an important cross-country freight link from the East Coast Main Line to the West Coast Main Line and is used as a diversionary route for Edinburgh-Newcastle passenger trains when engineering work closes the East Coast Main Line.

ROUTE DESCRIPTION
Haltwhistle to Carlisle

Like Hexham, Haltwhistle has lost its status as a junction, but features one of the line's classic Tudor-style station buildings, a decorative water tower dating back to the steam era and an ornate footbridge. Beyond Haltwhistle, the railway parts company with the South Tyne – whose source lies in the Pennines to the south – and continues the gentle climb to the watershed before beginning a steeper descent through Brampton and Wetheral towards Carlisle. At Gilsland the railway twice crosses Hadrian's Wall and, between the stations at Brampton and Wetheral, runs across three more modern but no less imposing structures – the Gelt, Corby and Wetheral viaducts, all built in stone in the 1830s to the design of Francis Giles.

West of Brampton, through wooded countryside, there are long southwards panoramas to the fells and immediately before Wetheral the five-arch viaduct of the same name (over the River Eden) affords excellent views of the village and nearby Corby Castle. The gradient eases as the line reaches the flood plain of the River Eden on the approaches to Carlisle and there then follows a sequence of junctions as the line penetrates the city's still-complex railway geography – Petteril Bridge Junction (Settle & Carlisle line), London Road Junction (chord line to the southbound West Coast Main Line) and Carlisle South Junction, where the line from Newcastle joins the West Coast Main Line immediately south of the city's eight-platform Citadel station.

Carlisle is still a great railway centre, with train services fanning out over four of the other scenic railways featured in this book, plus the important cross-country route to Glasgow via Dumfries and Kilmarnock – which also has its own picturesque qualities. The railway from Newcastle to Carlisle is not often mentioned in the same breath as the classic scenic routes of England, but it traverses some beautiful countryside and deserves greater attention for its wonderful array of engineering and architectural features dating from the earliest days of Britain's railways.

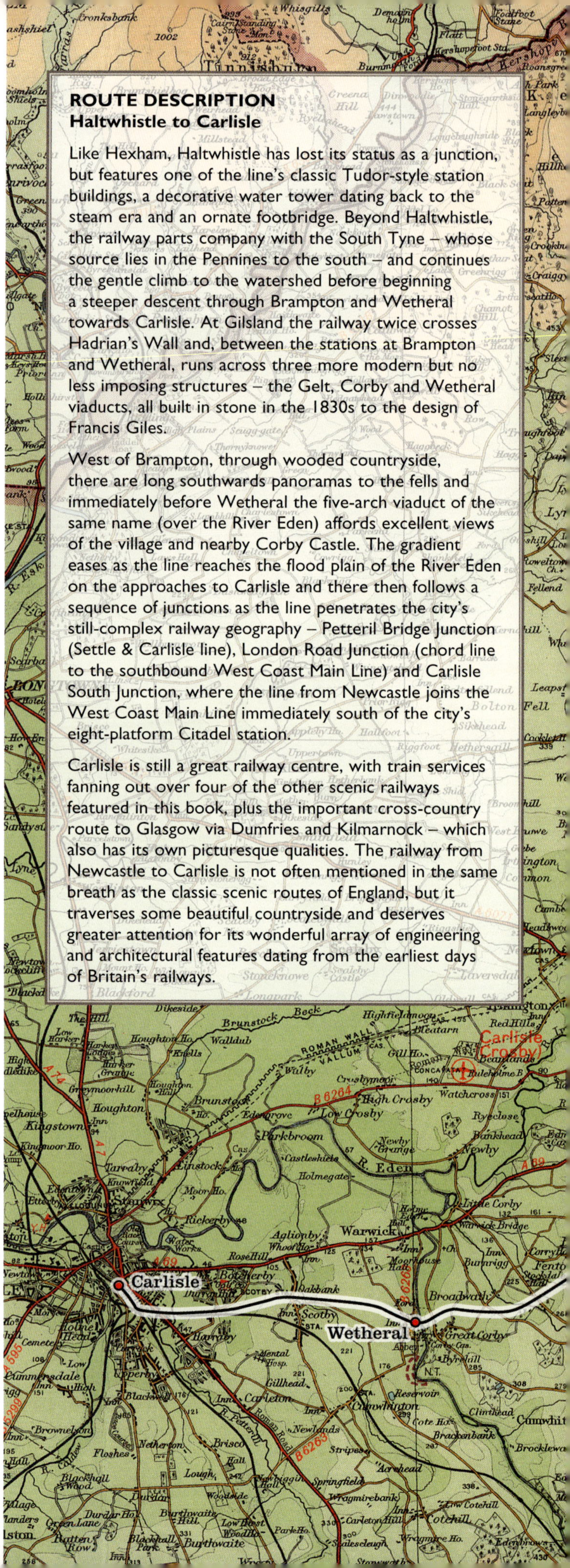

A diesel unit on a Newcastle to Carlisle service crosses Wetheral Viaduct 100 ft above the River Eden, in September 2010.

The graceful lines of Haltwhistle's distinctive signal box are seen to great effect on the approach from the South Tyne Walkway.

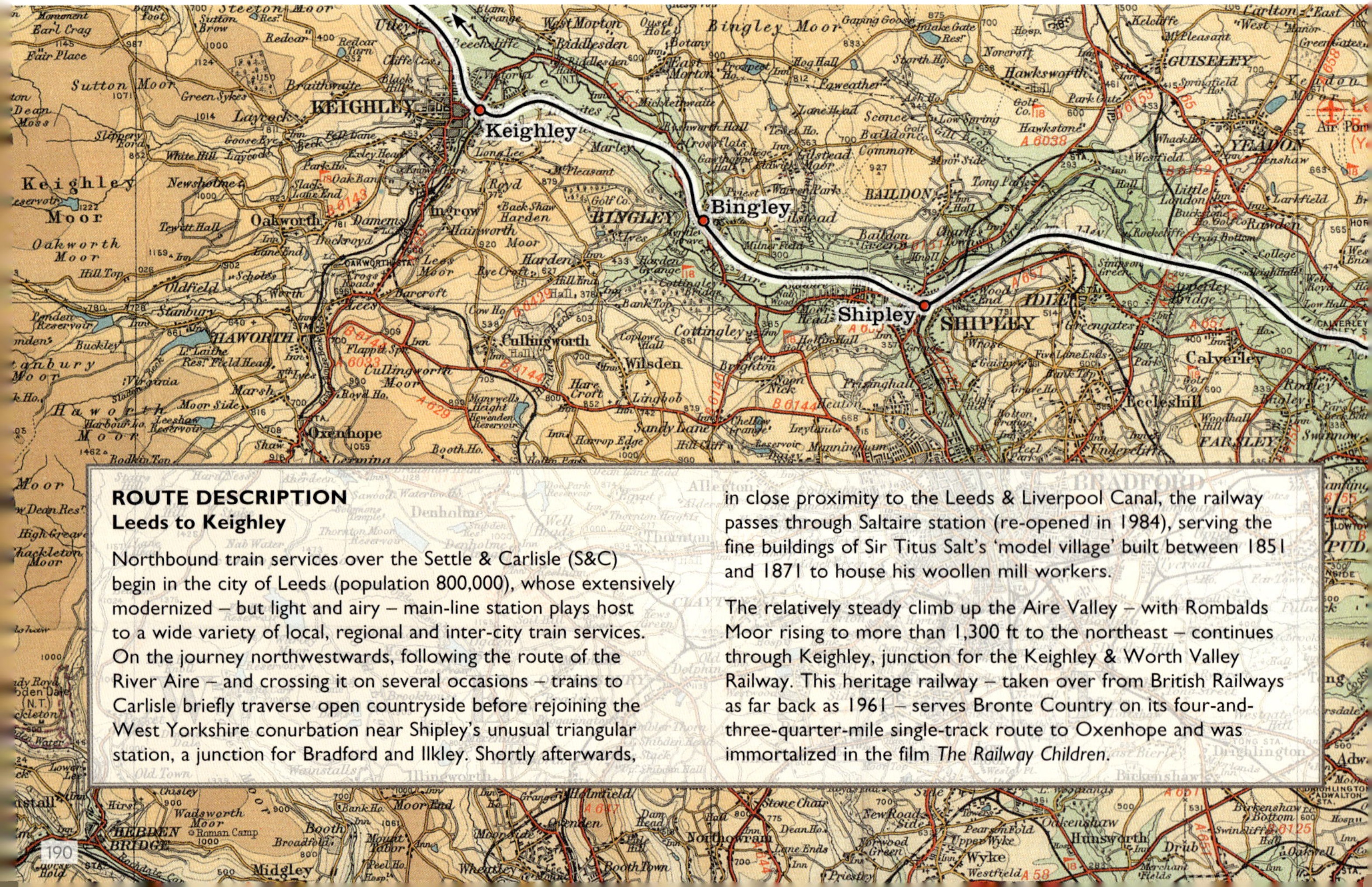

ROUTE DESCRIPTION

Leeds to Keighley

Northbound train services over the Settle & Carlisle (S&C) begin in the city of Leeds (population 800,000), whose extensively modernized – but light and airy – main-line station plays host to a wide variety of local, regional and inter-city train services. On the journey northwestwards, following the route of the River Aire – and crossing it on several occasions – trains to Carlisle briefly traverse open countryside before rejoining the West Yorkshire conurbation near Shipley's unusual triangular station, a junction for Bradford and Ilkley. Shortly afterwards, in close proximity to the Leeds & Liverpool Canal, the railway passes through Saltaire station (re-opened in 1984), serving the fine buildings of Sir Titus Salt's 'model village' built between 1851 and 1871 to house his woollen mill workers.

The relatively steady climb up the Aire Valley – with Rombalds Moor rising to more than 1,300 ft to the northeast – continues through Keighley, junction for the Keighley & Worth Valley Railway. This heritage railway – taken over from British Railways as far back as 1961 – serves Bronte Country on its four-and-three-quarter-mile single-track route to Oxenhope and was immortalized in the film *The Railway Children*.

Jubilee No. 5690 'Leander' heads an excursion from Leeds to Carlisle up 'the Long Drag' between Horton-in-Ribblesdale and Ribblehead in November 2010.

LEEDS TO CARLISLE

VIA SETTLE & CARLISLE LINE

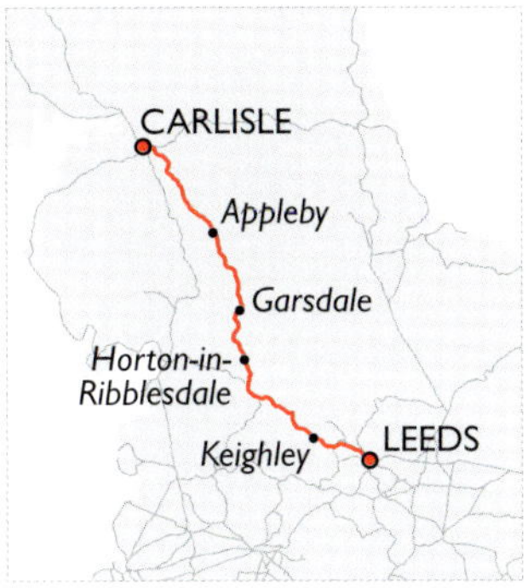

Superlatives rightly abound when it comes to describing England's most dramatically scenic railway, carved seventy-two miles across the Pennines without the aid of mechanical shovels – and later surviving two closure threats to become a major tourist attraction.

The Settle & Carlisle Line owes its existence to the determination of the mighty Midland Railway Company that it should have its own route from London to Carlisle in order to tap the lucrative Anglo-Scottish rail travel market. Denied running powers from its northern outpost at Ingleton (at the northwestern extremity of the Yorkshire Dales) over the linked route of its London & North Western (LNWR) rival to Low Gill on the West Coast Main Line, the Midland in 1866 secured an Act to construct a seventy-two-mile line from Settle Junction – fifteen miles southeast of Ingleton – to Carlisle, across some of the most difficult terrain in England. The Midland was supported in this initiative by putative through-route partners in the shape of the Lancashire & Yorkshire (LYR), Glasgow & South Western (GSWR) and North British (NBR) railway companies – the latter being particularly anxious to secure a 'friendly' southern outlet for its Waverley Route from Edinburgh, extended from Hawick in the Scottish Borders to Carlisle in 1862.

Alarmed by this competitive prospect, the LNWR in 1868 concluded an agreement with the Midland – which by then had had second thoughts about such an expensive undertaking through such inhospitable countryside – for the previously denied running powers, from Ingleton through Low Gill to Carlisle. Unfortunately for the Midland, the LYR and NBR both petitioned against the Bill for abandonment of the planned route and by 1869 construction had begun. The six-plus years it took for this double-track main line to reach Carlisle involved some of the most heroic construction work in British railway history – a workforce of up to 6,000, predominantly 'navvies' (short for 'navigators' who built the earlier canals), and £3.5 million were needed to drive the line over two summits of more than 1,000 ft and traverse countless rivers and streams en route. Lacking the mechanized equipment, which would soon transform railway building, the construction team under engineer-in-chief John Crossley faced extremes of both weather and topography, WR Mitchell and David Joy recording in *Settle-Carlisle Railway* (1979):

> *The terrain was difficult, for the glaciers that carved the northern dales left many slopes liberally coated with boulder clay and rubble. Disturb that clay in wet weather, and it took*

A Leeds-Carlisle service in February 2009, just north of the town of Settle, nestling in the valley of the River Ribble between high fells.

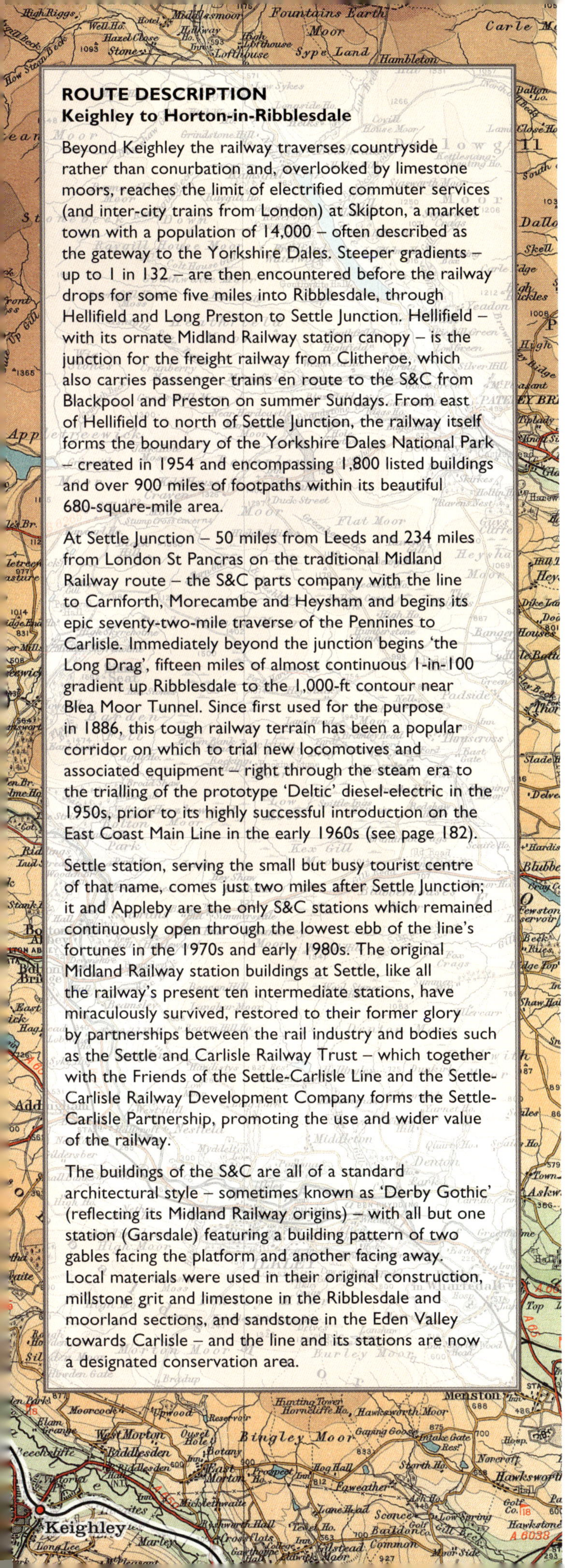

ROUTE DESCRIPTION

Keighley to Horton-in-Ribblesdale

Beyond Keighley the railway traverses countryside rather than conurbation and, overlooked by limestone moors, reaches the limit of electrified commuter services (and inter-city trains from London) at Skipton, a market town with a population of 14,000 – often described as the gateway to the Yorkshire Dales. Steeper gradients – up to 1 in 132 – are then encountered before the railway drops for some five miles into Ribblesdale, through Hellifield and Long Preston to Settle Junction. Hellifield – with its ornate Midland Railway station canopy – is the junction for the freight railway from Clitheroe, which also carries passenger trains en route to the S&C from Blackpool and Preston on summer Sundays. From east of Hellifield to north of Settle Junction, the railway itself forms the boundary of the Yorkshire Dales National Park – created in 1954 and encompassing 1,800 listed buildings and over 900 miles of footpaths within its beautiful 680-square-mile area.

At Settle Junction – 50 miles from Leeds and 234 miles from London St Pancras on the traditional Midland Railway route – the S&C parts company with the line to Carnforth, Morecambe and Heysham and begins its epic seventy-two-mile traverse of the Pennines to Carlisle. Immediately beyond the junction begins 'the Long Drag', fifteen miles of almost continuous 1-in-100 gradient up Ribblesdale to the 1,000-ft contour near Blea Moor Tunnel. Since first used for the purpose in 1886, this tough railway terrain has been a popular corridor on which to trial new locomotives and associated equipment – right through the steam era to the trialling of the prototype 'Deltic' diesel-electric in the 1950s, prior to its highly successful introduction on the East Coast Main Line in the early 1960s (see page 182).

Settle station, serving the small but busy tourist centre of that name, comes just two miles after Settle Junction; it and Appleby are the only S&C stations which remained continuously open through the lowest ebb of the line's fortunes in the 1970s and early 1980s. The original Midland Railway station buildings at Settle, like all the railway's present ten intermediate stations, have miraculously survived, restored to their former glory by partnerships between the rail industry and bodies such as the Settle and Carlisle Railway Trust – which together with the Friends of the Settle-Carlisle Line and the Settle-Carlisle Railway Development Company forms the Settle-Carlisle Partnership, promoting the use and wider value of the railway.

The buildings of the S&C are all of a standard architectural style – sometimes known as 'Derby Gothic' (reflecting its Midland Railway origins) – with all but one station (Garsdale) featuring a building pattern of two gables facing the platform and another facing away. Local materials were used in their original construction, millstone grit and limestone in the Ribblesdale and moorland sections, and sandstone in the Eden Valley towards Carlisle – and the line and its stations are now a designated conservation area.

on the consistency of Yorkshire pudding mixture, whereas in hot, dry weather it hardened until it had the stubborn nature of concrete; it might then bend a pickhead. A frustrated engineer said: 'I have know [sic] the men to blast the boulder clay like rock and within a few hours have to ladle out the same stuff from the same spot like soup in buckets.

This epic feat of engineering – including fourteen tunnels and twenty viaducts – was completed in 1875, but passenger traffic did not commence until the following year. Anticipating the line's future reputation for very high standards of comfort (with which, in conjunction with the GSWR and NBR, it sought to compete with the faster East Coast and West Coast routes from London to Edinburgh and Glasgow), the first train featured carriages with upholstered seats in Third Class – then a rare luxury for passengers used to wooden seating.

The line's heyday – until the modern freight revival from the 1990s onwards – came in the Edwardian years of the early twentieth century. Three day-time expresses from the Midland's grand St Pancras terminus in London all carried through carriages for Edinburgh (via the Waverley Route) and Glasgow (via the GSWR's Dumfries and Kilmarnock route) on a daily basis, while each night the line played host to three trains from London conveying sleeping cars variously to Aberdeen, Dundee, Edinburgh, Glasgow, Perth and Stranraer.

The First World War brought a major reduction in services and the 'Grouping' of 1923 placed the Midland in the same large new company – the London, Midland & Scottish Railway (LMS) – as its arch rival the LNWR, thereby rendering much of the previous competition redundant. The GSWR and its great Caledonian Railway rival also became part of the LMS, but the Waverley Route (as part of the NBR) was lumped in with the London & North Eastern Railway (LNER), where it found itself out on a limb without the link to a 'friendly' outlet to the south at Carlisle.

A Leeds-Carlisle service calls at lonely but scenic Garsdale station in September 2010 – formerly the junction for Wensleydale (see pages 170–171).

66 108 hauls a train of imported coal from Hunterston on the Forth of Clyde to Drax power station in Yorkshire across Dent Head Viaduct in August 2009.

ROUTE DESCRIPTION

Horton-in-Ribblesdale to Garsdale

The station at the village of Horton-in-Ribblesdale was one of eight on the S&C to be fully re-opened in 1986, allowing visitor access to a wide range of hill walking opportunities. Horton is well placed for exploring the 'Three Peaks', comprising Pen y Ghent (2,272 ft) across the Ribble Valley, Ingleborough (2,373 ft) to the northwest and the more distant Whernside (2,418 ft) to the west of Ribblehead Viaduct.

Continuing north at a 1-in-100 gradient, the railway reaches Ribblehead station – whose sidings are now used for loading timber trains – and the magnificent viaduct, which once threatened to cause the line's closure. To reduce wear and tear, the track is now singled across this twenty-four-arch structure, which has come to symbolize the whole Settle & Carlisle line. North of the viaduct is the first of the seven intermediate signal boxes on the line at lonely Blea Moor, just south of the one-mile 869-yd tunnel of that name. The S&C is still controlled by traditional manual signal boxes, but Network Rail has undertaken extensive modernization of the route's infrastructure to accommodate increasing freight traffic – notably coal from Scottish open cast mines and imported coal from the port of Hunterston to English power stations.

The line reaches its first summit within Blea Moor Tunnel – at a height of 1,151 ft – and over the next ten miles to the railway's highest point at Ais Gill (1,169 ft) gradients are less severe as the S&C swings around the steep fell slopes, over the watersheds between the Ribble, Dee and Eden rivers. Dent is England's highest main-line station at 1,150 ft, but it is four miles from Dent village and the access road involves a climb of 600 ft over gradients of 1 in 5! Garsdale was formerly the junction for the Midland Railway branch to Hawes in Wensleydale (see pages 170–171) but is now a lonely and spectacular spot, popular with walkers.

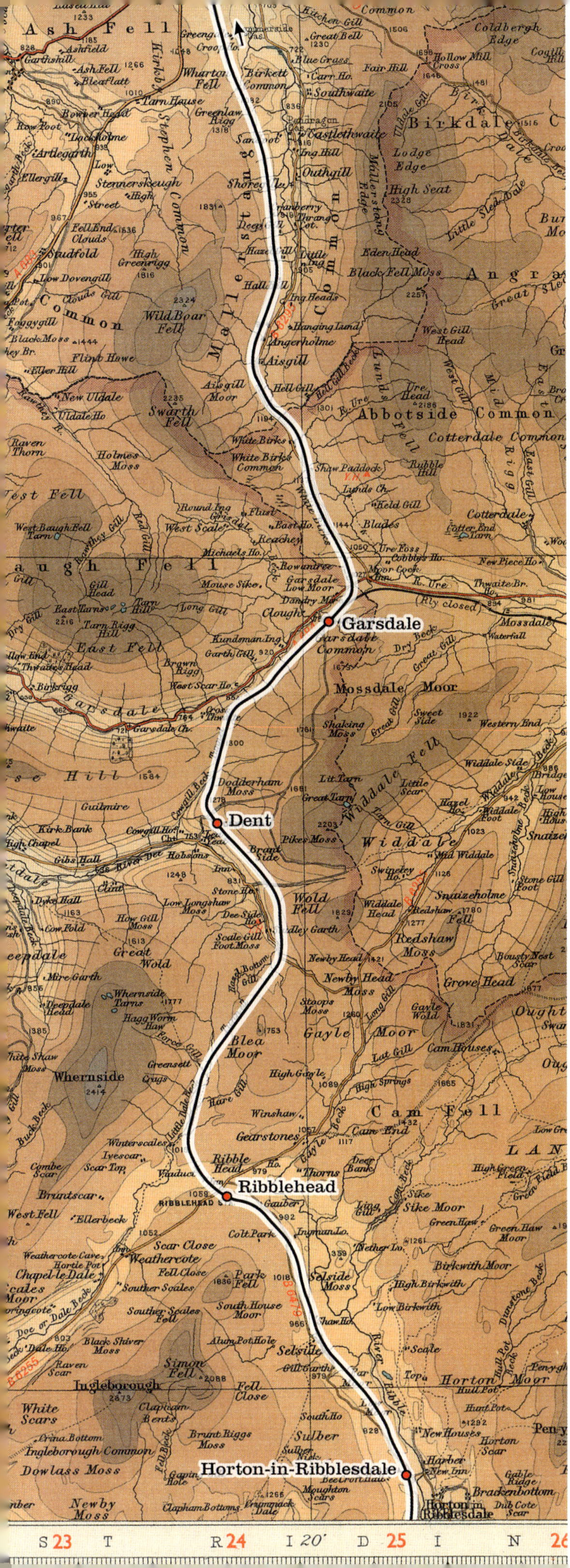

However, in line with the wider push for speed and luxury on the premier trunk routes, the 1930s saw the re-establishment of three day-time expresses on the Settle & Carlisle, with the morning trains to and from Glasgow and Edinburgh being named respectively 'The Thames-Clyde' and 'The Thames-Forth' (later 'The Waverley'). The latter, however, disappeared completely following the 1969 closure of the entire ninety-eight and a quarter miles of the Waverley Route – probably the worst of all the 'Beeching' closures – and in the early 1970s the Settle & Carlisle (S&C) reached its nadir. The line had earlier won a reprieve from the 'Beeching' threat, but all the intermediate stations except Settle and Appleby were closed in 1970, and campaigners protested about a passenger service run-down, which many foresaw would lead to a renewed plan for complete closure. The line however remained busy with through freight traffic and continued to have a key diversionary role, usually on Sundays when the West Coast Main Line was undergoing engineering maintenance.

Fortunately, summer 1975 brought the beginning of a revival of the line's passenger fortunes, when the 'Dales Rail' initiative, led by the Yorkshire Dales National Park, saw a diesel unit chartered for five weekends serving five temporarily re-opened intermediate stations as well as Settle and Appleby. The line also handled a good number of steam-hauled special trains during the late 1970s and the 1980s. The threat to the railway was not

In December 2008 a southbound diesel unit has just crossed Ribblehead Viaduct prior to reaching Ribblehead station.

6201 'Princess Elizabeth' climbs south of Kirkby Stephen at Birkett Common, at the head of the Cumbrian Mountain Express charter train, in August 2009.

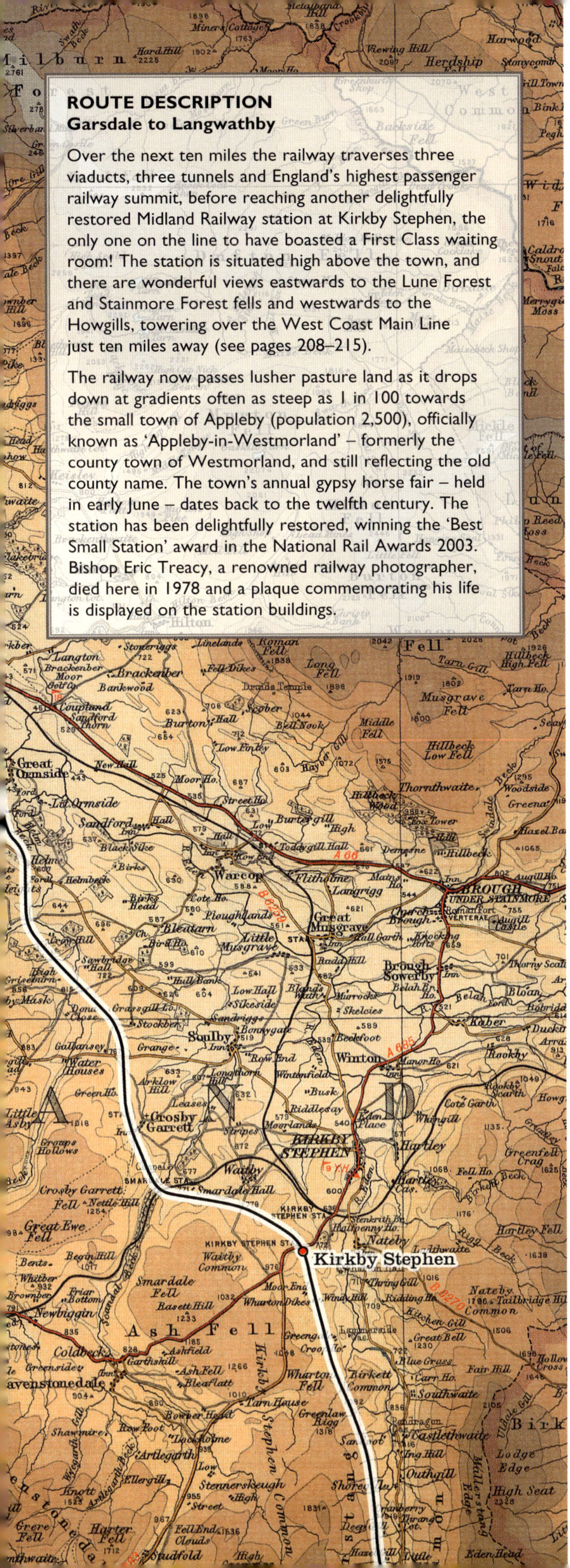

ROUTE DESCRIPTION
Garsdale to Langwathby

Over the next ten miles the railway traverses three viaducts, three tunnels and England's highest passenger railway summit, before reaching another delightfully restored Midland Railway station at Kirkby Stephen, the only one on the line to have boasted a First Class waiting room! The station is situated high above the town, and there are wonderful views eastwards to the Lune Forest and Stainmore Forest fells and westwards to the Howgills, towering over the West Coast Main Line just ten miles away (see pages 208–215).

The railway now passes lusher pasture land as it drops down at gradients often as steep as 1 in 100 towards the small town of Appleby (population 2,500), officially known as 'Appleby-in-Westmorland' – formerly the county town of Westmorland, and still reflecting the old county name. The town's annual gypsy horse fair – held in early June – dates back to the twelfth century. The station has been delightfully restored, winning the 'Best Small Station' award in the National Rail Awards 2003. Bishop Eric Treacy, a renowned railway photographer, died here in 1978 and a plaque commemorating his life is displayed on the station buildings.

lifted however, and, citing the cost of maintaining the many tunnels and viaducts – notably the twenty-four-arch Ribblehead Viaduct – in 1984 British Rail (BR) issued notice that it proposed to withdraw all passenger services.

However, a line support campaign, the Friends of the Settle-Carlisle line had been established in 1981, and together with local authorities and communities along the line, a major campaign against closure was mounted. Ironically BR, who in 1983 had appointed a proven but maverick manager, Ron Cotton, to manage the closure of the line, unwittingly aided this ultimately successful campaign. Cotton had introduced local branding and innovative marketing on Merseyside and had pioneered the use of off-peak Saver tickets nationally, transforming the railway's ability to compete with road and air transport. He quickly brought his marketing flair to bear on the very special circumstances of the S&C, as recorded by Stan Abbott and Alan Whitehouse in *The Line that Refused to Die* (1994):

> *And so Mr Cotton began his career on Settle-Carlisle doing what he was best at. He marketed the line according to the philosophy he had evolved at Liverpool and elsewhere: simplified fares and reduced fares to match demand and capacity. He was given a virtual free hand to milk the S&C for all it was worth before closure…*

By the end of 1984, revenue on the Settle & Carlisle was eighty per cent up on 1982, and although the Government made a belated attempt to sell the line to the private sector, consent to closure was finally refused in 1989. The logic of the S&C as an important tourist, freight and diversionary route had finally been accepted, and the railway has since gone from strength to strength, with increased regular services operated by Northern, a burgeoning passenger charter market (including steam-hauled specials) and more freight (primarily coal from Scotland to England) than it has seen for many years.

Diesel units on a Leeds to Carlisle service cross the 117-ft-high Arten Gill Viaduct, between Ribblehead and Dent, in July 2008.

6201 'Princess Elizabeth' heads the Cumbrian Mountain Express over Long Meg Viaduct, between Langwathby and Lazonby and Kirkoswald, in July 2011.

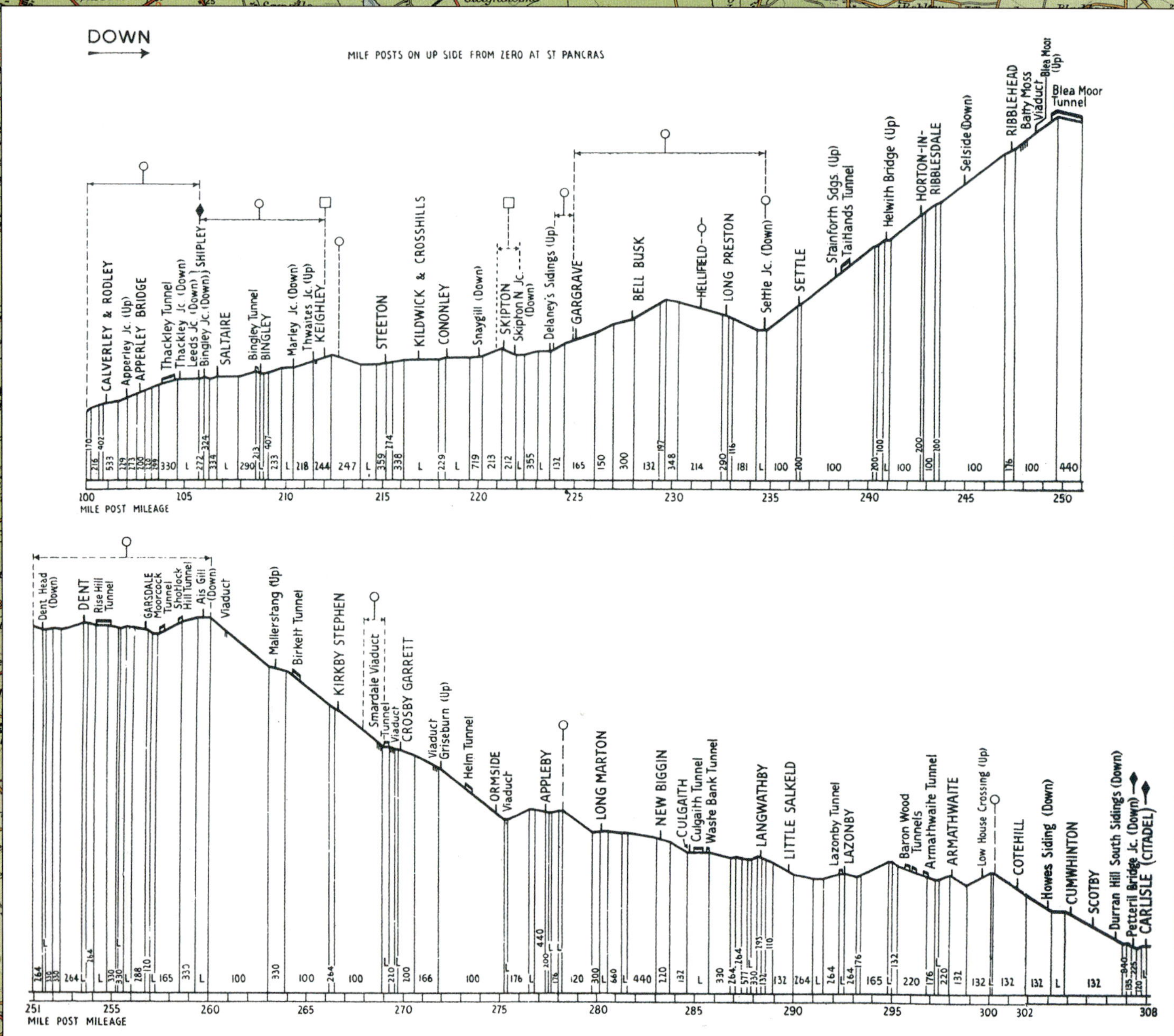

'The Long Drag' from Settle Junction to the railway's summits, and the southbound equivalent, are the key operational challenges of the Settle & Carlisle line.

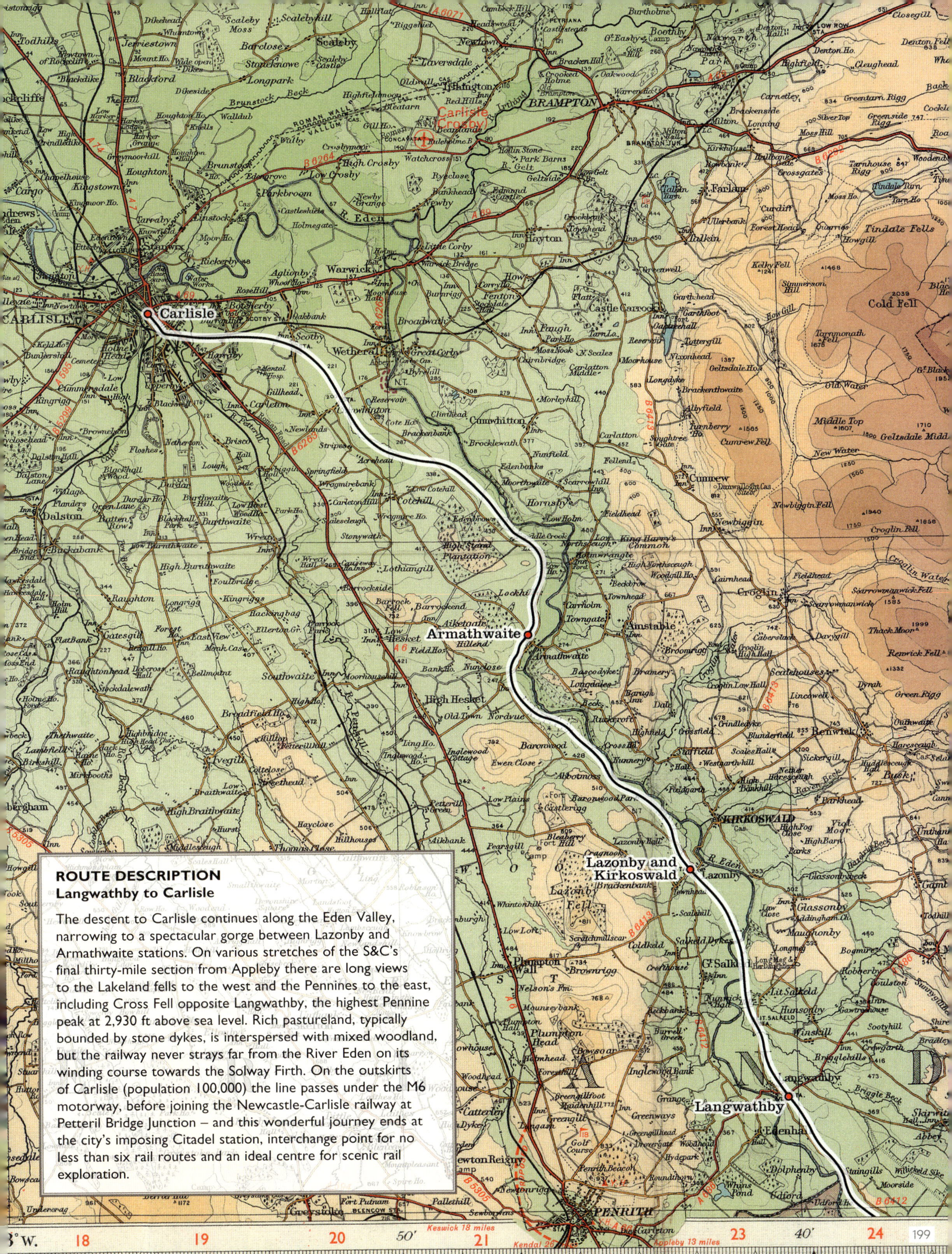

ROUTE DESCRIPTION

Langwathby to Carlisle

The descent to Carlisle continues along the Eden Valley, narrowing to a spectacular gorge between Lazonby and Armathwaite stations. On various stretches of the S&C's final thirty-mile section from Appleby there are long views to the Lakeland fells to the west and the Pennines to the east, including Cross Fell opposite Langwathby, the highest Pennine peak at 2,930 ft above sea level. Rich pastureland, typically bounded by stone dykes, is interspersed with mixed woodland, but the railway never strays far from the River Eden on its winding course towards the Solway Firth. On the outskirts of Carlisle (population 100,000) the line passes under the M6 motorway, before joining the Newcastle-Carlisle railway at Petteril Bridge Junction – and this wonderful journey ends at the city's imposing Citadel station, interchange point for no less than six rail routes and an ideal centre for scenic rail exploration.

Royal Scot No. 46115 Scots Guardsman, on a Ravenglass-Carnforth charter train, has a 'brief encounter' with a Class 156 on Kent Viaduct near Arnside in June 2010.

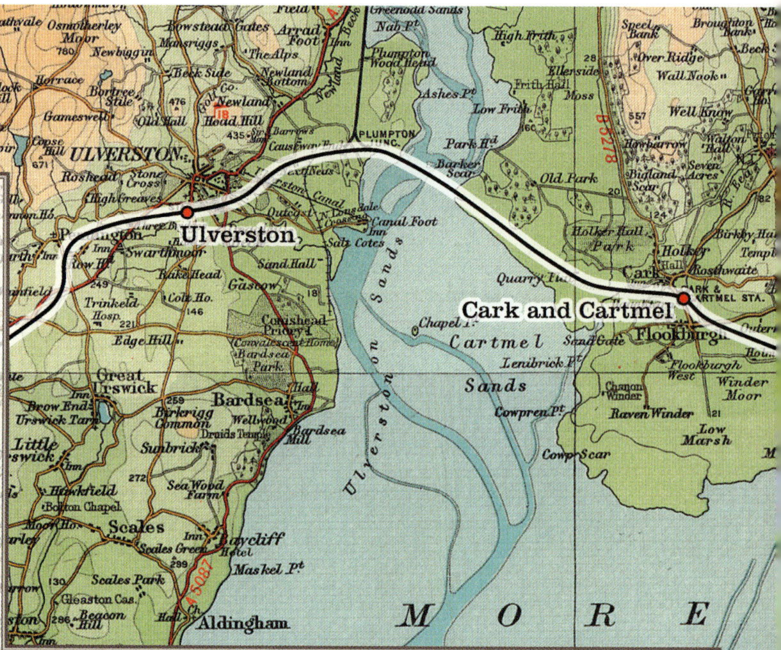

ROUTE DESCRIPTION

Lancaster to Ulverston

From Lancaster, trains for the Cumbrian Coast follow the West Coast Main Line (see pages 208–215) as far as the junction in the town of Carnforth (population 5,000). The platforms are adjacent to the station's *Brief Encounter* visitor centre, celebrating David Lean's classic romantic film – starring Trevor Howard and Celia Johnson – which was filmed here in 1945. Rather less well known is that the most recent part of the station complex features what was the longest unsupported single-piece concrete roof in Britain when constructed in 1937!

Leaving behind the complex of lines serving West Coast Railways' operational base for locomotives and coaches providing heritage steam and diesel charter trains across Britain – and the gigantic last-surviving cast-concrete coaling tower in the country – the railway soon encounters the first of the many sea defence works erected and reinforced over the years to protect the line infrastructure. For much of the next seventeen miles to Ulverston the line skirts (and in places crosses) Morecambe Bay, whose wide expanses of tidal reach have proved fatal to careless walkers and winkle pickers using the traditional low-tide routes across the sands. The seven miles of sea embankments have protected up to 20,000 acres of previously unusable land, much of which was transformed into agricultural use, aided in turn by the availability of the railway to transport produce and animals to more distant markets.

Two of the line's most impressive structures – the Kent and Leven Viaducts over the respective river estuaries – still give the railway a much more direct route to the towns of Grange-over-Sands and Ulverston than is possible by road. Both viaducts – which have been refurbished and strengthened in recent years by Network Rail – involved the first British use of a novel form of piling using high-pressure jets of water to force the intervening sand upwards. The Kent Viaduct is 1,300 ft long and links Arnside with Grange-over-Sands, which was transformed by the arrival of the railway in 1857 from a small fishing village to a popular but stylish seaside resort and 'hydropathic' health centre.

The railway clings to the coastline for three miles beyond Grange, then briefly swings inland through the charmingly-named Cark and Cartmel station, serving Cartmel Priory (founded in 1189) and Holker Hall, the home of Lord Cavendish of Furness, for whom special waiting rooms were provided for the dukes and their guests. Like many of the stations between Carnforth and Carlisle, the remaining building – although unstaffed – retains a number of its original features. The sylvan coastal countryside and attractive towns and villages of the line between Carnforth and Ulverston offer a delightfully uncrowded alternative to some of the extremes of popularity in the Lake District to the north.

The line soon returns to the coastline and strides across the impressive Leven Viaduct, a forty-nine span of 2,400-ft length, leading to the small town of Ulverston, whose station provided (and still provides) perhaps the architectural apotheosis of the Furness Railway's reputation for fine station buildings, as described by David Joy in *A Regional History of the Railways of Great Britain*: "With its Italianate flamboyancy of scale, tall clocktower incorporating corner urns, elaborate roof and richly monogrammed ironwork, it symbolized the company's pinnacle of achievement."

LANCASTER TO CARLISLE VIA BARROW-IN-FURNESS

CUMBRIAN COAST LINE

This fascinating but little-known rural railway hugs the Irish Sea coastline for much of its 115-mile route around western Lakeland, passing the sites of some of Victorian Britain's most meteoric coal and steel-generated industrial growth.

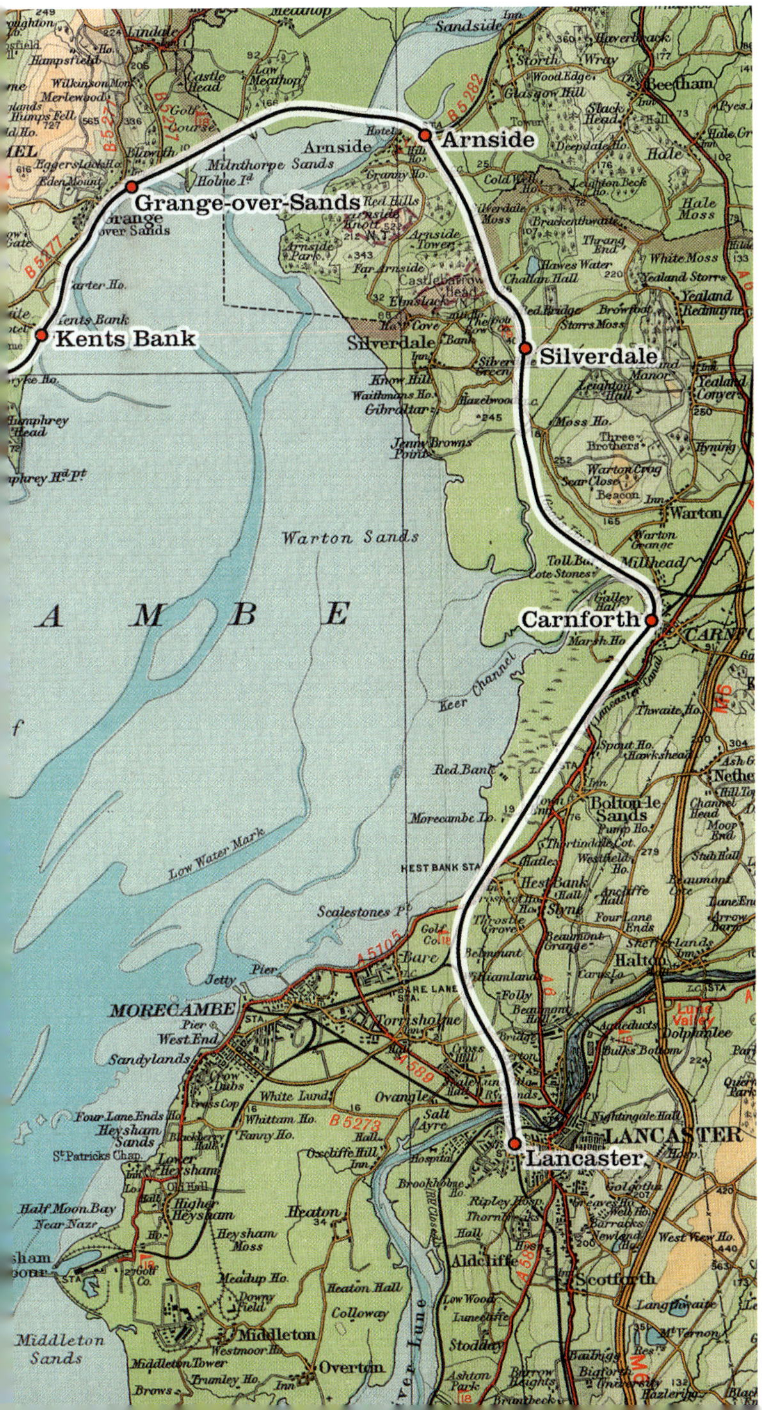

The Cumbrian Coast Line comprises surviving sections of five originally independent railways from mid-Victorian times:

- the Maryport & Carlisle Railway (opened in 1845)
- the Furness Railway from Kirkby-in-Furness to Barrow-in-Furness (1846)
- the Whitehaven Junction Railway from Whitehaven to Maryport (1847)
- the Whitehaven & Furness Junction Railway (1850)
- the Ulverston & Lancaster Railway (1857)

With the renowned George Stephenson as engineer, the Maryport & Carlisle Railway (M&CR) was built in stages, primarily for the movement of coal traffic, and opened throughout in 1845. Even with the addition of branch lines, the M&CR was always a small railway but it was generally highly profitable, based on the core business of moving hundreds of thousands of tons of coal annually from pit to port. It survived as an independent operation right through to the 'Grouping' of 1923.

The origins of today's surviving section of the Furness Railway (from Dalton through Barrow to Kirkby-in-Furness) lie partly in the drive to improve access to markets for local supplies of haematite iron ore. The Furness Railway, from quarry to coast through the town of Barrow, opened in 1846, and the company's network was extended eastwards to Ulverston (in 1854) and to Whitehaven to the north through the take-over of the Whitehaven & Furness Junction Railway in 1865. The latter had opened as an independent passenger and freight operation in 1850, three years after Whitehaven secured its first direct access to the national rail network with the opening of the Whitehaven Junction Railway northwards to Maryport.

Back in the 1830s the railway engineer George Stephenson had proposed routeing the planned Lancaster & Carlisle Railway via the Cumbrian coast to avoid steep inland gradients. When the Parliamentary Commissioners rejected the coastal route in 1843, an alternative scheme to link the developing local railway in Furness to the national network was promoted in the Ulverston & Lancaster Railway. This nineteen-mile scheme around the northern coast of Morecambe Bay received the Royal Assent in 1851, but construction was not easy – involving major viaducts across the tidal (and sandy) estuaries of the Rivers Kent and Leven – and the railway was not opened until 1857.

Thus was the Furness Railway connected to the national rail network, and just five years later it took over the Ulverston & Lancaster so that Carnforth became its southern boundary. The latter was controlled by the LNWR (which had taken over the Lancaster & Carlisle), and by 1866 the larger company controlled both this southern frontier of the Furness Railway and its northern frontier, following the takeover of the Whitehaven Junction Railway in the same year.

ROUTE DESCRIPTION
Ulverston to Ravenglass

Inland from Ulverston the railway tackles some of the toughest terrain of the entire route, aided by two tunnels but with gradients as steep as 1 in 80, before reaching Dalton Junction and one of the more unusual features of the rail network in Cumbria. A loop line diverges southwards to penetrate the heart of Barrow town, while the 'Dalton Loop' strikes northwards to Park South Junction where it is rejoined by the line from Barrow. All scheduled passenger trains take the line via the town, while the Dalton Loop is used only by freight and passenger charter services. The direct rail distance from Dalton Junction to Park South Junction is less than a mile, but passengers heading for points beyond Barrow must put up with eight and a half additional miles of travel – not, it must be said, on the most scenic stretch of the Cumbrian Coast line.

Barrow-in-Furness, with a population of 59,000, is the largest town in Cumbria. It saw astronomical growth in the second half of the nineteenth century, with the population swelling from just 4,684 in 1851 to 58,172 in 1881, and by the end of the century it was home to the largest steel works in the world. Steel making is now long gone, but a large military shipyard continues to be a key source of local manufacturing employment, and the adjacent commercial port of Barrow, which is still served by rail, handles exports and imports, including nuclear fuels and waste.

North of Barrow station, the railway drops to single track before re-joining the double track 'avoiding line' at Park South Junction. The peaks of the south west Lake District can be seen ahead, and the railway is soon taking a long loop around the deep inlet formed by Duddon Sands, which it was originally intended to bridge near Askam, avoiding the long detour which both rail and road face today. On the north side of the inlet is the small town of Millom, where there is much evidence of the line's community rail partnership. The station – one of the classic designs for the Furness Railway by the Lancaster architects Paley & Austin in the mid- to late-nineteenth century – is staffed by a community group and houses a local history museum.

For much of the next seventeen miles to Seascale, the railway runs around a mile inland over flat coastal country – but with frequent glimpses of the Lakeland peaks to the east, as well as the wide expanse of the Irish Sea to the west. Just before Ravenglass, the line crosses the River Esk close to where it flows into the sea, and then reaches the junction station – for the Ravenglass & Eskdale narrow-gauge railway (see page 204) – where the station is now the Ratty Arms pub owned by the railway company.

A Barrow-Carlisle service crosses the River Esk at Eskmeals south of Ravenglass in July 2011.

B1 No. 1306 and V2 No. 4771 'Green Arrow' head a northbound charter train round the Millom curve in June 1975.

Ulverston's classic Furness Railway station sports the 'double arrow' symbol, launched by British Rail in 1965, which is still the generic sign for a railway station.

As Furness (around Barrow and Millom) and West Cumberland were transformed into one of the leading industrial regions of the British Isles, based around iron, steel, coal and shipbuilding, so the railways – which were integral to these developments – prospered and expanded. The boom years of Barrow, beginning in the 1860s, were in fact relatively short lived – by the end of the 1870s new methods of steel making ended the dependence on the local haematite ore. In 1882 both local ore production and the number of vessels entering the port of Barrow peaked – at 1.4 million tons and 2,132 vessels respectively – and, as described by David Joy in *A Regional History of the Railways of Great Britain* (1990):

> *Thereafter it was a story of gradual economic decline, with railway expansion in the region virtually at an end, as it became all too apparent that the attempt to create a new industrial centre of world importance based on Barrow had failed. The town was just too far away from the manufacturing centres of Lancashire and Yorkshire to become a great port for general commerce, its position being further undermined by construction of the Manchester Ship Canal.*

However, railway passenger traffic was coming into its own, with the Furness Railway opening a tourism-inspired branch line from Ulverston to Windermere Lake Side in 1869 and enthusiastically promoting this and the earlier Coniston branch (completed in 1859) with cheap fares, package tours linked to horse bus and steamer services, and colourful publicity. Another line to become a key tourist route, as it also provided cross-country connections, was the link from Workington to Penrith through the heart of the Lake District by way of Cockermouth and Keswick. This opened in phases, initially from the port to Cockermouth and then throughout to Penrith in 1865, and was worked by the LNWR from the outset. Tourists did not generate anything like the profits of industrial freight traffic, but they became a key part of railway business in the Lake Counties, peaking in the halcyon Edwardian years just before the First World War.

Like rail routes throughout Britain, the lines in Furness and west Cumbria emerged from the war in a run-down state and in need of investment. The war had also demonstrated the benefits of unified control of the railways, and in 1923 the remaining

ROUTE DESCRIPTION
Ravenglass to Whitehaven

Two miles beyond this seeming rural idyll come the first signs of the area's economic dependence on nuclear power, at the rail-served Drigg waste depository. A further two miles north the line once again swings close to the coastline, which it will hug for most of the route north as far as Maryport. Immediately beyond Sellafield station and the massive adjacent nuclear complex, the line shrinks from double track to single track – which extends the eleven miles to Whitehaven, broken by a crossing loop at St Bees station which serves the well-known nearby public school. Here the railway turns inland along the valley of the Pow Beck to avoid some very difficult coastal terrain and then, after Corkickle station, passes through the 1,283-yd Whitehaven Tunnel to reach Whitehaven (population 25,000). The town was for a brief period in the eighteenth century the third largest in the north of England. The town centre was built on a classic Georgian grid pattern, and grew into a major coal-mining centre and associated port – but the pits have long gone, as has much of the commercial port activity, replaced by pleasure craft use.

RAVENGLASS & ESKDALE RAILWAY
Ravenglass to Dalegarth

This long-established outpost of miniature railway operation began life in 1875 as a 3-ft narrow gauge line built to convey iron ore to the coast, although passengers were eventually carried too. Once the mines closed, the railway was converted to 15-in gauge by the model railway engineer WJ Bassett-Lowke as a test bed for his locomotives, and resumed carrying passengers (and granite) in 1915.

After the quarries closed, local people and railway enthusiasts formed the Ravenglass & Eskdale Railway Preservation Society in 1960 to save the line, now known as *La'al Ratty*. The organizational structure, with the railway owned and operated by a private company, backed by the preservation society, is still in place. Today the line's steam and diesel hauled trains are a major tourist draw, attracting around 120,000 passengers annually. Services run for most of the year, with only six weeks in mid-winter when there are no trains.

Unusually for a 'heritage' railway, the Ratty uses a radio signalling system, which predates adoption of similar principles for operation of rural railways on the national rail network. Key features include centralized control, spring-loaded points at loops, and on-train equipment rather than fixed equipment at remote locations. Over seven miles of single-track route there are five request halts and three intermediate crossing loops, and the line climbs at gradients as steep as 1 in 55. In that short distance, the scenery changes from tidal estuary through attractive pastoral country to craggy hillsides, past abandoned quarries to the line's terminus of Dalegarth, at the foot of England's highest mountains, with the Scafell range (3,209 ft) at their peak.

three owners of the coast line (the Furness Railway, the LNWR and the Maryport & Carlisle Railway) were all grouped into the massive London Midland and Scottish Railway.

A new Barrow Central station – built in an unusual rustic style for a heavily industrialized area – had opened in 1882, but was severely damaged, like much of the surrounding town, by German bombing in 1941. Peacetime brought nationalization of the railways and in due course further closures of iron and steel works in Furness and West Cumbria, but the area received a major boost economically with the 1956 opening of Britain's first nuclear power station at Calder Hall and the adjacent Windscale nuclear reprocessing plant.

A rail connection to the latter in due course brought in trains of irradiated nuclear elements from power stations across Britain – this and passenger traffic generated by workers commuting to the site almost certainly saved the central section of the Cumbrian Coast line, from Whitehaven to Barrow, from threatened closure in the mid-1960s. The site became highly controversial after a fire on 10 October 1957 – the worst nuclear accident in UK history, releasing radioactive material that spread as far as mainland Europe. Part of the site was later renamed Sellafield in response to the notoriety of the name Windscale, but trains have continued to carry nuclear material without incident throughout the life of the nuclear complex.

Industrial contraction and the rise of the car and the lorry did however bring a steady erosion of all the standard-gauge freight and passenger branch lines that fed into the 'main line' through Furness and West Cumberland. The Coniston branch – which involved a circuitous route from the main population centres – closed completely in 1962, and the 'Beeching Axe' closed the Windermere Lake Side branch in 1965. Fortunately part of the latter has survived as the Lakeside and Haverthwaite Railway, a heritage steam operation, which is now the last surviving Furness Railway branch line. The final passenger branch to go – in 1966 – was the highly scenic line from Workington through Cockermouth to Keswick, which could and should have continued as a sustainable alternative to the excessive car traffic and new roads which have blighted the most popular parts of the Lake District. The one branch line survivor is the much-loved narrow-gauge Ravenglass & Eskdale Railway, which runs for seven miles inland from Ravenglass on the Cumbrian Coast line to Dalegarth for Boot (see opposite).

Today's regular passenger train services along the coast are operated by Northern, while south of Barrow their local trains to and from Lancaster are supplemented by First Transpennine limited-stop services to Preston and Manchester. Between Barrow and Carlisle local trains run virtually hourly during the day, but there are no evening services on the more sparsely populated section between Millom and Whitehaven. The Cumbrian Coast Line – still operated by traditional manual signal boxes, is a popular destination for charter trains such as the *Northern Belle*, and the scenic delights of this north-western outpost of the English rail network are rightly becoming better known to a much wider national audience.

Davey Paxman 2-8-2 'River Esk' heads a Dalegarth-Ravenglass narrow-gauge train past The Green in September 2011.

Classic coastal terrain – a Class 156 on a Carlisle-Barrow service near Whitehaven in May 2011.

The distinctive and well-proportioned station building at Aspatria dates back to the original nineteenth-century Maryport & Carlisle Railway.

ROUTE DESCRIPTION

Whitehaven to Carlisle

Immediately beyond Whitehaven's utilitarian modern station, the railway returns to the coast and the first of the sea defences that line much of the next seven dramatic miles to Workington, part of which is just single track. The town of Workington (population 24,000) was formerly a major steel-making centre, manufacturing rails for export across the world. It still has the busiest port in Cumbria – with its own rail connection – handling forest products and containers and, increasingly, components for the offshore wind industry, as well as their operations and maintenance bases.

North of Workington station is the site of one of the rail industry's greatest achievements of modern times in Cumbria. On the night of 19 November 2009, torrential rain led to the main road bridge from the north into Workington being swept away by flood waters – but the railway bridge survived, and within days Network Rail had constructed a temporary station north of the River Derwent. With the help of the UK Government and local authorities, the Carlisle-headquartered Direct Rail Services (a freight haulage subsidiary of the Nuclear Decommissioning Authority) operated a regular special train service linking the two stations through to October 2010 when a new road bridge was opened.

Five miles north of Workington the railway reaches the former industrial town of Maryport and then swings inland and eastward on the final section of the line towards Carlisle. Now passing through more pastoral gently rolling country, the railway has attractive original station buildings still surviving at Aspatria and at Dalston, where daily trains from the Grangemouth refinery in central Scotland bring oil for local distribution by road.

The final approaches to Carlisle station reflect the town's long and complex railway history – passing the site of the former Glasgow & South Western Railway's Currock locomotive shed (more than nine miles from the company's nearest tracks) which survived as a wagon workshop into the early twenty-first century, and then crossing by bridge the freight lines connecting the Cumbrian Coast line with the routes to Newcastle and Settle. Carlisle's Citadel station provides a suitably grand end to the spectacular 115-mile journey from Lancaster. The train takes over three and a half hours, calling at up to thirty-four intermediate stations (almost half of which are request stops), compared to as little as fifty-one minutes via the West Coast Main Line – but this is a journey to be savoured and an intriguing rail corridor to be explored at leisure.

LANCASTER TO CARLISLE VIA OXENHOLME

WEST COAST MAIN LINE

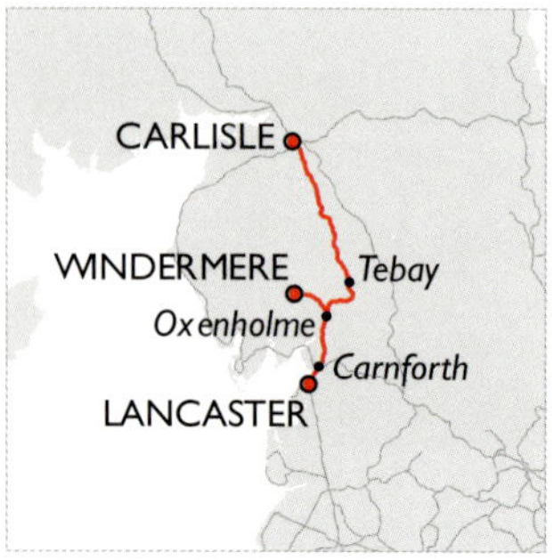

This mountainous section of the West Coast Main Line climbs steeply through the Lake District to one of Britain's highest railway summits at Shap before making a long descent to the Border town and major junction of Carlisle.

The origins of the West Coast Main Line through Cumbria lie in the mid-1830s, when a railway to Scotland was beginning to be seen as a logical extension of the early trunk routes advancing north from London – either the line under construction to Preston or the envisaged 'East Coast' line from London through York to Newcastle. Years of deliberation about alternative inland or coastal routes north of Lancaster ensued and despite a strong steer from parliamentary commissioners towards the route finally selected, it was not in fact until 1843 that sufficient funds were raised to begin promotion of the Lancaster & Carlisle Railway.

Joseph Locke – who had surveyed the route on behalf of the Grand Junction Railway back in 1836-7 – was appointed Engineer-in-Chief, leading what had become a race against the East Coast route which, by 1843, was advancing towards Newcastle with plans to push onwards to Berwick. The Royal Assent in 1844 provided for £900,000 capital, comprising both local investors and 'southern' partners in the shape of the London & Birmingham Railway, the Grand Junction Railway, the North Union Railway and the Lancaster & Preston Junction Railway – a clear indication of how this sixty-nine-mile railway was seen from the very beginning as fulfilling a key Anglo-Scottish trunk route function.

The engineering contract was the largest placed to date for a railway – of necessity a massive undertaking across hard terrain in potentially difficult weather. Work began shortly after the Royal Assent and although the project over-ran, it was no mean achievement to have opened for passenger service through to Carlisle by late 1847 – tribute to the work of nearly 10,000 navvies from England, Scotland and Ireland, who also interspersed their work on the job with regular bouts of drinking, fighting and/or striking!

The route originally envisaged by parliamentary commissioners was indeed followed, but with the crucial difference that instead of tunnelling through the limestone hills north of Tebay, Joseph

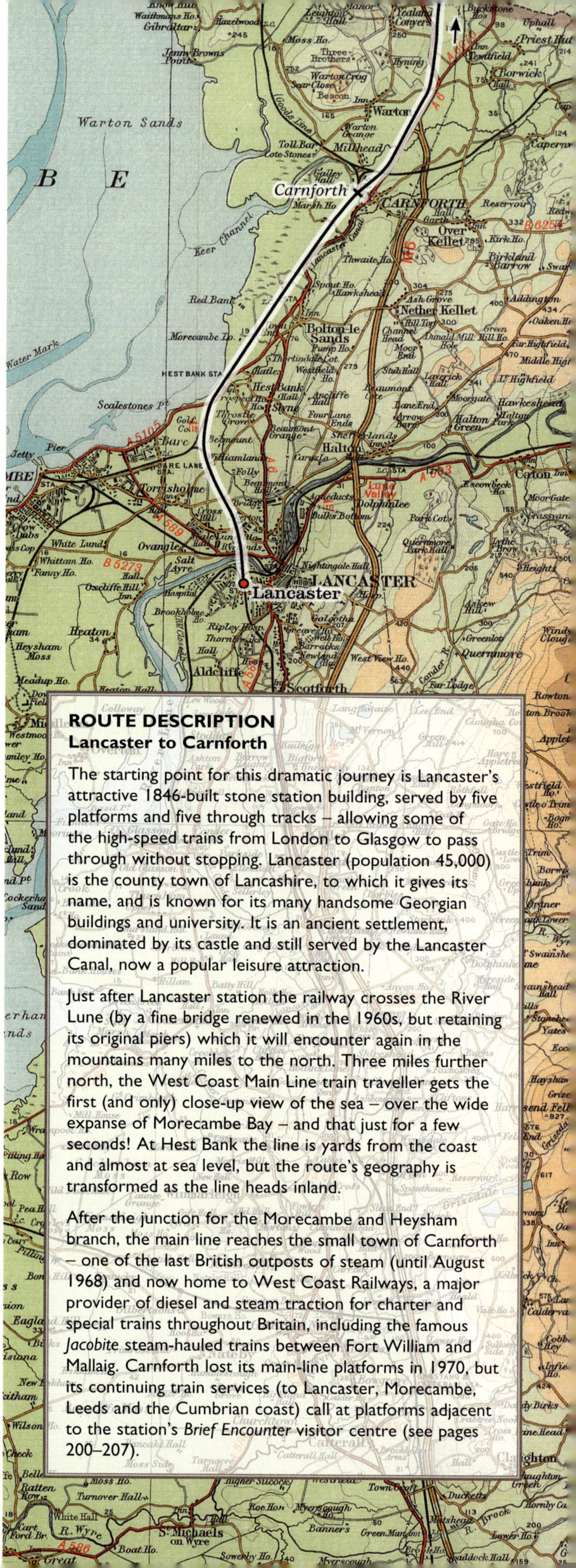

ROUTE DESCRIPTION
Lancaster to Carnforth

The starting point for this dramatic journey is Lancaster's attractive 1846-built stone station building, served by five platforms and five through tracks – allowing some of the high-speed trains from London to Glasgow to pass through without stopping. Lancaster (population 45,000) is the county town of Lancashire, to which it gives its name, and is known for its many handsome Georgian buildings and university. It is an ancient settlement, dominated by its castle and still served by the Lancaster Canal, now a popular leisure attraction.

Just after Lancaster station the railway crosses the River Lune (by a fine bridge renewed in the 1960s, but retaining its original piers) which it will encounter again in the mountains many miles to the north. Three miles further north, the West Coast Main Line train traveller gets the first (and only) close-up view of the sea – over the wide expanse of Morecambe Bay – and that just for a few seconds! At Hest Bank the line is yards from the coast and almost at sea level, but the route's geography is transformed as the line heads inland.

After the junction for the Morecambe and Heysham branch, the main line reaches the small town of Carnforth – one of the last British outposts of steam (until August 1968) and now home to West Coast Railways, a major provider of diesel and steam traction for charter and special trains throughout Britain, including the famous *Jacobite* steam-hauled trains between Fort William and Mallaig. Carnforth lost its main-line platforms in 1970, but its continuing train services (to Lancaster, Morecambe, Leeds and the Cumbrian coast) call at platforms adjacent to the station's *Brief Encounter* visitor centre (see pages 200–207).

Overlooked by the Howgills, a Virgin Voyager heads past Greenholme, north of Tebay (between Oxenholme and Penrith) with a Penzance-Glasgow train in August 2005.

A Royal Mail electric multiple unit heads south through the Lune Gorge on a Motherwell-Warrington postal service in May 2007.

Locke opted to save costs and time by following a surface route. For more than a century this bequeathed a major handicap to railway operation, with 'banking' engines (based at Tebay) required to assist trains in the rear over the four gruelling miles of 1-in-75 gradient to Shap Summit until the end of steam on the route in 1967.

With the completion of the Caledonian Railway from Carlisle to Glasgow and Edinburgh in 1848, the West Coast partners established the first continuous railway from London to Scotland's big two cities more than two years ahead of their East Coast rivals. However, the rivalry was by no means over and has extended right into the modern era of electric traction and high-speed trains.

In 1879 the Lancaster & Carlisle Railway finally became part of the giant London & North Western Railway, which had been created back in 1846. From the earliest days the main line through the Lake District had been a catalyst for the construction of connecting cross-country routes and branch lines, the first (see right) being the Kendal & Windermere Railway from Kendal Junction (now called Oxenholme). As on the other new railways, the Lancaster-Carlisle line widened the economic reach of people, goods and services – and while much of the terrain could not support any great expansion of economic activity, limestone, granite and milk for urban markets became key local freight traffics.

The railway was co-host to the famous 'railway races' of the late nineteenth century. The shorter and more easily graded East Coast route from London to Edinburgh (operated by the Great Northern, North British and North Eastern companies) was originally much faster than the West Coast route (operated by the London & North Western and Caledonian) – nine hours against ten. But over a period of two months from early June 1888, the schedules on both routes got faster and faster, until the East Coast train completed the journey in just seven hours and twenty-seven minutes on 13 August. The next day both sets

Deep in the Lake District, a service from Manchester Airport nears the end of the line at Windermere in August 2004.

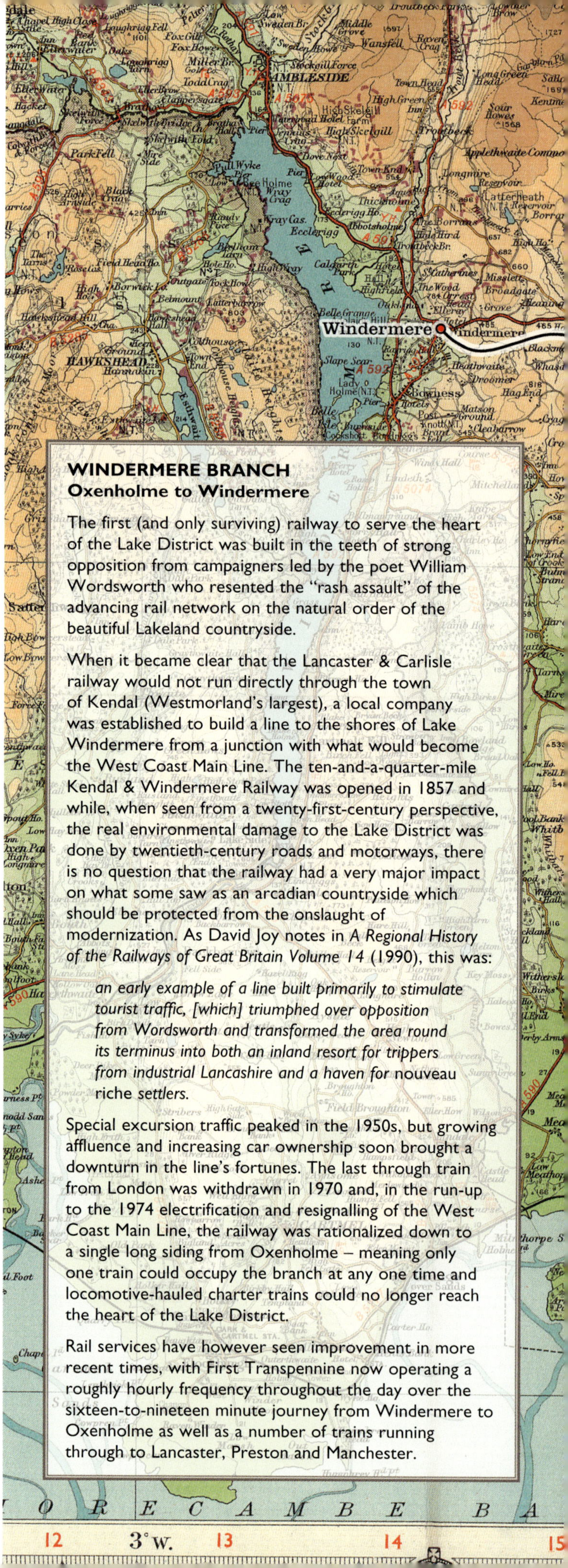

WINDERMERE BRANCH
Oxenholme to Windermere

The first (and only surviving) railway to serve the heart of the Lake District was built in the teeth of strong opposition from campaigners led by the poet William Wordsworth who resented the "rash assault" of the advancing rail network on the natural order of the beautiful Lakeland countryside.

When it became clear that the Lancaster & Carlisle railway would not run directly through the town of Kendal (Westmorland's largest), a local company was established to build a line to the shores of Lake Windermere from a junction with what would become the West Coast Main Line. The ten-and-a-quarter-mile Kendal & Windermere Railway was opened in 1857 and while, when seen from a twenty-first-century perspective, the real environmental damage to the Lake District was done by twentieth-century roads and motorways, there is no question that the railway had a very major impact on what some saw as an arcadian countryside which should be protected from the onslaught of modernization. As David Joy notes in *A Regional History of the Railways of Great Britain Volume 14* (1990), this was:

> *an early example of a line built primarily to stimulate tourist traffic, [which] triumphed over opposition from Wordsworth and transformed the area round its terminus into both an inland resort for trippers from industrial Lancashire and a haven for nouveau riche settlers.*

Special excursion traffic peaked in the 1950s, but growing affluence and increasing car ownership soon brought a downturn in the line's fortunes. The last through train from London was withdrawn in 1970 and, in the run-up to the 1974 electrification and resignalling of the West Coast Main Line, the railway was rationalized down to a single long siding from Oxenholme – meaning only one train could occupy the branch at any one time and locomotive-hauled charter trains could no longer reach the heart of the Lake District.

Rail services have however seen improvement in more recent times, with First Transpennine now operating a roughly hourly frequency throughout the day over the sixteen-to-nineteen minute journey from Windermere to Oxenholme as well as a number of trains running through to Lancaster, Preston and Manchester.

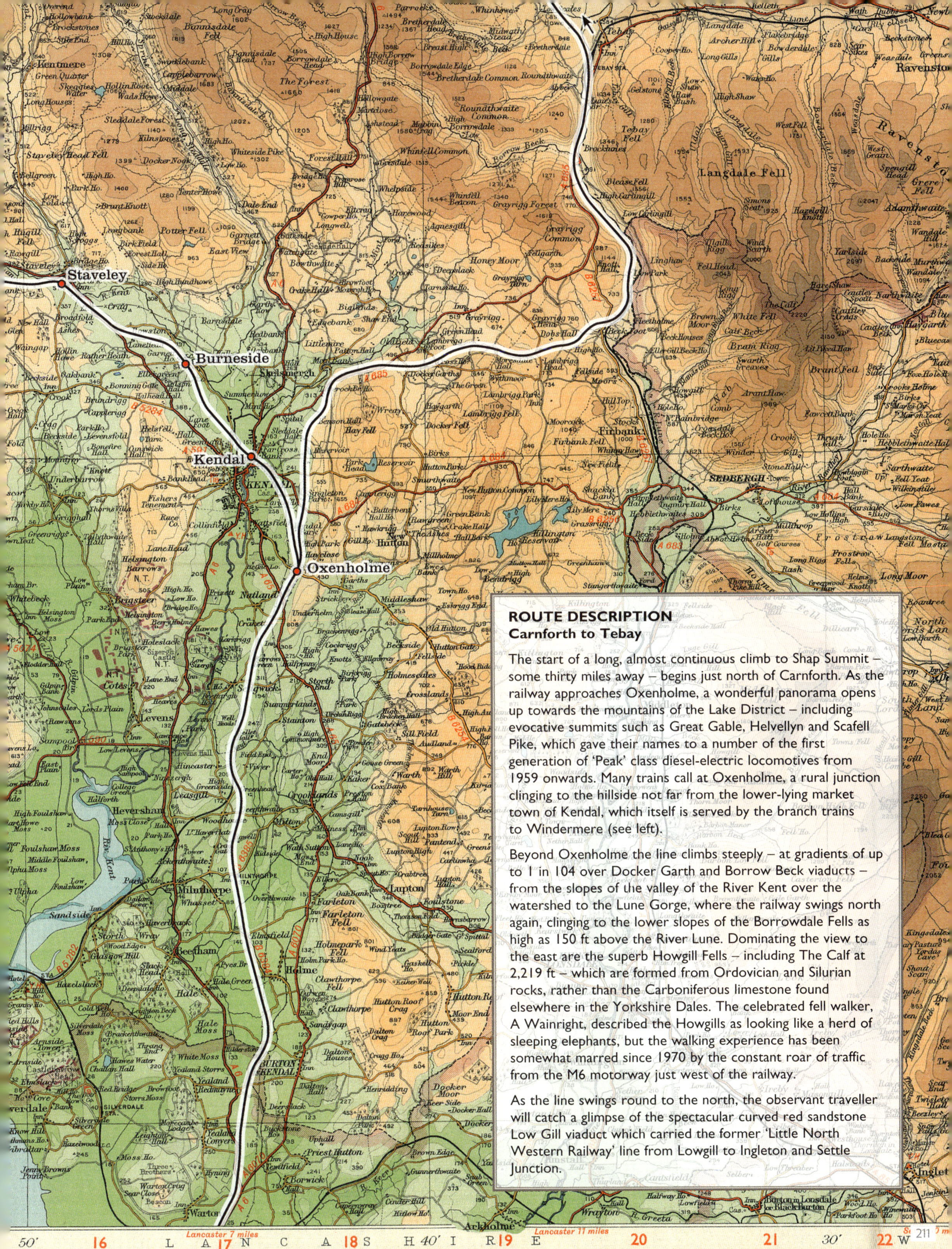

ROUTE DESCRIPTION

Carnforth to Tebay

The start of a long, almost continuous climb to Shap Summit – some thirty miles away – begins just north of Carnforth. As the railway approaches Oxenholme, a wonderful panorama opens up towards the mountains of the Lake District – including evocative summits such as Great Gable, Helvellyn and Scafell Pike, which gave their names to a number of the first generation of 'Peak' class diesel-electric locomotives from 1959 onwards. Many trains call at Oxenholme, a rural junction clinging to the hillside not far from the lower-lying market town of Kendal, which itself is served by the branch trains to Windermere (see left).

Beyond Oxenholme the line climbs steeply – at gradients of up to 1 in 104 over Docker Garth and Borrow Beck viaducts – from the slopes of the valley of the River Kent over the watershed to the Lune Gorge, where the railway swings north again, clinging to the lower slopes of the Borrowdale Fells as high as 150 ft above the River Lune. Dominating the view to the east are the superb Howgill Fells – including The Calf at 2,219 ft – which are formed from Ordovician and Silurian rocks, rather than the Carboniferous limestone found elsewhere in the Yorkshire Dales. The celebrated fell walker, A Wainright, described the Howgills as looking like a herd of sleeping elephants, but the walking experience has been somewhat marred since 1970 by the constant roar of traffic from the M6 motorway just west of the railway.

As the line swings round to the north, the observant traveller will catch a glimpse of the spectacular curved red sandstone Low Gill viaduct which carried the former 'Little North Western Railway' line from Lowgill to Ingleton and Settle Junction.

of companies agreed to minimum train times and the costly races stopped.

By 1895, the focus had moved further north with the opening of the Forth Bridge and Aberdeen was the goal. These mid-summer races lasted just five weeks (see pages 230–241) before the companies agreed that train journeys on both routes should take a minimum of eight and a quarter hours from London to Edinburgh, an agreement that was to last until 1932.

Following the 1923 'Grouping', which put the West Coast Main Line under the sole control of the London, Midland & Scottish Railway (LMS) and the East Coast under the London & North Eastern Railway (LNER), the big new companies began developing more powerful steam locomotives and modern coaching stock. The quest for speed (and luxury) – and its colourful promotion in the classic poster designs of that 'Golden Age' – reached its apotheosis in the mid-1930s, when the LMS and LNER vied for premier position between London and Scotland. The West Coast response to the 1935 introduction of the streamlined 'Silver Jubilee' express on the East Coast came with the streamlined 'Coronation Scot', which was scheduled to take just six and a half hours for its journey.

The Second World War brought this competition to an end, and following nationalization – and the growing threat from the car and the lorry – rail retrenchment began in the 1950s with the withdrawal of services over connecting routes and branch lines. Steam ended in late 1967, with the closure of the Tebay and Carlisle Kingmoor locomotive sheds – and all the intermediate stations on the main line, except Oxenholme and Penrith, had gone by 1970. Two years later, with the closure of the Penrith-Keswick branch, the Lancaster-Carlisle main line lost all but one of its intermediate junctions between Carnforth (for the cross-country routes to the Cumbrian coast and Leeds) and Carlisle, only Oxenholme (for Windermere) remaining.

More positively, electrification and resignalling of the entire northern half of the West Coast Main Line – from Crewe to Glasgow – was completed in 1974. This major project transformed Anglo-Scottish travel – Shap ceased to be an obstacle, with passenger trains reaching the summit at speeds of up to 95 mph. The Lancaster-Carlisle line has since gone from strength to strength, with the introduction of 125-mph electric 'Pendolino' and diesel 'Super Voyager' trains variously serving Oxenholme and Penrith intermediately. The growth of intermodal (containerized) rail freight has had a particular impact on the West Coast Main Line, as the premier British freight route – with trains helping locally to reduce the number of heavy lorries on the M6 through Cumbria and contributing to national carbon emissions reduction through their greater energy efficiency.

Princess Coronation Pacific No. 6233 'Duchess of Sutherland' passes Greenholme on the climb to Shap summit with a rail tour in April 2007.

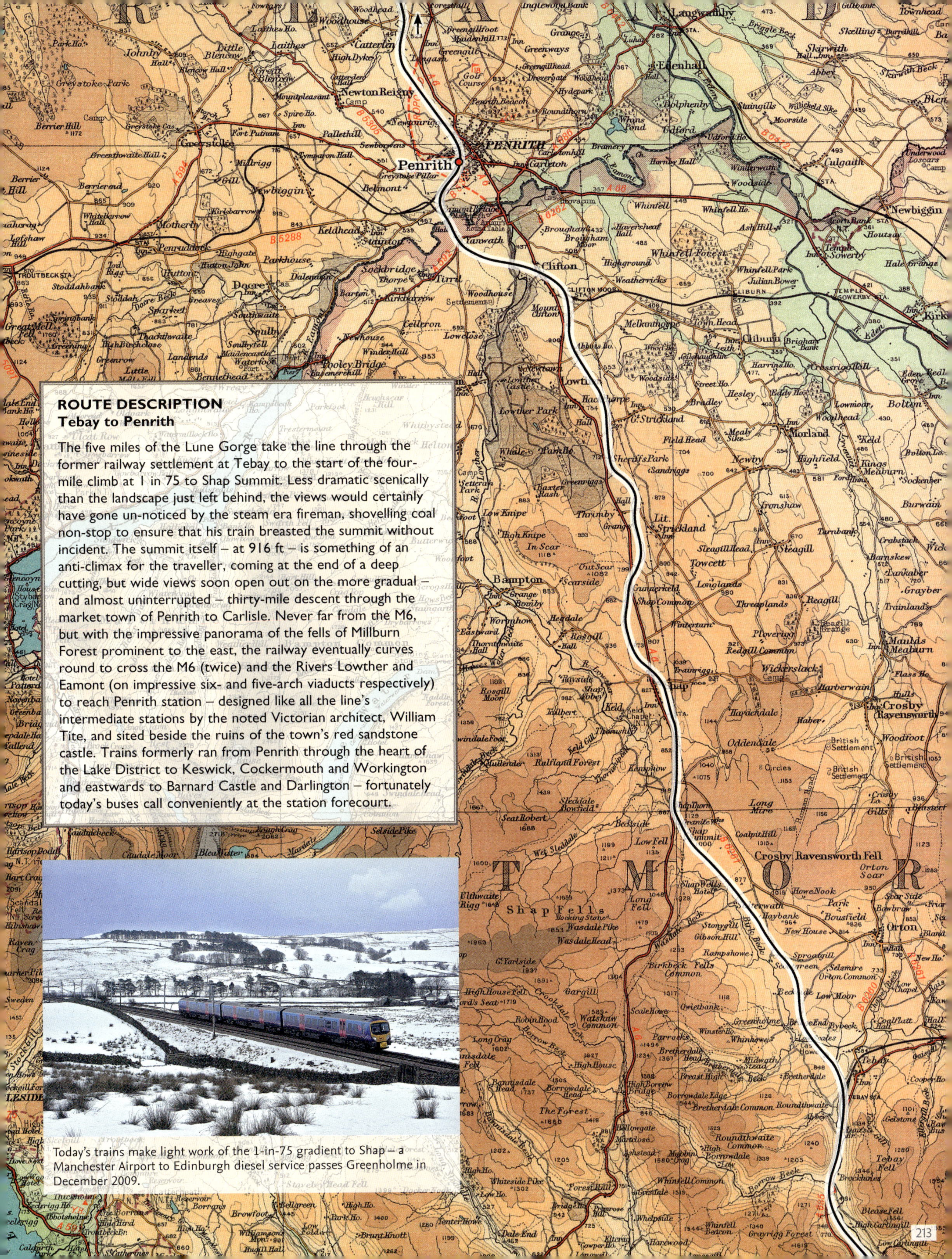

ROUTE DESCRIPTION
Tebay to Penrith

The five miles of the Lune Gorge take the line through the former railway settlement at Tebay to the start of the four-mile climb at 1 in 75 to Shap Summit. Less dramatic scenically than the landscape just left behind, the views would certainly have gone un-noticed by the steam era fireman, shovelling coal non-stop to ensure that his train breasted the summit without incident. The summit itself – at 916 ft – is something of an anti-climax for the traveller, coming at the end of a deep cutting, but wide views soon open out on the more gradual – and almost uninterrupted – thirty-mile descent through the market town of Penrith to Carlisle. Never far from the M6 but with the impressive panorama of the fells of Millburn Forest prominent to the east, the railway eventually curves round to cross the M6 (twice) and the Rivers Lowther and Eamont (on impressive six- and five-arch viaducts respectively) to reach Penrith station – designed like all the line's intermediate stations by the noted Victorian architect, William Tite, and sited beside the ruins of the town's red sandstone castle. Trains formerly ran from Penrith through the heart of the Lake District to Keswick, Cockermouth and Workington and eastwards to Barnard Castle and Darlington – fortunately today's buses call conveniently at the station forecourt.

Today's trains make light work of the 1-in-75 gradient to Shap – a Manchester Airport to Edinburgh diesel service passes Greenholme in December 2009.

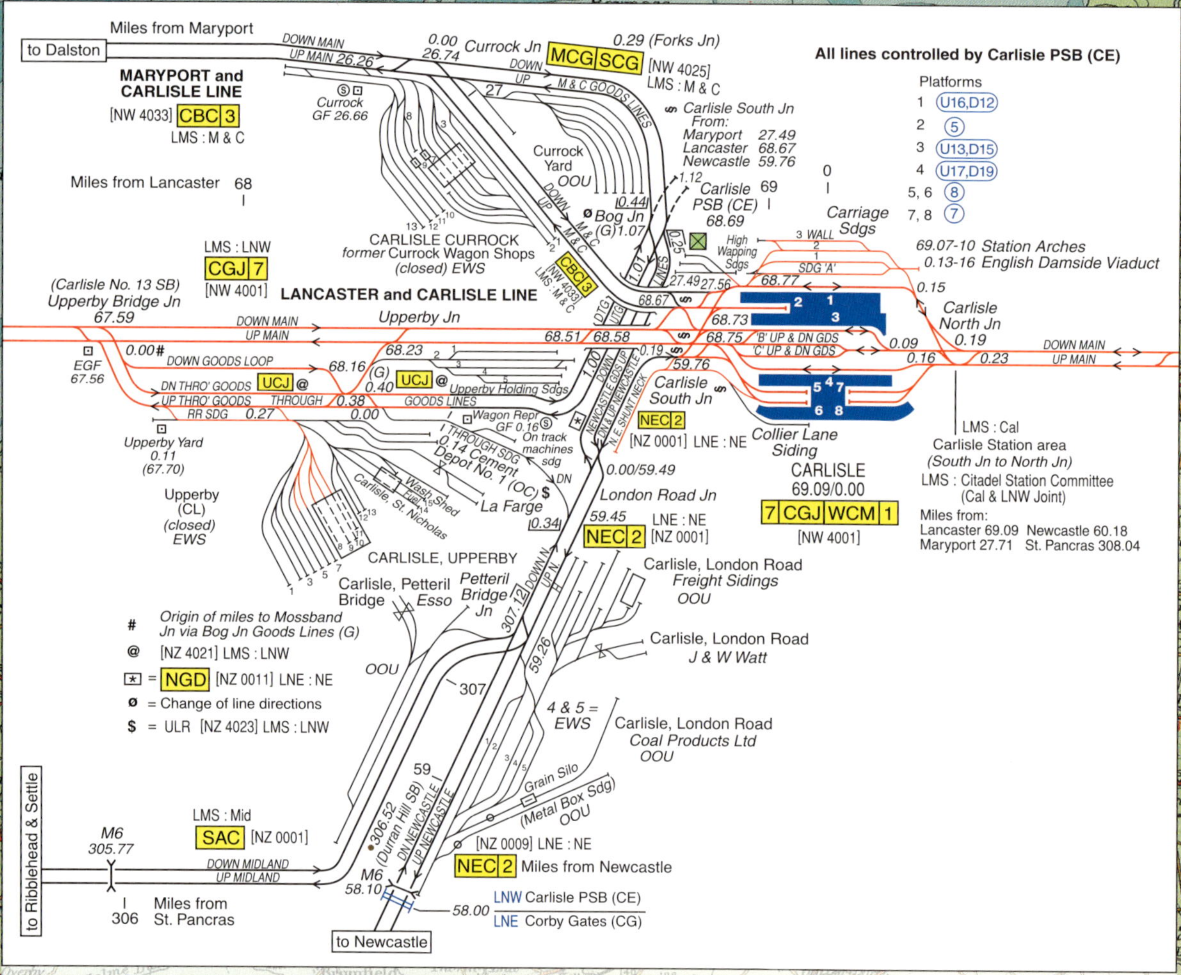

The still-complex railway geography of Carlisle is well illustrated by this track diagram showing the convergence of five different routes within the city's boundaries.

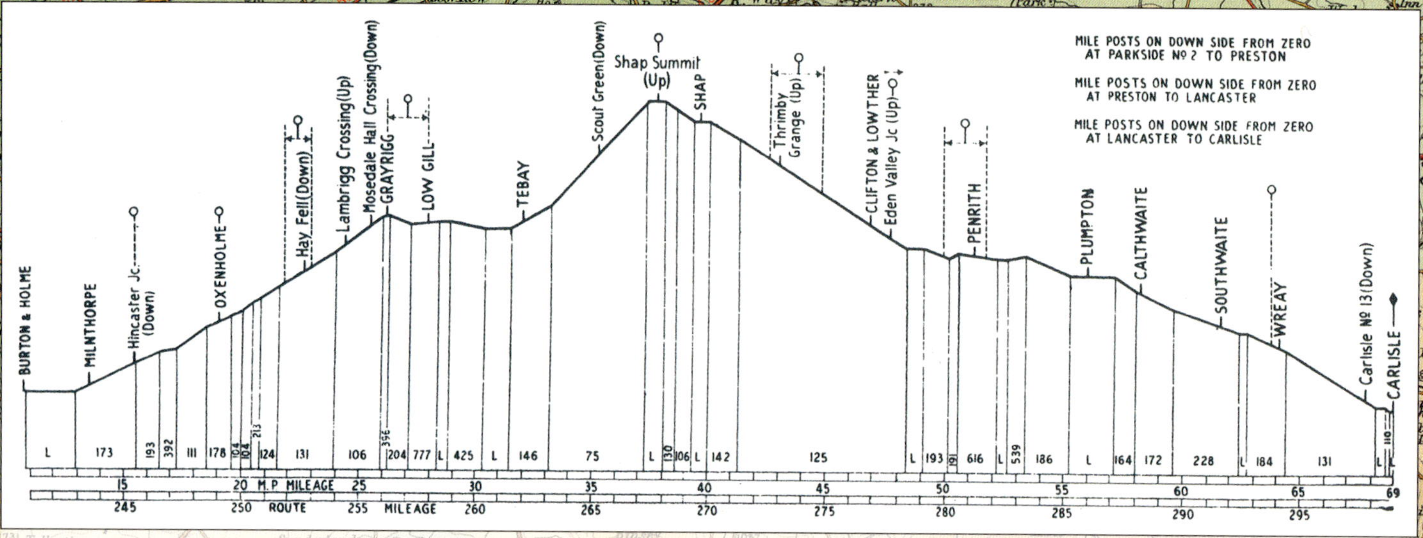

The long northbound and southbound climbs to England's fourth highest railway summit stand out in this gradient profile from north of Carnforth to Carlisle.

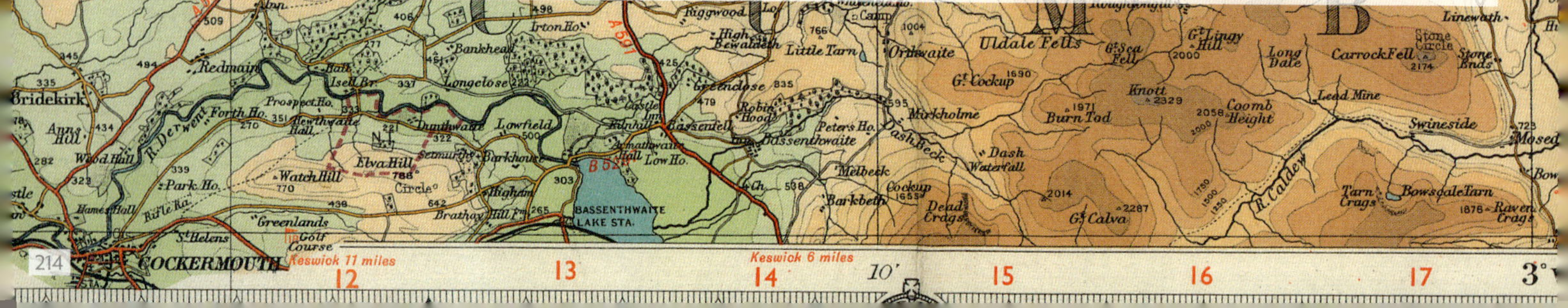

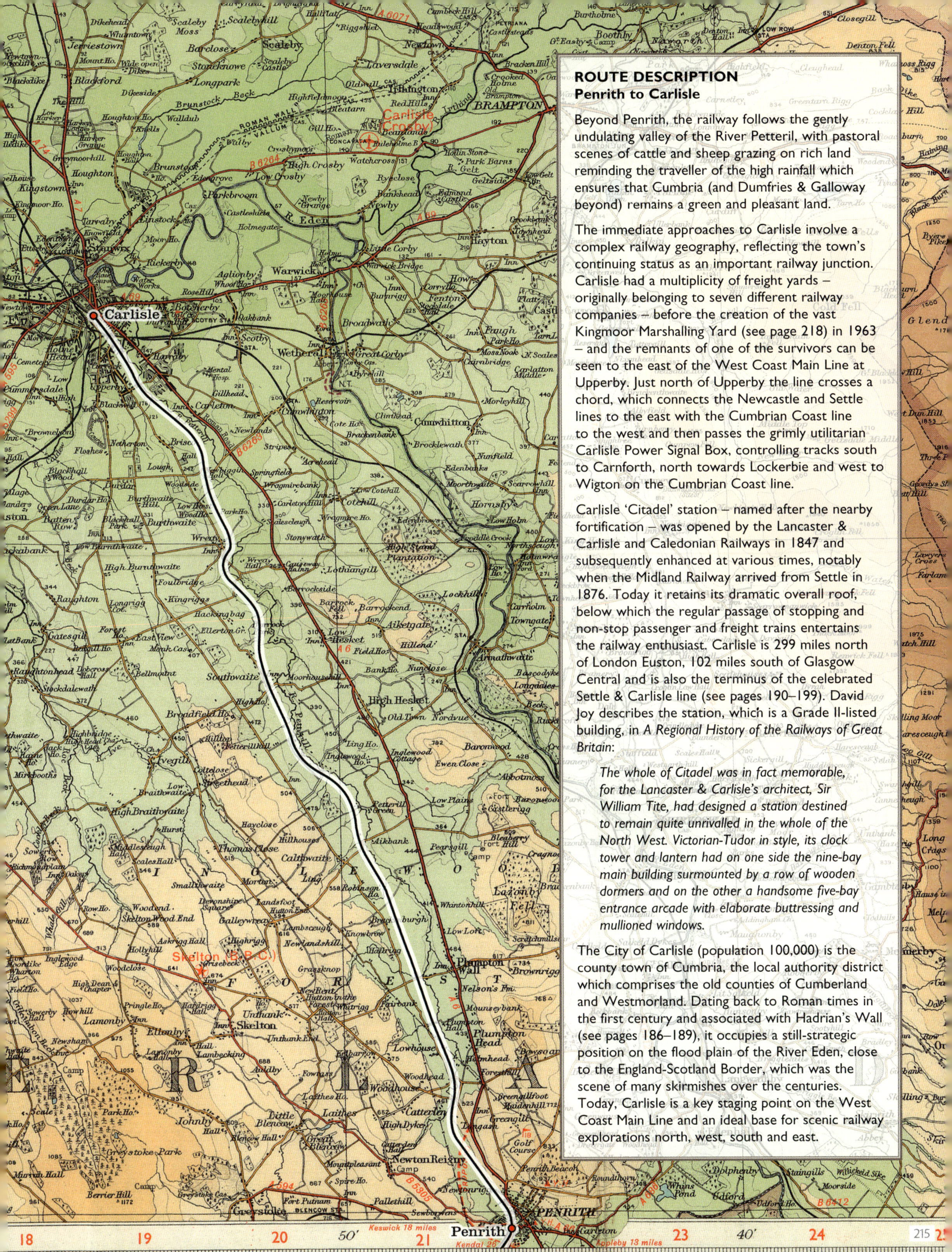

ROUTE DESCRIPTION
Penrith to Carlisle

Beyond Penrith, the railway follows the gently undulating valley of the River Petteril, with pastoral scenes of cattle and sheep grazing on rich land reminding the traveller of the high rainfall which ensures that Cumbria (and Dumfries & Galloway beyond) remains a green and pleasant land.

The immediate approaches to Carlisle involve a complex railway geography, reflecting the town's continuing status as an important railway junction. Carlisle had a multiplicity of freight yards – originally belonging to seven different railway companies – before the creation of the vast Kingmoor Marshalling Yard (see page 218) in 1963 – and the remnants of one of the survivors can be seen to the east of the West Coast Main Line at Upperby. Just north of Upperby the line crosses a chord, which connects the Newcastle and Settle lines to the east with the Cumbrian Coast line to the west and then passes the grimly utilitarian Carlisle Power Signal Box, controlling tracks south to Carnforth, north towards Lockerbie and west to Wigton on the Cumbrian Coast line.

Carlisle 'Citadel' station – named after the nearby fortification – was opened by the Lancaster & Carlisle and Caledonian Railways in 1847 and subsequently enhanced at various times, notably when the Midland Railway arrived from Settle in 1876. Today it retains its dramatic overall roof, below which the regular passage of stopping and non-stop passenger and freight trains entertains the railway enthusiast. Carlisle is 299 miles north of London Euston, 102 miles south of Glasgow Central and is also the terminus of the celebrated Settle & Carlisle line (see pages 190–199). David Joy describes the station, which is a Grade II-listed building, in *A Regional History of the Railways of Great Britain*:

> *The whole of Citadel was in fact memorable, for the Lancaster & Carlisle's architect, Sir William Tite, had designed a station destined to remain quite unrivalled in the whole of the North West. Victorian-Tudor in style, its clock tower and lantern had on one side the nine-bay main building surmounted by a row of wooden dormers and on the other a handsome five-bay entrance arcade with elaborate buttressing and mullioned windows.*

The City of Carlisle (population 100,000) is the county town of Cumbria, the local authority district which comprises the old counties of Cumberland and Westmorland. Dating back to Roman times in the first century and associated with Hadrian's Wall (see pages 186–189), it occupies a still-strategic position on the flood plain of the River Eden, close to the England-Scotland Border, which was the scene of many skirmishes over the centuries. Today, Carlisle is a key staging point on the West Coast Main Line and an ideal base for scenic railway explorations north, west, south and east.

SCOTLAND

The apotheosis of steam travel in the grandest of British scenery – late afternoon sun catches a Mallaig-bound train skirting the south side of Loch Eilt, between Glenfinnan and Lochailort.

ROUTE DESCRIPTION

Carlisle to Lockerbie

Emerging from the impressive 'train shed' (overall roof) of Carlisle station – formerly known as Carlisle Citadel – the line heads northwest and within the first five miles is, unusually, crossed by two other railways. The first carries the remaining short freight stump of what was once the ninety-eight-mile Waverley Route to Edinburgh via Hawick and Galashiels – probably the worst of all the 'Beeching cuts', leaving the Scottish Borders as the only mainland region of Britain without a train service. The second rail overbridge is a flyover carrying freight trains from Scotland over the West Coast Main Line and into Kingmoor Yard, one of the major marshalling yards constructed as part of British Railways' 1955 Modernisation Plan. By the time Kingmoor was opened in 1963, much of the traffic it was intended to carry had already been lost to road haulage and the yard never fulfilled its potential. Its extensive remaining sidings are still, however, an important stabling and exchange point for Anglo-Scottish freight trains.

The railway then runs over the low-lying estuary of the River Esk, with the Solway Firth stretching far to the west, and crosses the England-Scotland border at the River Sark, where it runs close to the village of Gretna Green, formerly famed as the host for 'runaway marriages'. At Gretna Junction, the former Glasgow & South Western Railway route to Glasgow via Dumfries and Kilmarnock – now an important cross-country line for passenger and coal traffic – diverges north westwards. Just a mile beyond the junction are the loops at Quintinshill, the site of the worst accident in British railway history, in 1915. Due to a signalman's negligence, an express passenger service and a special troop train – both travelling at speed – collided with two stationary goods trains and a local passenger service. The collision and the ensuing inferno killed twelve civilians and 215 officers and men of the Royal Scots.

Ironically, just fifteen miles north of Quintinshill, the railway passes through the town of Lockerbie, scene of the worst air disaster in British history – the bombing of Pan Am Flight 103 on 21 December 1988.

A stylish tariff from the restaurant car of the 'Caledonian' named day train which linked London Euston and Glasgow Central in the 1950s and 60s.

A Pendolino tilting train from Glasgow Central to London Euston speeds across the flat border lands north of Carlisle in December 2010.

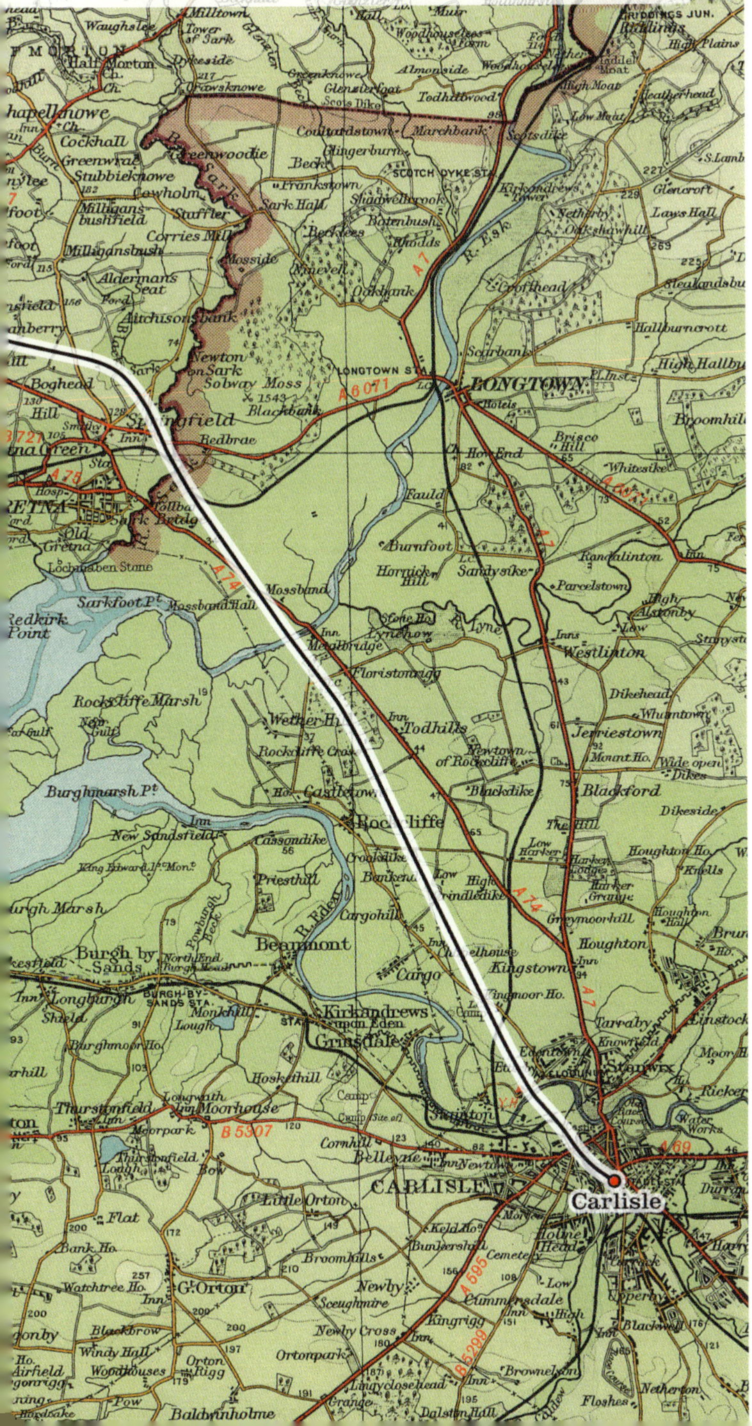

CARLISLE TO CARSTAIRS

WEST COAST MAIN LINE

The final upland stretch of the London-Glasgow railway takes the line across the England-Scotland Border, up Annandale and over Beattock Summit before dropping down through Clydesdale to Carstairs, the junction for Glasgow and Edinburgh.

The first railways in Central Scotland were opened in the 1830s, physically independent of the developing network in the north of England, but by 1840 putative railway promoters were showing intense interest in constructing cross-Border routes through or round the Southern Uplands. However, an 1841 Royal Commission – little realizing the traffic-generating potential of new trunk railways – concluded that Anglo-Scottish business justified the construction of just one line, and that:

> *so far as regards the interest of the traveller, both with respect to the economy of his time and of his purse, the preferable route for the railway communication to Edinburgh and Glasgow would be by the proposed Carlisle and Lockerbie line with a branch from Symington.*

The 'Caledonian Railway' – soon to extend its territory across central Scotland and north to Aberdeen and Oban – was authorized in 1845 and opened throughout from Carlisle to Glasgow and Edinburgh in 1848. Substantial engineering works – including twenty-three viaducts, but surprisingly no tunnels – were required to take the line through the Southern Uplands to the junction at Carstairs. The biggest operating challenge was the long northbound slog of ten miles up to Beattock Summit, at gradients of 1 in 74 to 1 in 88. Throughout the steam era, trains required assistance at the rear from banking locomotives based at Beattock station.

The West Coast route was the first to offer a direct rail service from London to Scotland without a change of train. The railway had an immediate impact – its first passenger expresses took twelve and a half hours for the through journey and the Post Office soon switched its London-Edinburgh mail traffic from the East Coast to the West Coast. The market for cattle was transformed, with livestock trains from Scotland to London's Smithfield market delivering cattle quickly and in prime condition, compared to the traditional droving and shipping arrangements.

As described in the feature on the Lancaster-Carlisle section (see pages 208–215), the West Coast Main Line has enjoyed a

ROUTE DESCRIPTION

Lockerbie to Beattock Summit

Beyond Lockerbie, the Southern Uplands become more prominent, with the Lowther Hills to the northwest and the Moffat Hills to the northeast of the village of Beattock.

The stiff climb to Beattock Summit begins just north of Beattock station, as famously recorded in *Night Mail*, the 1936 documentary film produced by the GPO Film Unit, with music by Benjamin Britten and a poem by WH Auden featuring the memorable opening lines: "This is the Night Mail crossing the border, Bringing the cheque and the postal order..." While letters are no longer sorted on railway 'Travelling Post Offices', London-Glasgow mail still travels by train, in fast purpose-designed electric units. The locomotive which starred in *Night Mail* – LMS Royal Scot No. 46115 'Scots Guardsman' – survived into preservation, and enjoyed much later fame on 20 June 2012 when it conveyed the Olympic Torch along the East Coast Main Line from York to Thirsk.

The train takes the strain – an electric-hauled intermodal service from Daventry to Mossend crosses the Clyde Viaduct at Crawford in May 2010.

regular association with speed, both during the 'railway races' of 1888 and 1895 and in the mid-1930s when the London, Midland & Scottish Railway and the London & North Eastern Railway – created in the 'Grouping' of 1923 – vied with each other to achieve the fastest and most luxurious journeys from London to Scotland.

The outbreak of the Second World War brought an end to the high-speed train services and the West Coast Main Line (WCML) reverted to a more prosaic – but crucial – role, supporting the war effort. After peace returned, the rail system was soon nationalized, but the first major change in the Carlisle-Carstairs operation did not come until the start of dieselization in the 1950s. By 1967 'bankers' were just history and so, by 1972, were all the intermediate stations – except Lockerbie – following the implementation of the 1963 'Beeching Report' proposals. This left the forty-eight miles between Lockerbie and Carstairs as by far the longest section of railway in Britain without an intermediate station, albeit along a sparsely populated corridor.

1974 saw the completion of WCML electrification from Crewe Junction to Glasgow, allowing a dramatic acceleration of train service. That same decade British Rail began developing the 'Advanced Passenger Train', designed to tilt on curves and thus be able to use the existing railway alignment to run at speeds of up to 155 mph. Three electric prototypes were built, but extensive trials between London and Glasgow were ultimately unsuccessful and the project was abandoned in 1986. Today's tilting 'Pendolino' trains – which operate at up to 125 mph and complete the 401-mile journey in as little as four hours and eight minutes – were developed in Italy, although final assembly was undertaken in Britain.

The upgraded WCML is now busier than ever, and the Carlisle Carstairs section sees more freight traffic than any other route in Scotland, with up to fifty trains every twenty-four hours conveying a wide variety of commodities from coal through import/export containers to supermarket supplies.

Despite passenger train speeds of up to 125 mph, the attractive views can still be fully appreciated from the train, as much of the dramatic scenery is at some distance from the railway. Carlisle to Carstairs provides a perfect foretaste of the scenic delights to come across Scotland's railway network.

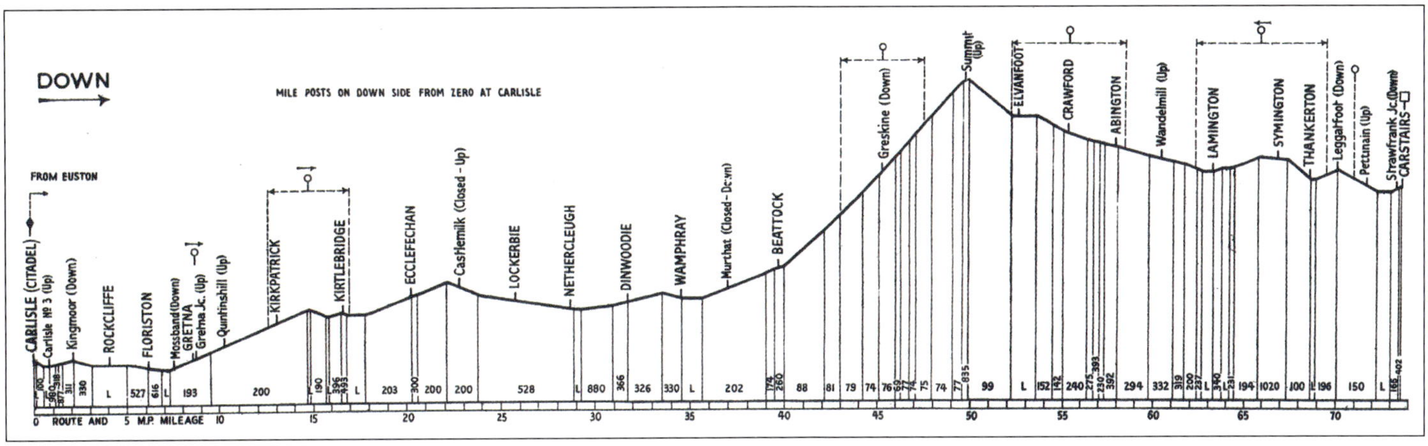

The contrast in severity of the northbound and southbound climbs to Beattock Summit is shown clearly in the Carlisle-Carstairs gradient profile.

A northbound Pendolino descends Clydesdale through Scotland's Southern Uplands in September 2005.

ROUTE DESCRIPTION
Beattock Summit to Carstairs

Today's electric Pendolino and diesel Super Voyager trains are so highly powered that the rail traveller now has little sense of the severity of the climb faced by steam engines over the first 120 years of the line's history. Beyond Beattock Summit (the fifth highest in Scotland, at 1,016 ft), a less severe gradient – 1 in 99 at its steepest, just after the summit – takes the railway twenty-three miles down Clydesdale to Carstairs. Over the fifty-eight miles from Carlisle to Abington the railway and the M74 motorway are rarely more than a mile apart and the contrast between the rail and road footprints is striking – the railway occupies the same double-track corridor as in Victorian times, but the M74 has six lanes and the immediately parallel 'all-purpose' B7076 / A702 road has a further two.

At Abington the motorway diverges northwestwards through the hills, while the railway continues northeastwards along the Clyde Valley, skirting around the impressive bulk of the Tinto Hills and its highest peak at Tinto (2,333 ft). The village of Symington sits just east of the Tinto Hills, and was formerly the junction for the Caledonian Railway line through Biggar to the Borders market town of Peebles.

Until 1990, Carstairs was an important operational hub on the railway – not just a junction, but also the point where trains from Edinburgh and Glasgow were joined before proceeding south to Manchester, Liverpool, Birmingham and southwest England. Now only the overnight sleeper between Edinburgh/Glasgow and London is split and joined at Carstairs. Trains from Carlisle to Edinburgh no longer pass through the station, swinging away eastward at Carstairs South Junction (formerly Strawfrank Junction) instead. Interestingly, although the station is now officially known as just 'Carstairs', the village that adjoins it is called 'Carstairs Junction', while 'Carstairs' village is a mile away to the west!

Beyond Carstairs the West Coast Main Line heads through urban-industrial Lanarkshire to Glasgow – the gateway to the West Highlands and an attraction in its own right, with many museums, art galleries, parks and a riverside walkway along the River Clyde. Most of the route to Edinburgh is sparsely populated, skirting north west of the Pentland Hills, before dropping down into Edinburgh – capital of Scotland, home of the Scottish Parliament, and a city that needs little introduction.

Pacific No. 71000 'Duke of Gloucester' – built by British Railways in 1954 – heads north through Abington with a charter train in April 2008.

Carluke 2 miles
Carluke 3 miles
West Calder 12 miles
Edinburgh 24 miles
21
22
23
24
25
26
27
40'
30'
CARSTAIRS
Carstairs
Carstairs Junction
Carstairs Mains
CARNWATH
Newbigging
Walston
WALSTON
DOLPHINTON
Elsrickle
LANARK
Lanark
Kirkfieldbank
Cleghorn
Lanark Moor
Hyndford
Pettinain
PETTINAIN
Libberton
LIBBERTON
Covington
Quothquan
Biggar Common
BIGGAR
Biggar
Skirling
SKIRLING
Thankerton
Carmichael
CARMICHAEL
COVINGTON
Symington
Tinto
TINTO 2335
Coulter
CULTER
BROUGHTON
GLENHOLM
KILBUCHO
Wiston
WISTON
Lamington
LAMINGTON
WANDEL
Roberton
ROBERTON
Douglas Water
Culter Fell
Abington
Crawfordjohn
Crawford
CRAWFORD
Leadhills
Wanlock
Harleburn Head
Clyde Law
Tweedshaws
River Tweed
Biggar Water
Crawfordjohn
Duneaton Water
Camps Water
Daer Water
Elvan Water
Louise Wood Law

BERWICK-UPON-TWEED TO EDINBURGH

EAST COAST MAIN LINE

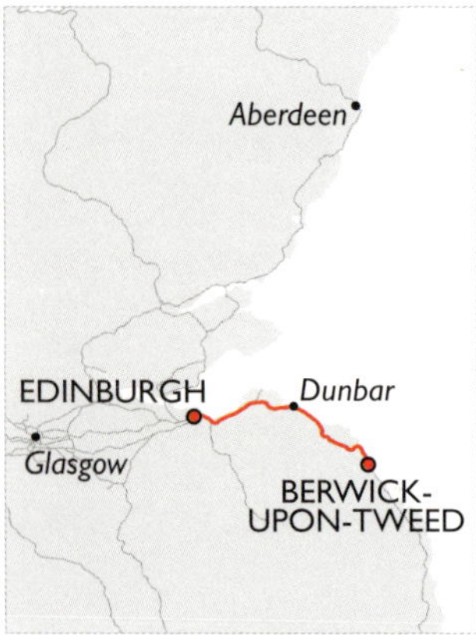

The final fifty-seven and a half miles of the East Coast Main Line from London to Edinburgh carry Anglo-Scottish express trains along cliff tops flanking the North Sea, crossing the Border before swinging inland skirting the rolling Lammermuir Hills and descending into the rich farming country of East Lothian.

The railway from Edinburgh to Berwick was planned in the early 1840s by the directors of the new North British Railway (NBR) as the first phase of a link to Newcastle and onwards to London. The NBR was authorized in 1844 and the line opened for traffic in 1846. On several occasions the York-based North Eastern Railway sought to take over the NBR, but was thwarted, only to succeed in securing 'running powers' in 1862 for their trains to operate over the NBR's tracks into Edinburgh. Overnight sleeper services were introduced between Edinburgh and London in 1873 and in 1893 dining cars appeared, with all coaches interlinked with corridor connections providing every passenger with access to the facilities.

The Berwick-Edinburgh line played host to one route of the famous – but short-lived – 'railway races' of 1888 and 1895, when trains from London to Edinburgh and Aberdeen respectively via the East and West Coast Main Lines competed at ever-faster speeds to win traffic from their rivals. After the last of the races in 1895 (see pages 230–241), agreement was reached between the rival companies on minimum journey times, remaining in place until as late as 1932. The quest for speed led to the introduction in 1937 of the 'Coronation' train, with its streamlined A4 locomotive, specially-built articulated coaches and a streamlined observation car at the rear – reducing the journey time from London to Edinburgh to just six hours.

In 1948, after the railways' exhausting war role, came nationalization. In August of that year, just months into the new era, the new British Railways Scottish Region faced a major challenge when unprecedented rainfall led to breaching of the railway by landslips and washouts at many locations between Berwick and Dunbar. Trains were diverted inland via Kelso and the Waverley Route, but after just eleven weeks the railway engineers and contractors had restored the line.

Further dislocation – and tragedy – struck the line in March 1979 when Penmanshiel Tunnel, south of Dunbar, collapsed during reconstruction work to accommodate higher containers. Two workers were entombed, the tunnel had to be sealed up and the line was then diverted around the site. Modernization work, which commenced in the 1970s, culminated in the East Coast Main Line's electrification in 1991, allowing the fastest train from Edinburgh to London to complete the journey in just three hours and fifty-nine minutes.

An Edinburgh-Birmingham 'Voyager' train speeds along the cliff tops south of Burnmouth in September 2004.

The East Coast Main Line between Berwick and Edinburgh is now busier passenger-wise than at any time in its 160-plus years of operation – primarily with Anglo-Scottish inter-city trains from as far afield as London, Birmingham, Bristol and Penzance, heading for Edinburgh, Glasgow, Dundee, Aberdeen and

Inverness. Local ScotRail services operate to Edinburgh from Dunbar and the branch terminus at North Berwick to Edinburgh, serving the surviving intermediate stations in East Lothian.

As a premier trunk route, the line is double track throughout, controlled by centralized signalling centres in Edinburgh and Tweedmouth (just south of Berwick). Line speeds are as high as 125 mph and the final approach to Edinburgh is one of the fastest to any city in Britain, with trains usually decelerating less than two miles from the terminus at Waverley. On the southbound climb to Penmanshiel, passengers on (very rare) occasions may experience the disorienting sensation of overtaking another southbound train on the parallel track, as both tracks here are 'bi-directional' to allow slow-moving freight services to be overtaken on the move.

This stretch of railway provides a fittingly impressive conclusion to the fast rail journey from London. Despite its high speed, the journey still provides ample opportunity to appreciate the subtleties of the landscape where Scotland and England meet, with its impressive coastal and inland scenery often bathed in sunshine. The express train delivers a tantalizing flavour of the countryside, towns and villages of Berwickshire and East Lothian, ripe for further exploration by foot, bike and local train.

The LNER's streamlined 'Coronation' train ran between London and Edinburgh from 1937 until the outbreak of the Second World War.

A northbound Virgin Voyager train speeds past the former Innerwick station in the lush lands of East Lothian in July 2006.

ROUTE DESCRIPTION

Berwick-upon-Tweed to Dunbar

From Berwick station the line heads north along the cliff tops overlooking the North Sea (known as the German Ocean before the First World War), soon passing the historic lineside indicators marking the England-Scotland boundary at Marshall Meadows. At several locations on this section of the route, the railway has had to be shifted tens of metres inland to avoid imminent cliff-top slippage.

Beyond Burnmouth, formerly the junction for a short branch to the fishing town of Eyemouth, the line sweeps inland to avoid some difficult coastal terrain, passing within sight of Ayton Castle and then – at somewhat slower speeds – threading its way through typical Borders hill country along the valley of the Eye Water, past the scenes of the epic floods of 1948. At Reston, there is little sign remaining of the cross-country line to Duns and St Boswells which was terminally severed by the floods, but engineers' sidings at Grantshouse illustrate the ongoing task of railway track maintenance, while long loops on either side of the main line provide 'refuge' for freight trains as they wait to be overtaken by their high-speed passenger counterparts.

From the Penmanshiel diversion, where trains have to slow to negotiate the post-1979 curve, the line runs downhill along the steeply sided valley of Penmanshiel Wood on the most severely graded section of track between London and Edinburgh – four miles at 1 in 96 towards the East Lothian coast.

Dunglass viaduct carries the railway above no less than four historic road bridges nearby, of which the newest – carrying the A1 trunk road – was built in 1992 to replace a faulty 1932 construction. Like so many Victorian railway structures, the railway viaduct was built to last. That it has lasted so long and so well – carrying heavier, faster and more frequent trains than ever anticipated – is a remarkable tribute to the companies, engineers, contractors and workmen who built them and those who have maintained them.

Skirting the coastline once more, the railway passes two of the few blemishes on the landscape of the Berwickshire and East Lothian section – the controversial Torness nuclear power station, which was completed in the 1990s, and the Oxwellmains cement plant. Both provide traffic for the railway – a relief for local roads – with cement trains running regularly to strategic railheads across Scotland.

Dunbar is the only station between Berwick and Edinburgh to be served by inter-city services. As the birthplace of the explorer, naturalist and conservationist John Muir – the creator of the national parks in the USA – the town is also the stepping-off point for the John Muir Way coastal path. Both the landscape and the weather conspire to make this ideal walking country – in this part of East Lothian, annual hours of sunshine are higher than almost anywhere else in Scotland and it records similar rainfall averages to the driest parts of England.

A northbound charter train hauled by A4 Pacific No. 60007 'Sir Nigel Gresley' skirts the cliff tops at Lamberton north of Berwick.

ROUTE DESCRIPTION
Dunbar to Edinburgh

Near Dunbar station, the distinctive mass of the Bass Rock comes into view to the north. This is a steep-sided volcanic rock, 350 ft high and home to a large colony of gannets. Currently uninhabited, it was the site of an important castle which was subsequently used as a prison before being abandoned to the birds.

Beyond Dunbar, striking inland to head directly towards Edinburgh, the railway soon comes within sight of the distinctive peak of North Berwick Law. This 613-ft-high volcanic survivor towers over the popular golfing resort of North Berwick, served by a four-mile branch line which narrowly escaped the 'Beeching axe' in 1969 and has since been electrified, seeing a huge growth in popularity amongst commuters and leisure travellers alike.

Past the tiny village of Drem, still served by trains from North Berwick, Anglo-Scottish expresses move swiftly on through the barley and wheat fields of the fertile East Lothian landscape, offering glorious views across the Forth, dotted with little islands, to the hills of Fife beyond. Past Longniddry, Prestonpans, Wallyford and Musselburgh, the suburbs of Edinburgh soon enclose the railway with the distinctive shape of Arthur's Seat just a mile to the south. As Meadowbank Stadium – host of the Commonwealth Games in 1970 and 1986 – comes into view, trains slow for the final approach through Calton Tunnel before reaching journey's end for most – but not all – rail passengers. Nestling beside the city's historic Old Town, Edinburgh's Waverley Station is watched over by its world-famous castle and the mighty Victorian bulk of the Balmoral Hotel, built by the North British Railway as the 'North British Hotel' to serve one of Britain's largest railway stations.

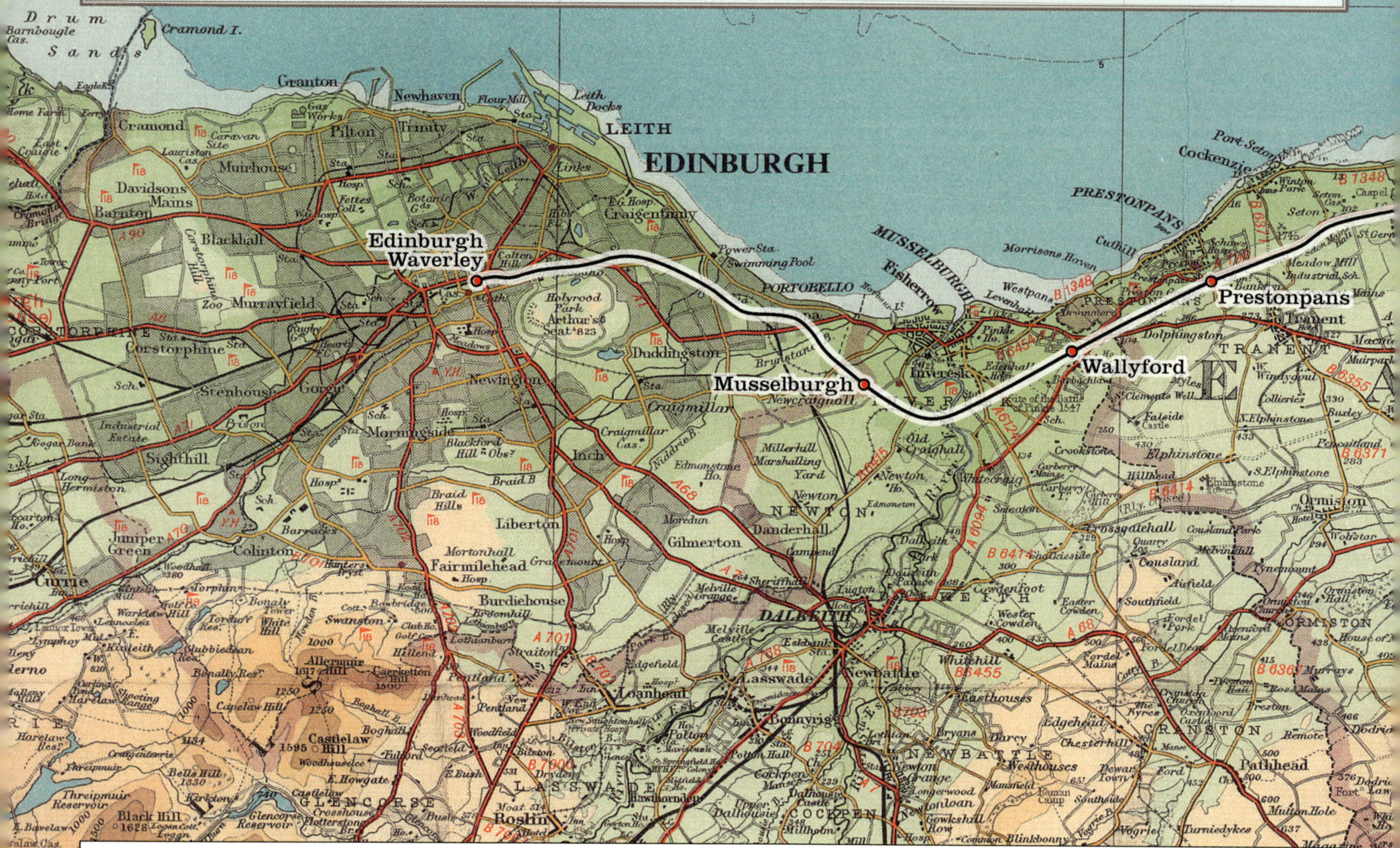

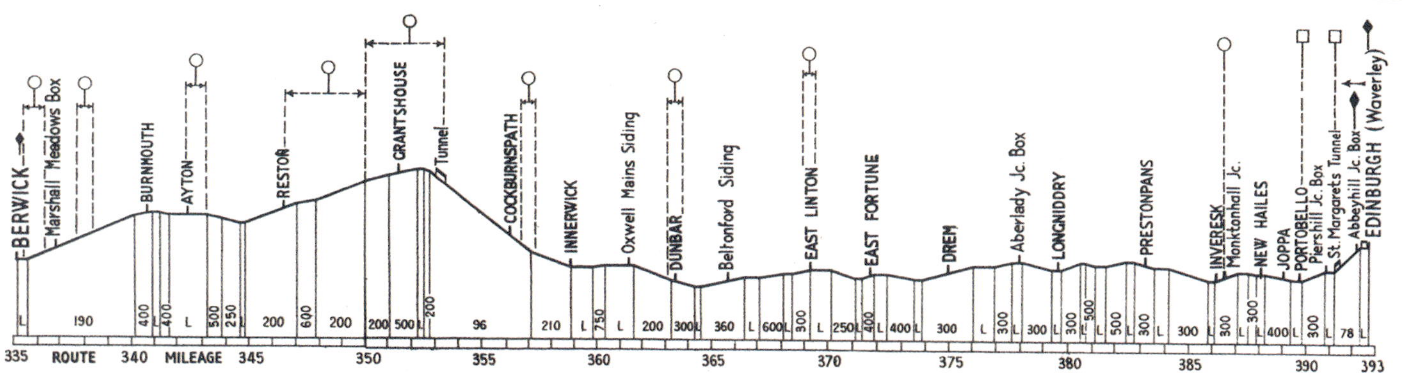

The stiff four-mile climb at 1 in 96 up Cockburnspath Bank stands out in the Edinburgh-Berwick gradient profile.

Journey's end from London – inter-city and local trains at the east end of Edinburgh Waverley station, overlooked by the Old Town and Edinburgh Castle.

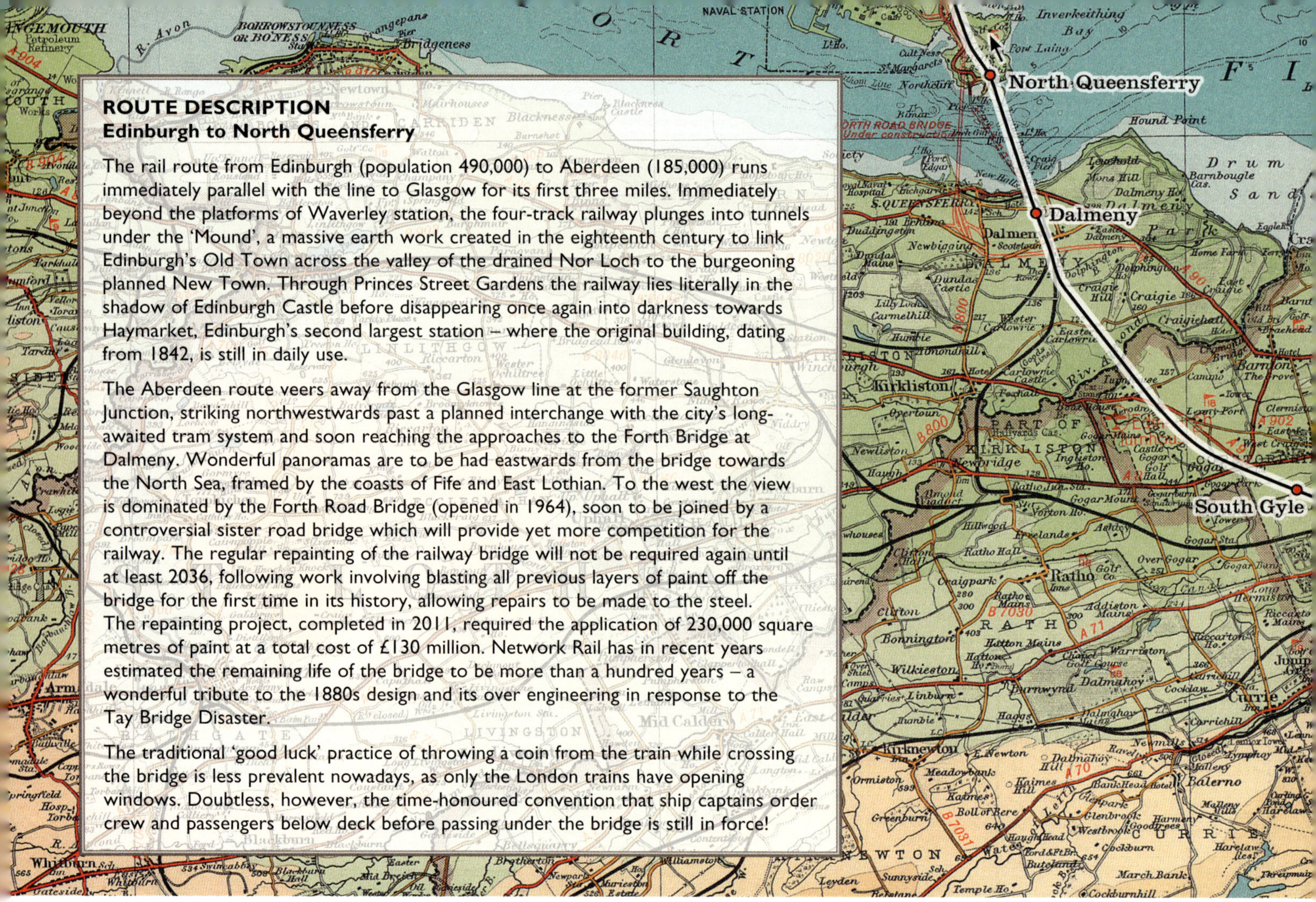

ROUTE DESCRIPTION
Edinburgh to North Queensferry

The rail route from Edinburgh (population 490,000) to Aberdeen (185,000) runs immediately parallel with the line to Glasgow for its first three miles. Immediately beyond the platforms of Waverley station, the four-track railway plunges into tunnels under the 'Mound', a massive earth work created in the eighteenth century to link Edinburgh's Old Town across the valley of the drained Nor Loch to the burgeoning planned New Town. Through Princes Street Gardens the railway lies literally in the shadow of Edinburgh Castle before disappearing once again into darkness towards Haymarket, Edinburgh's second largest station – where the original building, dating from 1842, is still in daily use.

The Aberdeen route veers away from the Glasgow line at the former Saughton Junction, striking northwestwards past a planned interchange with the city's long-awaited tram system and soon reaching the approaches to the Forth Bridge at Dalmeny. Wonderful panoramas are to be had eastwards from the bridge towards the North Sea, framed by the coasts of Fife and East Lothian. To the west the view is dominated by the Forth Road Bridge (opened in 1964), soon to be joined by a controversial sister road bridge which will provide yet more competition for the railway. The regular repainting of the railway bridge will not be required again until at least 2036, following work involving blasting all previous layers of paint off the bridge for the first time in its history, allowing repairs to be made to the steel. The repainting project, completed in 2011, required the application of 230,000 square metres of paint at a total cost of £130 million. Network Rail has in recent years estimated the remaining life of the bridge to be more than a hundred years – a wonderful tribute to the 1880s design and its over engineering in response to the Tay Bridge Disaster.

The traditional 'good luck' practice of throwing a coin from the train while crossing the bridge is less prevalent nowadays, as only the London trains have opening windows. Doubtless, however, the time-honoured convention that ship captains order crew and passengers below deck before passing under the bridge is still in force!

Morning sun bathes the mighty Forth Bridge as a Class 170 heads north towards Fife across 'the eighth wonder of the World'.

EDINBURGH TO ABERDEEN

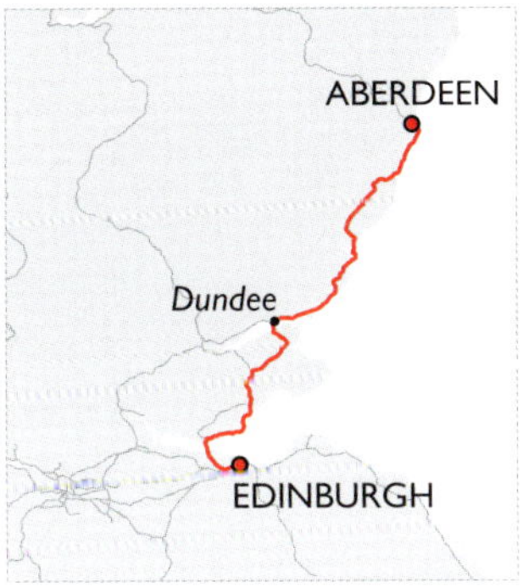

The 130-mile train journey from Edinburgh to Aberdeen takes the traveller from 'the Athens of the North' across two of the most famous railway bridges in the world – over the Firths of Forth and Tay – before culminating in a dramatic cliff-top approach to 'the Granite City'.

Partly through the constraints of physical geography, the railway between Edinburgh and Aberdeen was built in piecemeal fashion by a variety of companies, the first phase completed as early as 1839 and the last not finished until as late as 1890, with the opening of the Forth Bridge.

A railway from the rapidly growing industrial city of Dundee – famed for its 'jute, jam and journalism' – to the fishing port of Arbroath was authorized in 1836 and opened as a non-standard 5-ft 6-inch-gauge line in 1838. This was in fact the first town-to-town – as opposed to coal-carrying – railway to be authorized in Scotland, and initially operated in isolation from the rest of the rail network. It was converted to standard gauge in 1847.

Much of this route follows the coastline, seen here with Black Five No. 45407 'Lancashire Fusilier' heading south between Aberdeen and Stonehaven in April 2009.

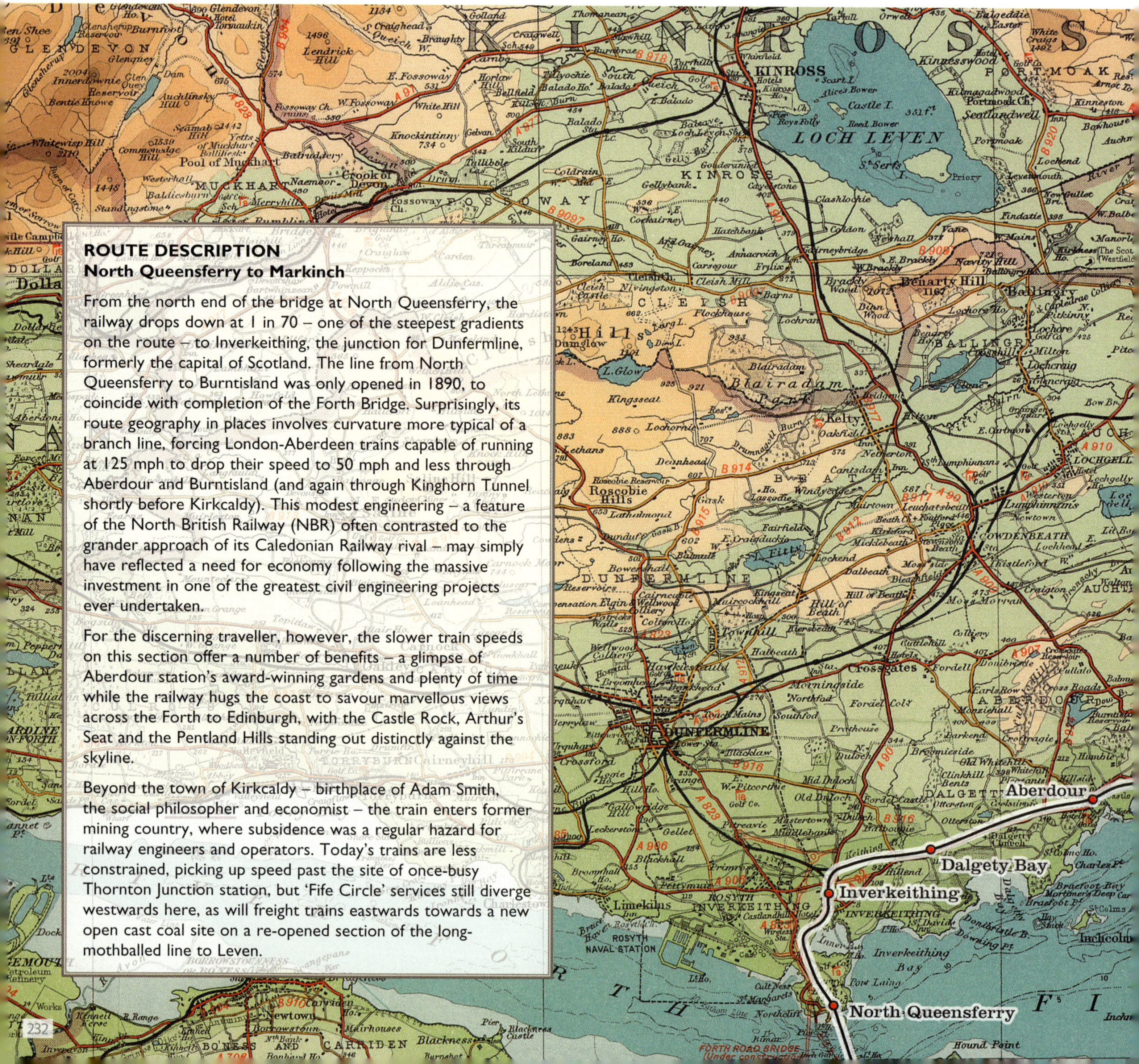

ROUTE DESCRIPTION
North Queensferry to Markinch

From the north end of the bridge at North Queensferry, the railway drops down at 1 in 70 – one of the steepest gradients on the route – to Inverkeithing, the junction for Dunfermline, formerly the capital of Scotland. The line from North Queensferry to Burntisland was only opened in 1890, to coincide with completion of the Forth Bridge. Surprisingly, its route geography in places involves curvature more typical of a branch line, forcing London-Aberdeen trains capable of running at 125 mph to drop their speed to 50 mph and less through Aberdour and Burntisland (and again through Kinghorn Tunnel shortly before Kirkcaldy). This modest engineering – a feature of the North British Railway (NBR) often contrasted to the grander approach of its Caledonian Railway rival – may simply have reflected a need for economy following the massive investment in one of the greatest civil engineering projects ever undertaken.

For the discerning traveller, however, the slower train speeds on this section offer a number of benefits – a glimpse of Aberdour station's award-winning gardens and plenty of time while the railway hugs the coast to savour marvellous views across the Forth to Edinburgh, with the Castle Rock, Arthur's Seat and the Pentland Hills standing out distinctly against the skyline.

Beyond the town of Kirkcaldy – birthplace of Adam Smith, the social philosopher and economist – the train enters former mining country, where subsidence was a regular hazard for railway engineers and operators. Today's trains are less constrained, picking up speed past the site of once-busy Thornton Junction station, but 'Fife Circle' services still diverge westwards here, as will freight trains eastwards towards a new open cast coal site on a re-opened section of the long-mothballed line to Leven.

Aberdour's award-winning station, seen here in summer 2010.

Further south in Fife, the railway age had originated in the late eighteenth century with wooden waggonways – modelled on the pioneering practice in Tyneside – conveying coal from colliery to coast. Robert Stevenson (of lighthouse fame) promoted the first standard-gauge railway proposed for Fife – from the shores of the Forth to Perth – in 1819, but the railway age was only just emerging and Stevenson hedged his bets by suggesting this might instead be a canal! The first line to be built in Fife was in practice the Edinburgh & Northern Railway, opened in 1850 from Burntisland (where ferries crossed from Granton in Edinburgh) to Tayport, where ferries plied to Broughty Ferry east of Dundee.

In the 'Railway Mania' of the mid-1840s, Perth- and Dundee-based rival rail companies set their sights on the conquest of the city of Aberdeen on the northeast coast, but the first phase to be opened by the Aberdeen Railway in early 1848 was just ten miles of what would become the Caledonian Railway main line, from Guthrie on the Arbroath & Forfar Railway (opened in 1839) north to Dubton, with branches to the towns of Brechin and Montrose. By 1850 the line was opened as far north as Portlethen, eight miles south of Aberdeen, and in 1854 the railway connection from Central Scotland via Perth and Forfar finally reached its Guild Street terminus in 'the Granite City'. As Thomas and Turnock note in *A Regional History of the Railways of Great Britain* (1993):

> *The railway had an immediate impact. Mail was carried by rail rather than mail coaches and the local post offices standardised their clocks by Greenwich Mean Time, involving a 'loss' of some eight minutes. The rest of the city soon fell into line and uniform time became the norm.*

Trainloads of Aberdeen Angus cattle were sent to the Smithfield market in London, and in the 1880s the North British Railway began operating a daily fish train from Aberdeen to London for the Billingsgate market, including wagons fed in by the Great North of Scotland Railway from key fishing ports such as Fraserburgh, Peterhead, Banff, Portsoy and Buckie. This was to be a lasting feature of the railway scene, with a daily Aberdeen-London fish train surviving until the 1979 Penmanshiel tunnel collapse (see page 224).

In 1849 the Edinburgh & Northern had amalgamated with the Edinburgh, Leith & Granton to form the Edinburgh, Perth & Dundee railway – but despite the name, it had no railway connecting these three places, due to the formidable physical barriers presented by the Firths of Forth and Tay. Connecting ferries were a time-consuming inconvenience, and even the introduction of the world's first train ferry in 1850 was no substitute for a direct rail route. It would not be until 1878 that the Tay was bridged, to a design of Carlisle-born Thomas Bouch, who had also designed the train ferry. The tragic story of the collapse of this bridge on the stormy night of the 28 December 1879 is well known – and it was not until 1887 that the replacement rail bridge was completed.

By 1866 the Aberdeen Railway had been absorbed by the larger Caledonian Railway, whose main line brought traffic from the

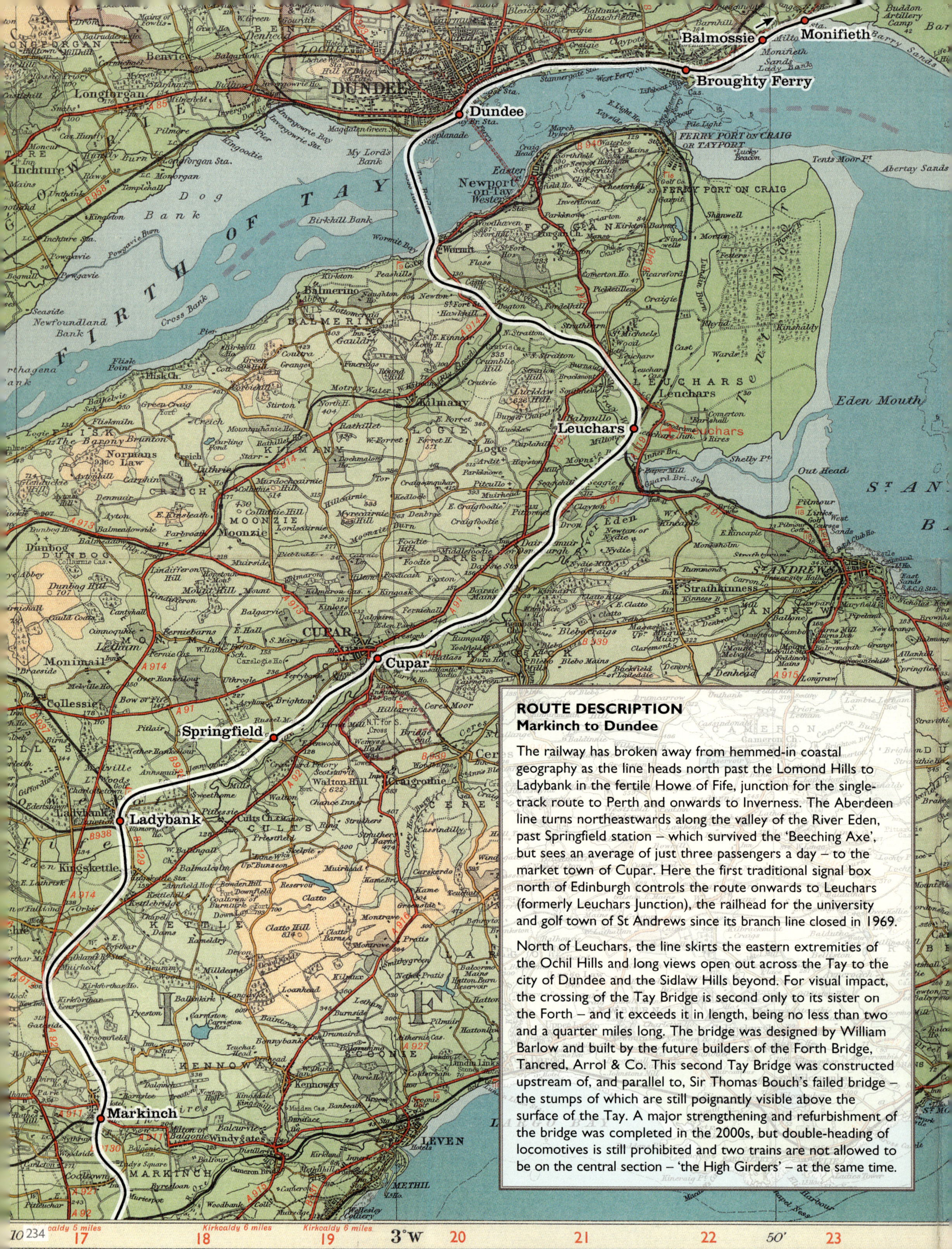

ROUTE DESCRIPTION

Markinch to Dundee

The railway has broken away from hemmed-in coastal geography as the line heads north past the Lomond Hills to Ladybank in the fertile Howe of Fife, junction for the single-track route to Perth and onwards to Inverness. The Aberdeen line turns northeastwards along the valley of the River Eden, past Springfield station – which survived the 'Beeching Axe', but sees an average of just three passengers a day – to the market town of Cupar. Here the first traditional signal box north of Edinburgh controls the route onwards to Leuchars (formerly Leuchars Junction), the railhead for the university and golf town of St Andrews since its branch line closed in 1969.

North of Leuchars, the line skirts the eastern extremities of the Ochil Hills and long views open out across the Tay to the city of Dundee and the Sidlaw Hills beyond. For visual impact, the crossing of the Tay Bridge is second only to its sister on the Forth – and it exceeds it in length, being no less than two and a quarter miles long. The bridge was designed by William Barlow and built by the future builders of the Forth Bridge, Tancred, Arrol & Co. This second Tay Bridge was constructed upstream of, and parallel to, Sir Thomas Bouch's failed bridge – the stumps of which are still poignantly visible above the surface of the Tay. A major strengthening and refurbishment of the bridge was completed in the 2000s, but double-heading of locomotives is still prohibited and two trains are not allowed to be on the central section – 'the High Girders' – at the same time.

Stanier Black Five loco No. 45407 is silhouetted against the evening sky on the Tay Bridge, hauling a southbound excursion to Edinburgh in April 2009.

south via Perth and Forfar. The penultimate link in the rival East Coast route came with the 1883 opening to through trains of the line from Arbroath to Kinnaber Junction – just north of Montrose on the existing Caledonian line – by the North British Railway, which had 'running powers ' over the Caledonian from Kinnaber to Aberdeen. By this time, the railway between Dundee and Arbroath had become a 'joint' operation of the Caledonian and North British.

In 1890 direct through journeys from London Kings Cross to Aberdeen via the East Coast Main Line were finally realized with the opening of the mighty Forth Bridge. Originally Thomas Bouch had been engaged to design this structure, but after the 1879 tragedy on the Tay, two other Englishmen replaced him, Sir John Fowler and Sir Benjamin Baker, designing a cantilevered structure built by Tancred, Arrol & Co between 1883 and 1890.

This was the first major steel bridge in the world and remains one of the greatest feats of bridge engineering anywhere. It is one and a half miles in length and weighs over 51,000 tonnes – as well as steel, the bridge's construction required 635,664 cubic ft of granite and 6.5 million rivets. At its peak, around 4,600 workers were employed in its construction, of whom nearly 100 died in accidents. The cost of the bridge was £30 million (£3 billion in today's prices), met not just by the North British Railway, which would benefit most directly from its opening (their share was 30%), but also by three English railway companies – the Midland Railway (32.5%), for whom this gave a link from London via Leeds, Carlisle and the Scottish Borders; the North Eastern Railway (18.75%); and the Great Northern Railway (18.75%): these companies joining in this venture to form the Forth Bridge Railway Company.

The bridge was formally completed on 4 March 1890 when HRH Edward Prince of Wales tapped a 'golden' rivet into place. The London-Aberdeen market now became the most symbolic target for competition between the East Coast Main Line and the West Coast Main Line. The East Coast group of companies (the Great Northern, the North Eastern and the North British) could now compete with the West Coast group (the London & North Western and the Caledonian) on better terms than ever before, with the journey distance reduced to 523½ miles compared to 540 via the West Coast.

This competition reached its climax in the summer of 1895 with the 'Races to the North', precipitated by the West Coast companies' concern that the arrival time of the overnight train from Euston at Aberdeen did not give enough leeway for the important connection to the Deeside line to Ballater, and accelerated their train by ten minutes. This led to conflicts with the North British at Kinnaber Junction, where the two routes converged, and the overnight trains from King's Cross and Euston were now due to pass within five minutes of each other. The West Coast acceleration started a tit-for-tat slashing of the timings, but their value probably lay more in the resulting publicity than any benefit for passengers, who ended up being deposited in Aberdeen at an unearthly hour in the morning. This acrimonious and potentially dangerous episode lasted just five weeks, with the various companies agreeing new minimum journey times to replace the "foolishness", as described by Thomas and Turnock:

> *The whole thing was an economic absurdity. It was the race* to the North. *There was no corresponding race* to the South. *The race trains were returned to London as that bane of the railway accountants' existence, empty coaching stock.*

ROUTE DESCRIPTION
Dundee to Montrose

The bridge curves eastwards into Scotland's fourth city, Dundee (population 155,000), formerly renowned for its 'jute, jam and journalism'. Dundee's station – traditionally known as Dundee Tay Bridge – is set in a deep cutting below sea level immediately before the tunnel which takes the railway east towards Arbroath. It was never a large station – and still has just four platforms – since the city previously had two termini, including the Caledonian Railway's architecturally impressive Dundee West, which was unfortunately demolished in the late 1960s. The line to Aberdeen climbs out of Dock Street Tunnel at a gradient of 1 in 60 – the steepest on the entire route – but the contrast could not be greater just a mile further east, where there begins fourteen miles of dead-level track, surprisingly the longest stretch on the entire London-Aberdeen route. Here the railway allows high-speed running past sand dunes, beaches, golf courses – and two of the least-used stations in Britain at Barry Links and the charmingly named Golf Street. With just one train a day provided in each direction, their unenviable status is perhaps no surprise!

On the approach to the fishing port of Arbroath – famed for its 'smokies' (smoked haddock) – the main line is briefly paralleled by Kerr's Miniature Railway, Scotland's oldest miniature line, which has been carrying holidaymakers along the sea front since 1935. Beyond Arbroath the railway turns briefly inland but returns to the coast with a striking panorama of Lunan Bay and one of the most dramatic beaches in Scotland. As the line drops down to Montrose, trains traversing the only section of single line between King's Cross and Aberdeen, the immediate view is ever changing – the railway runs right beside and over the entrance to the tidal Montrose Basin, host to large numbers of wildfowl, waders and up to 60,000 migrating geese each year.

The sweeping curve and distinctive sections of the Tay Bridge are captured in this scene with a Voyager heading for distant Penzance in September 2003.

Two of the delightful locomotives on Kerr's Miniature Railway at Arbroath.

After the 'Grouping' of 1923, the Edinburgh-Kinnaber Junction section of the route became part of the London & North Eastern Railway (LNER), with running powers over the London, Midland & Scottish Railway (LMS) from Kinnaber to Aberdeen. On their premier routes, the new companies vied with each other using both style and speed – in 1927, the overnight LMS express from London to Aberdeen and Inverness became the 'Royal Highlander', and the LNER christened its overnight train to Aberdeen as the 'Aberdonian'.

Steam reigned supreme on Edinburgh-Aberdeen trains until the late 1950s, but by 1961 virtually all expresses were in the hands of diesels. However, the modern motive power was not an unqualified success, and the classic A4 Pacific locomotives – which had been displaced by diesels on the East Coast Main Line – were drafted in for a steam swan song on Aberdeen-Glasgow trains between 1962 and 1966. A year after steam's final hurrah on the 'Strathmore Route', that line from Kinnaber Junction through Forfar to Perth fell under the 'Beeching Axe' – and all Glasgow trains were diverted to follow the same route as those for Edinburgh as far as Dundee, continuing south through the Carse of Gowrie to rejoin the original route at Perth.

In 1979, London-Aberdeen services were taken over by the new Inter City 125 High Speed Trains – at long last exceeding the record-breaking journey time made by steam in the Race to the North of 1895! Today, the re-engined 125s continue to give sterling service on the London trains, while ScotRail express services between Edinburgh and Aberdeen are operated by modern diesel units, many making no stops between Haymarket (in the west end of Edinburgh) and Leuchars in northeast Fife. The route is double track throughout, with the continuing exception of the one and a half miles of single track between Usan and Montrose, where the expense of rock blasting and building a new bridge over the River South Esk has always deterred easing this bottleneck – not a situation which would be tolerated on the parallel road network.

The southern section of the route between Edinburgh and Cupar is now under the control of a modern signalling centre sited beside the capital's Waverley station, but from Cupar northwards – other than in the immediate vicinity of Dundee and Aberdeen – the line still features traditional electro-mechanical signalling, controlled from signal boxes sited along the lineside.

The Edinburgh-Aberdeen line traverses a different Scotland from the famed scenic lines of the Highlands but it deserves equal billing, with its dramatic coastal views and the unforgettable experience of crossing two of the most famous railway bridges in the world.

Laurencekirk station, re-opened in 2009, is unusual in having retained its original (listed) building, in which a waiting room celebrates the town's heritage.

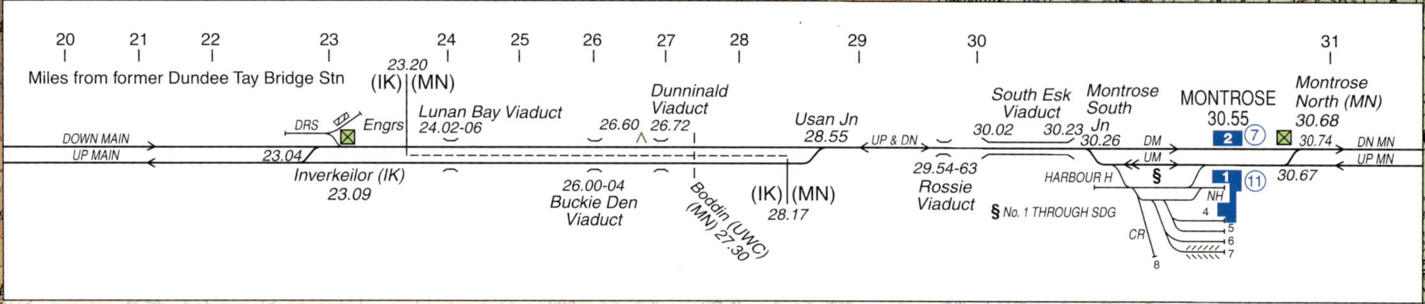

The only stretch of single-track railway on the 525 miles from London to Aberdeen is the one mile fifty-one chains (just over one and a half miles) from Usan Junction to Montrose South Junction.

ROUTE DESCRIPTION

Montrose to Stonehaven

The railway turns inland again, and two and a half miles north of Montrose passes the unrecognisable remains of the once-renowned Kinnaber Junction, the 'finishing post' of the 1895 Races to the North. The line is now entering the Howe of the Mearns, a gently rolling basin of fertile farmland bound by coastal hills and the mountains of the Angus glens to the north – the setting for Lewis Grassic Gibbon's classic novel, *Sunset Song*. The district's principal market town is Laurencekirk, whose station closed in 1967 but re-opened in 2009, with the original station building sympathetically restored. Onwards towards the coastal town of Stonehaven, the railway becomes more hemmed in by forested hillside – but then straddles the Cowie Water, high above the town on the fine Glenury Viaduct.

A symbol of the Granite City – the striking granite structure of Aberdeen University's Marischal College, much of which now serves as the headquarters of Aberdeen City Council.

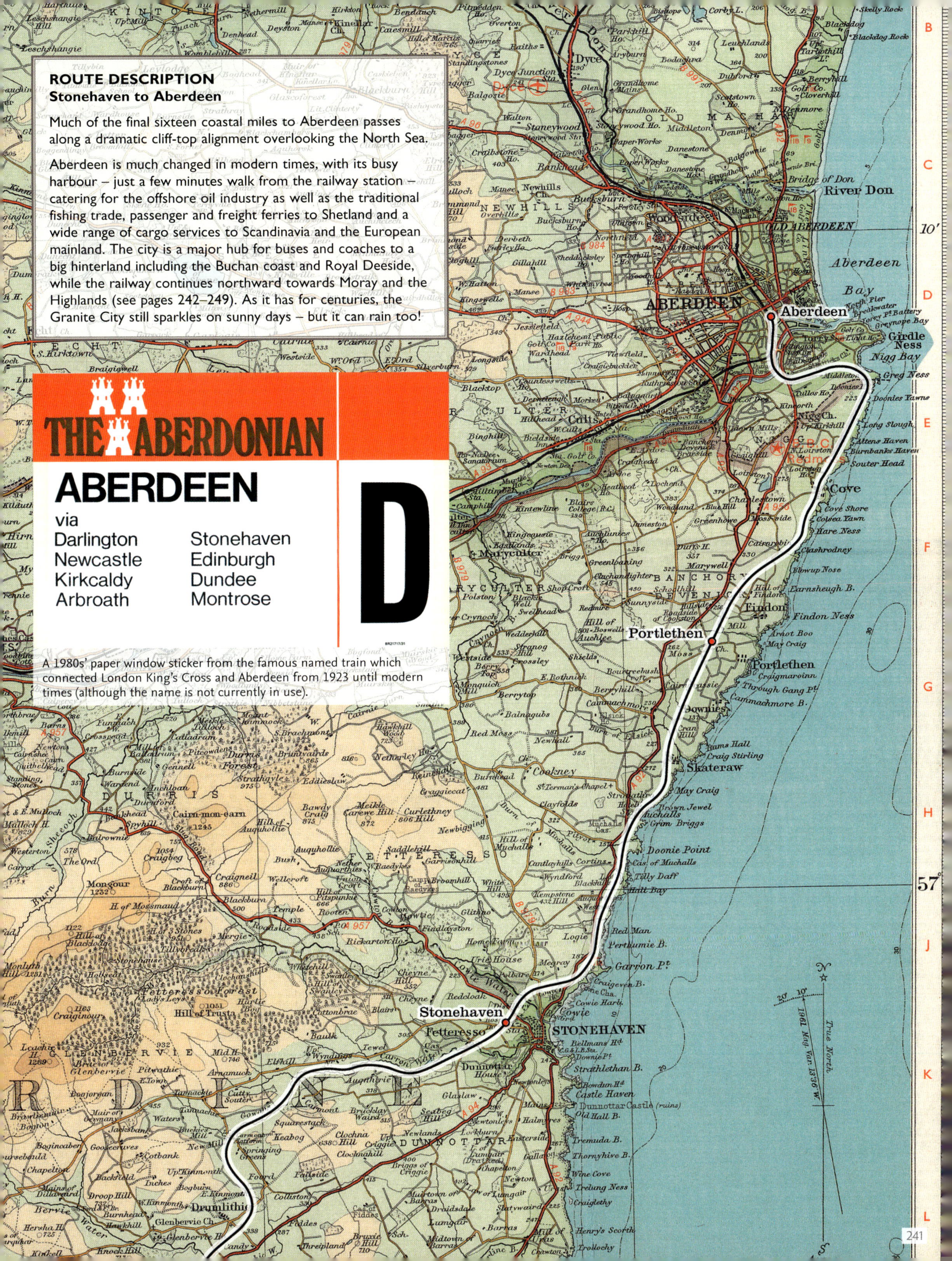

ROUTE DESCRIPTION
Stonehaven to Aberdeen

Much of the final sixteen coastal miles to Aberdeen passes along a dramatic cliff-top alignment overlooking the North Sea.

Aberdeen is much changed in modern times, with its busy harbour – just a few minutes walk from the railway station – catering for the offshore oil industry as well as the traditional fishing trade, passenger and freight ferries to Shetland and a wide range of cargo services to Scandinavia and the European mainland. The city is a major hub for buses and coaches to a big hinterland including the Buchan coast and Royal Deeside, while the railway continues northward towards Moray and the Highlands (see pages 242–249). As it has for centuries, the Granite City still sparkles on sunny days – but it can rain too!

A 1980s' paper window sticker from the famous named train which connected London King's Cross and Aberdeen from 1923 until modern times (although the name is not currently in use).

ABERDEEN TO INVERNESS

The 108-mile Aberdeen-Inverness line provides Scotland's most northerly cross-country rail link through largely unspoilt agricultural and forested country, serving a string of characterful small towns en route from the Granite City to the Highland Capital.

As construction of the Aberdeen Railway from the south edged closer to the city in the 1840s, so interest grew in the potential for a line striking north west from the Granite City to Elgin and Inverness. The first (unsuccessful) prospectus was issued in 1844, but it was not until 1846 that the 'Great North of Scotland Railway' (GNoSR) was authorized as far as Huntly – and the first trains did not run until 1854. The railway was built largely on the bed of the short-lived Aberdeenshire Canal between Aberdeen and Port Elphinstone (Inverurie), and was constructed as single track with crossing loops.

The advance of the new line prompted, by the fear of competition, the construction of the Inverness & Nairn Railway, which opened in 1855, and then the GNoSR was extended from Huntly to Keith the following year. Meantime the Inverness & Aberdeen Junction Railway was pushing east from Nairn and making an end-on connection with the GNoSR at Keith in 1858 – "a remarkable piece of railway building" as noted by Thomas and Turnock in *A Regional History of the Railways of Great Britain Volume 15* (1993).

The lack of a direct connection between the GNoSR's Aberdeen terminus at Kittybrewster and the Aberdeen Railway Company's Guild Street station in the city centre was a constant inconvenience, but it was not until 1867 that disagreements between the two companies were overcome and a new double-track railway linked their systems at the newly-opened Aberdeen (Joint) station. The fifty-three-mile GNoSR main line was doubled progressively, the second track finally reaching Keith in 1898.

Beyond Keith the western half of the route to Inverness was the Highland Railway's single line, only the last few miles from Dalcross into Inverness being double-track operation. Relations between the Highland and the GNoSR were strained for many years, in part due to the aspirations of each company to extend their territory into the heartlands of their rivals. The Highland was particularly sensitive about protecting its Inverness base against incursions from the east – in response opening the Aviemore-Inverness 'direct line' in 1898 – and matters were not helped when the GNoSR completed its rival routes from Keith to Elgin via Dufftown in 1863 and then via Buckie (the 'Moray Coast line') in 1886.

The GNoSR steadily enlarged its network with many branch lines throughout northeast Scotland, all but one of them (the Ballater branch) radiating from the Aberdeen-Keith-Elgin main line. While the Highland network could be described as 'all main line', the GNoSR network in contrast lent itself to the description 'all branch lines' – reflecting the dispersed pattern of agricultural settlements and small market or fishing towns. Despite the inter-company tensions, arrangements were soon made for through trains to operate seamlessly between Aberdeen and Inverness, without any change of locomotive at Keith.

As in the Highlands, the railway had a major impact on the regional economy, Thomas and Turnock recording that: "Thanks to the railway taking fat animals to market and bringing in the store cattle and the fertilizers, the region became one great beef factory." In the 1880s the North British Railway began operating a daily fish train from Aberdeen to London for the Billingsgate market; this would convey wagons fed in by rail from across the region, including traffic moved over the Moray Coast route at Banff, Buckie and Portsoy.

The new railways gave whisky distilling a substantial boost, and many of the new and expanded distilleries were in the Banffshire and Morayshire heartland, served by the GNoSR's route from Elgin to Keith via Craigellachie and Dufftown and the associated 'Speyside line' from Craigellachie to Boat of Garten. From the Victorian era through to the 1960s, virtually every distillery in this area had either its own rail sidings or even branch line,

Royal Scot No. 46115 'Scots Guardsman' heads the northbound 'Great Britain' rail tour away from the Spey Viaduct between Keith and Elgin in April 2011.

or was located a short horse-and-cart ride away from a railway goods yard. Much of the traffic – barley and coal in from the south and whisky out – was routed via the major junction and goods yard at Keith, then south to Aberdeen and Central Scotland.

Passenger transport was also transformed by the coming of the railway – with new travel opportunities and faster journeys. While the vast majority of rail passengers were 'local' within the region – with Aberdeen by far the most popular destination – longer-distance links were not neglected. In 1869 through coaches were briefly introduced between Elgin and London (Euston) via Craigellachie, and a through sleeping car was operated by the London & North Eastern Railway between Lossiemouth and London (King's Cross) between 1923 and 1939.

After the Second World War, weekly through coaches from Glasgow (Buchanan Street) to Keith and Elgin survived until the eve of the infamous 'Beeching Report' of 1963, but, as elsewhere in the country, competition from road transport was making major inroads on rail traffic despite the introduction of diesel locomotives in the late 1950s. Dr Beeching's prescription for the former GNoSR network was as bad as it could be, short of complete obliteration – all the branch lines were to go and all stopping services were to be withdrawn along the main line between Aberdeen and Inverness. The Grampian region was one of the relatively few parts of Britain where all Beeching's route closure proposals were implemented, the sole reprieve being for Insch station mid-way between Inverurie and Huntly.

Other than a short section of double track from Insch to Kennethmont, the Aberdeen-Keith route was singled by 1971, but the majority of the route remained controlled (as it does today) by traditional electro-mechanical signal boxes. Whisky traffic from Keith to Central Scotland lingered on until 1992 and thereafter new timber flows were attracted to the railway at Elgin and Keith, but it is now some years since the Aberdeen-Inverness line has seen regular freight traffic north of Dyce. Passenger traffic has grown substantially in recent decades – particularly commuters at the eastern and western ends of the line – but the limited number of crossing loops constrains the introduction of additional train services. For some twenty-five years there has been talk of enhancing the line's infrastructure, but despite lack of progress on the ground there has fortunately been enough capacity available for the route to remain a regular feature of 'The Royal Scotsman' luxury land cruise itinerary.

Many of the inter-city trains that reach Aberdeen from Edinburgh and Glasgow are now extended through to Dyce or Inverurie, eliminating the need to change trains. Aberdeen's station has seen a substantial refurbishment in association with a new retail development on land formerly occupied by a railway freight depot, now relocated to Dyce.

The Aberdeen-Inverness line is an under-rated link in the classic Scottish scenic rail itinerary, with faster and more frequent trains than at any time in its history, traversing a range of picturesque lowland and upland landscapes. And as Michael Pearson reflects in *Iron Road to Whisky Country* (2002), "Its stations serve a necklace of intriguingly historic towns yet to be overpowered by the kind of tourism which can suffocate the Highlands proper."

Birds alight above Ardmore Distillery in late 2005, as an Aberdeen-bound train heads past Kennethmont signal box on the line's only section of double track.

ROUTE DESCRIPTION
Aberdeen to Huntly

The north end of Aberdeen station – substantially rebuilt during the First World War – still conveys a flavour of the once-busy network of Great North of Scotland Railway (GNoSR) lines north of the city. There remain signs of the four former bay platforms and four former through platforms – now reduced to just two through platforms, but one of which (Platform 6) is the longest in Scotland, at some 1640 ft in length. Shortly after leaving the station, trains for Inverness encounter the steepest climb of the entire northbound route, at a gradient of 1 in 72 up the Denburn valley and through the only two tunnels on the line.

Six miles from Aberdeen (population 185,000) the line reaches its first intermediate station at Dyce – closed when stopping services from Aberdeen to Keith (and onwards to Elgin via Dufftown and Buckie) were withdrawn in 1968, but re-opened by British Rail in 1984 to serve local housing, industrial developments and the nearby airport.

Beyond Dyce the railway carefully skirts the flood plain of the River Don, crossing it just south of the market town and growing commuter settlement of Inverurie, where the wood-panelled interior of the original GNoSR station building has been tastefully restored. North of Inverurie – past the remains of the old GNoSR locomotive works, whose name survives courtesy of the local football team, Inverurie Loco Works FC – the line continues its gradual climb through the increasingly hilly and forested terrain of the district known as Garioch (pronounced Geeree), briefly following the route of the River Urie, a tributary of the Don. The distinctive peak of Bennachie (1,733 ft) stands out to the south.

The small town of Insch stands on the River Shevock, its traditional signal box controlling the first part of the five and a half miles of double track onwards to Kennethmont, the 590-ft summit of the long climb from Aberdeen. Trains rush past Ardmore Distillery down Strathbogie towards the next stop at Huntly, the terminus of the line from 1854 to 1856.

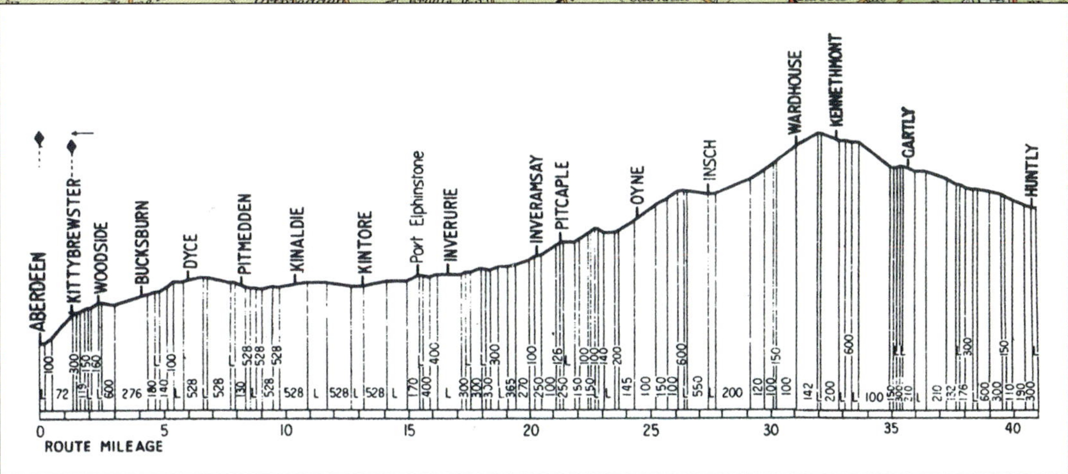

The first section of the line involves a long and almost continuous climb from Aberdeen to Kennethmont, but at relatively easy gradients.

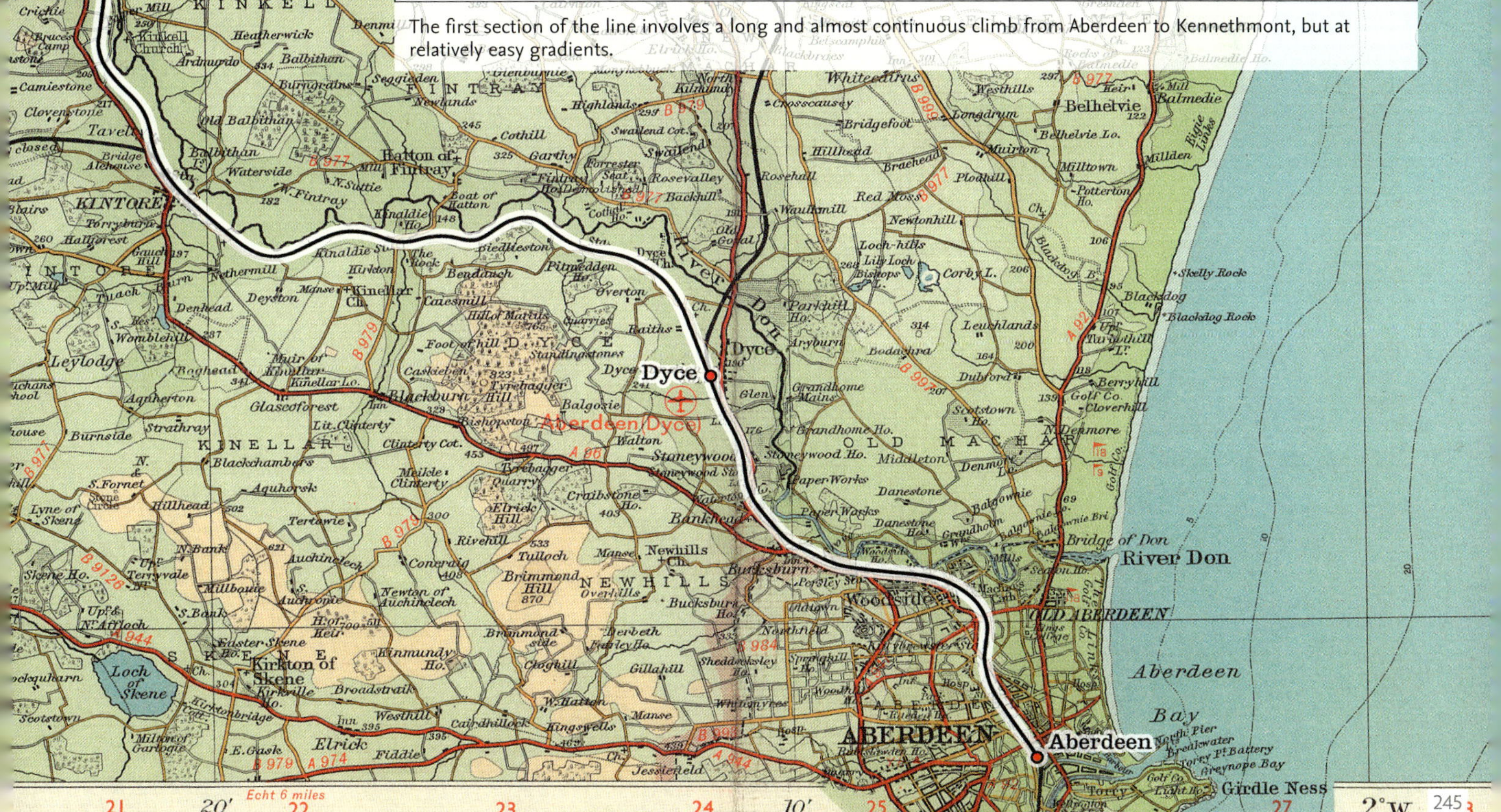

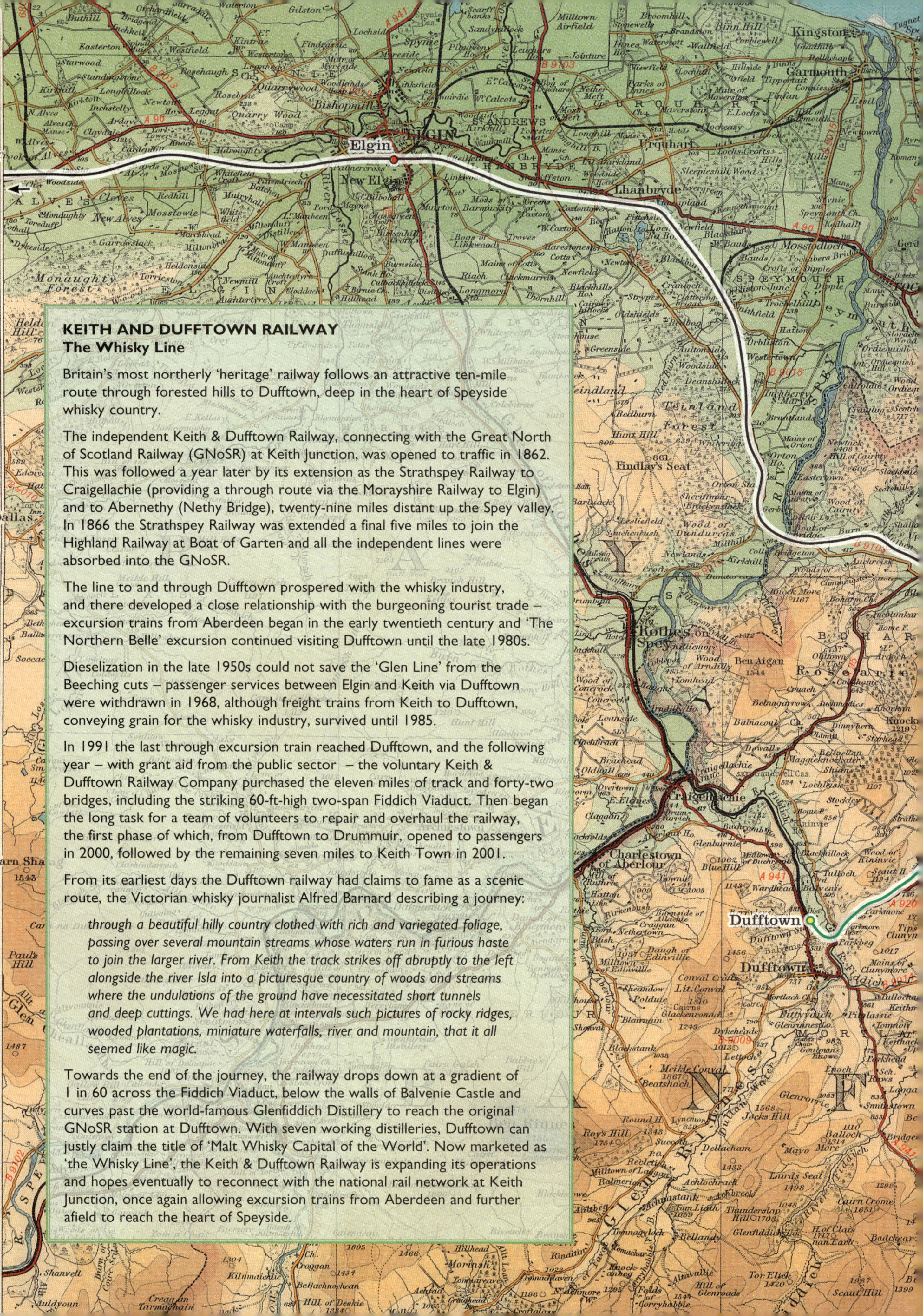

KEITH AND DUFFTOWN RAILWAY
The Whisky Line

Britain's most northerly 'heritage' railway follows an attractive ten-mile route through forested hills to Dufftown, deep in the heart of Speyside whisky country.

The independent Keith & Dufftown Railway, connecting with the Great North of Scotland Railway (GNoSR) at Keith Junction, was opened to traffic in 1862. This was followed a year later by its extension as the Strathspey Railway to Craigellachie (providing a through route via the Morayshire Railway to Elgin) and to Abernethy (Nethy Bridge), twenty-nine miles distant up the Spey valley. In 1866 the Strathspey Railway was extended a final five miles to join the Highland Railway at Boat of Garten and all the independent lines were absorbed into the GNoSR.

The line to and through Dufftown prospered with the whisky industry, and there developed a close relationship with the burgeoning tourist trade – excursion trains from Aberdeen began in the early twentieth century and 'The Northern Belle' excursion continued visiting Dufftown until the late 1980s.

Dieselization in the late 1950s could not save the 'Glen Line' from the Beeching cuts – passenger services between Elgin and Keith via Dufftown were withdrawn in 1968, although freight trains from Keith to Dufftown, conveying grain for the whisky industry, survived until 1985.

In 1991 the last through excursion train reached Dufftown, and the following year – with grant aid from the public sector – the voluntary Keith & Dufftown Railway Company purchased the eleven miles of track and forty-two bridges, including the striking 60-ft-high two-span Fiddich Viaduct. Then began the long task for a team of volunteers to repair and overhaul the railway, the first phase of which, from Dufftown to Drummuir, opened to passengers in 2000, followed by the remaining seven miles to Keith Town in 2001.

From its earliest days the Dufftown railway had claims to fame as a scenic route, the Victorian whisky journalist Alfred Barnard describing a journey:

> *through a beautiful hilly country clothed with rich and variegated foliage, passing over several mountain streams whose waters run in furious haste to join the larger river. From Keith the track strikes off abruptly to the left alongside the river Isla into a picturesque country of woods and streams where the undulations of the ground have necessitated short tunnels and deep cuttings. We had here at intervals such pictures of rocky ridges, wooded plantations, miniature waterfalls, river and mountain, that it all seemed like magic.*

Towards the end of the journey, the railway drops down at a gradient of 1 in 60 across the Fiddich Viaduct, below the walls of Balvenie Castle and curves past the world-famous Glenfiddich Distillery to reach the original GNoSR station at Dufftown. With seven working distilleries, Dufftown can justly claim the title of 'Malt Whisky Capital of the World'. Now marketed as 'the Whisky Line', the Keith & Dufftown Railway is expanding its operations and hopes eventually to reconnect with the national rail network at Keith Junction, once again allowing excursion trains from Aberdeen and further afield to reach the heart of Speyside.

ROUTE DESCRIPTION

Huntly to Elgin

Over much of this section extensive forests fringe the railway, an important source of freight traffic for the line in the 1990s.

At Keith, the station has been substantially rebuilt in recent years – formerly known as Keith Junction, it is now the railhead for a large tract of country stretching to the Moray Coast. The 'branch platform' for the route inland to Dufftown is still in place, but there is a gap of around 1640 ft between Network Rail's track and the independent operation of the Keith & Dufftown Railway at Keith Town station (see opposite).

The next stretch – the former Highland Railway line to Elgin via Mulben – is operationally distinctive. It has the longest single-track section – eighteen miles – and the fiercest southbound gradient (at 1 in 60) of the entire 108-mile route. Modern passenger trains take this in their stride, but for the occasional freight trains that use this route the maximum southbound load is just 1,230 tonnes, compared to 2,700 tonnes northbound. Five miles beyond Keith the railway plunges downhill to cross the River Spey on the impressive steel-girdered Spey Viaduct, whose construction in the late 1850s involved the most expensive individual work of engineering on the line. A distinctive feature of the entire route is that it is amongst the last in Britain to be accompanied by the traditional telegraph poles carrying telegraphic wires for communication between signal boxes.

At Elgin, the railway passes close to the former GNoSR station – a striking 'Scottish baronial' pile built in 1902 which gives a taste of the many impressive Victorian and Edwardian buildings which still grace this important regional centre (population 26,000). Today's railway has a rather more utilitarian structure on the nearby site of the former Highland Railway station.

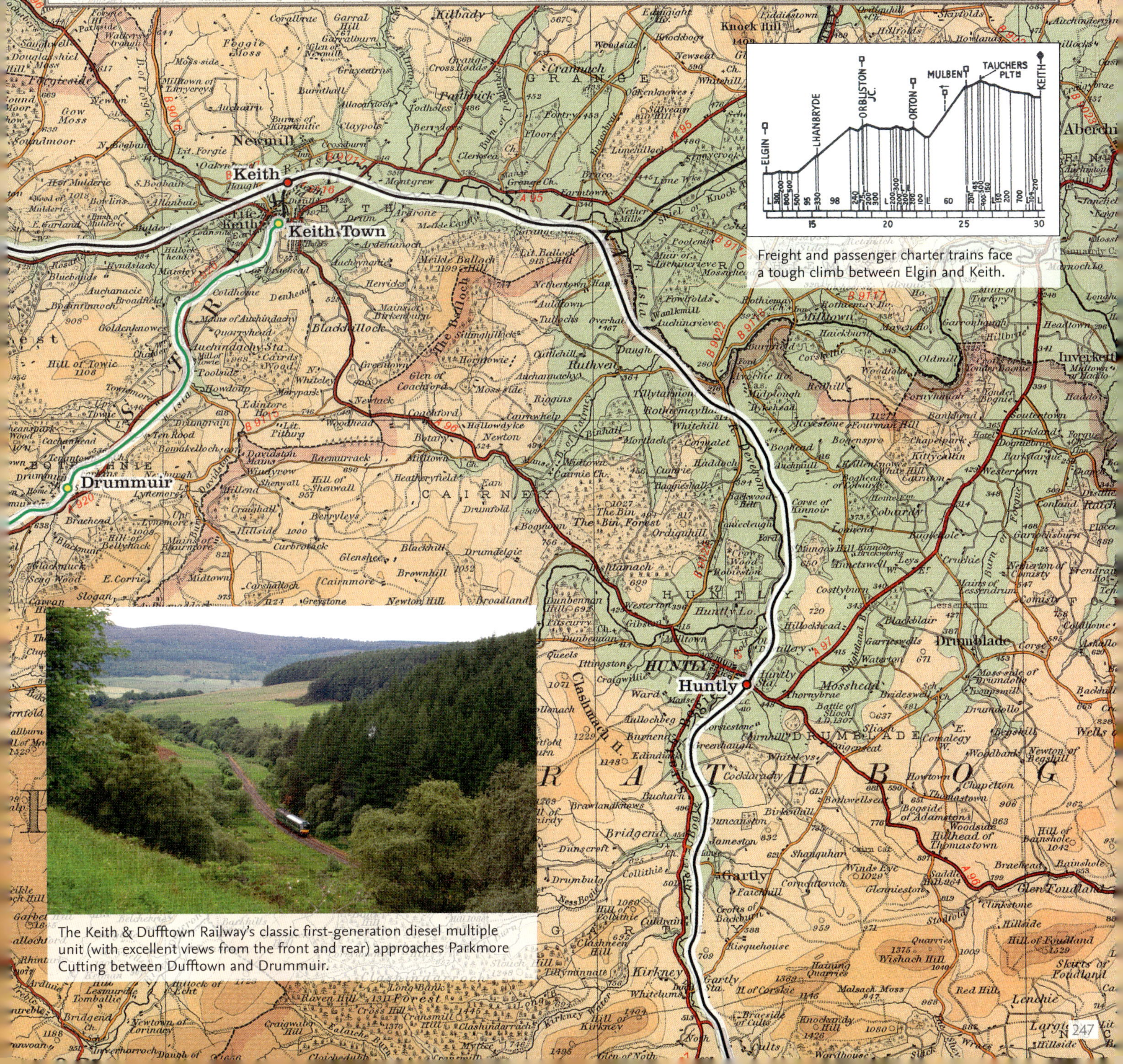

Freight and passenger charter trains face a tough climb between Elgin and Keith.

The Keith & Dufftown Railway's classic first-generation diesel multiple unit (with excellent views from the front and rear) approaches Parkmore Cutting between Dufftown and Drummuir.

ROUTE DESCRIPTION

Elgin to Inverness

From Elgin to Inverness, the railway is generally gently graded across the fertile lands of the Laigh of Moray, although large bridges were required to take the line over the Rivers Lossie, Findhorn and Nairn close to their outlets to the Moray Firth. Just west of Elgin at Alves a mothballed branch line – with rail track still intact – heads off towards Scotland's most modern malt whisky distillery at Roseisle. With fourteen copper stills and an annual production capacity of ten million litres, it is also one of Scotland's largest malt distilleries – and there are hopes that whisky will once again move south by rail from Speyside.

At Forres, the Aberdeen-Inverness line meets the former junction with the original Highland Main Line to Perth via Dava Moor and Grantown-on-Spey. Closed in 1965, it left Forres with reduced railway importance, and the former five platforms built on a triangle have been reduced to just one today – but a distinctive 1950s station building survives.

Both platforms at Nairn feature attractive Highland Railway buildings dating from 1885, which make for a welcoming gateway to this small holiday and retirement town where Sir Charlie Chaplin spent holidays in the later years of his life. Beyond Nairn the railway passes the Inverness airport at Dalcross, where a new station is planned to serve local housing and industrial developments. On the approaches to Inverness there are attractive views across the Inner Moray Firth towards the Black Isle and the more distant mountains of Ross-shire. At Inverness station – the rail crossroads of the Highlands – trains from Aberdeen make connections with services to destinations as widely scattered as Wick, Thurso, Kyle of Lochalsh, Glasgow, Edinburgh and London.

With the mountains of Ross-shire in the distance, an Inverness-Aberdeen service passes Gollanfield between Inverness and Nairn in December 2009.

PERTH TO INVERNESS

HIGHLAND MAIN LINE

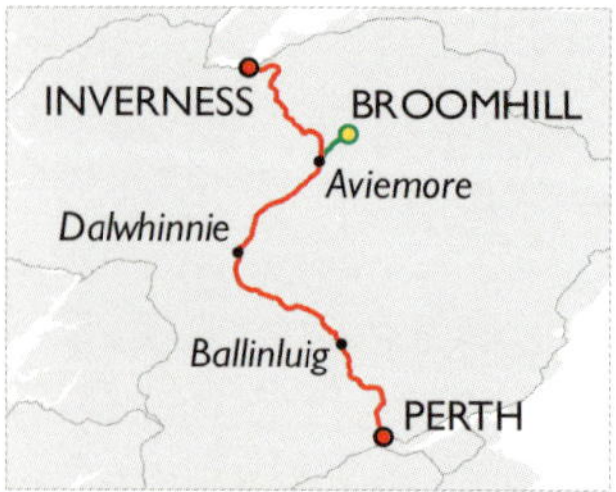

The 118 railway miles across the Grampians traverse fertile straths, wild heather moorland, bleakly dramatic mountain country – and include Britain's highest standard-gauge railway summit.

The Scottish Midland Junction Railway was the first line to push north from Perth, opening to Forfar in lowland Strathmore in 1848, but the first physical investigations to assess the challenge of driving a railway all the way through the Grampian Mountains to Inverness had been undertaken three years earlier. As John Thomas and David Turnock record in *A Regional History of the Railways of Great Britain* (1993):

> *The peripheral position of the [Highland] region meant that railway developments would inevitably be delayed: as with other technological innovations progress spread gradually from the south. But the time lag was exaggerated by the logistical difficulties of carrying out surveys and constructing railways in a mountainous environment. When Joseph Mitchell organised the survey of a route for the Perth-Inverness line in 1845 there were no ordnance surveys to provide initial guidance.*

The independent Perth & Dunkeld Railway opened north from Stanley Junction on the Scottish Midland to the small town of Dunkeld in 1856, but could never hope to be viable in isolation. Then, pushed ahead by a wish to improve on the circuitous route from Inverness to the south via Aberdeen, construction work on the Inverness & Perth Junction Railway – linking the existing railways at Forres and Dunkeld – began in 1861 and was completed just two years later, despite the challenging terrain involved. The engineer was Joseph Mitchell, who had trained under Thomas Telford, the famous road, bridge and canal builder, and under his control was built a 104-mile-long railway, surmounting summits of 1,052 ft above sea level at Dava and 1,484 ft at Druimuachdar, and yet which had no gradient steeper than 1 in 70 and only two short tunnels north of Dunkeld.

The line was opened in 1863, and in 1865 the Inverness & Perth Junction Railway merged with the Inverness & Aberdeen Junction Railway to become the Highland Railway, one of the five large companies to dominate the Scottish railway system until the 'Grouping' of 1923.

The railway was initially single track throughout, but doubling from Blair Atholl to Dalwhinnie was completed in 1909 – and with additional crossing loops the average length of single-track sections, on what had become known as the Highland Main Line, dropped to just three miles by 1911. In 1898 – alarmed by the prospect of a competing railway reaching Inverness via the Great Glen or from the Great North of Scotland Railway territory to the east – the Highland had opened the 'Direct Line' from Inverness to Aviemore via the 1,315-ft Slochd summit, including double track on the steep climb over the first ten miles south from Inverness.

The evening sun catches Findhorn Viaduct at Tomatin in May 2012, as 67 017 hauls the Inverness-London sleeper towards Slochd Summit .

The Perth-Inverness line was always difficult to operate. Much of the seventeen-and-a-half-mile northbound climb from Blair Atholl to Druimuachdar Summit involves gradients of 1 in 70, while the price of the direct line over Slochd was the most severe gradients of the entire route – 1 in 60 northbound and southbound. Double-heading of trains, with 'banking' engines in the rear, became a regular feature of Highland Main Line operation until the end of steam haulage, not least in the halcyon era of the railway when the summer 'season' brought tourists and hunting, shooting and fishing enthusiasts in their thousands to the Highlands. An 1888 account of a train which had become the apotheosis of this traffic – the early morning service from Perth to Inverness – recorded that on 7 August it comprised thirty-seven carriages from ten different railway companies!

Despite a sparse population and fragile economy, railways had come relatively quickly to the Highlands, reflecting the support of landowners who could see how the railway would help the development of their estates. Not only tourism was transformed – the concentration of operations and engineering activities at the Highland Railway's headquarters led to the steady growth of Inverness towards the status of Highland capital; fish landed in the Highlands could now reach distant markets in England quickly; and the cattle trade was revolutionized by transits of as little as a day to key markets instead of weeks of hard driving of the animals over rough drove roads.

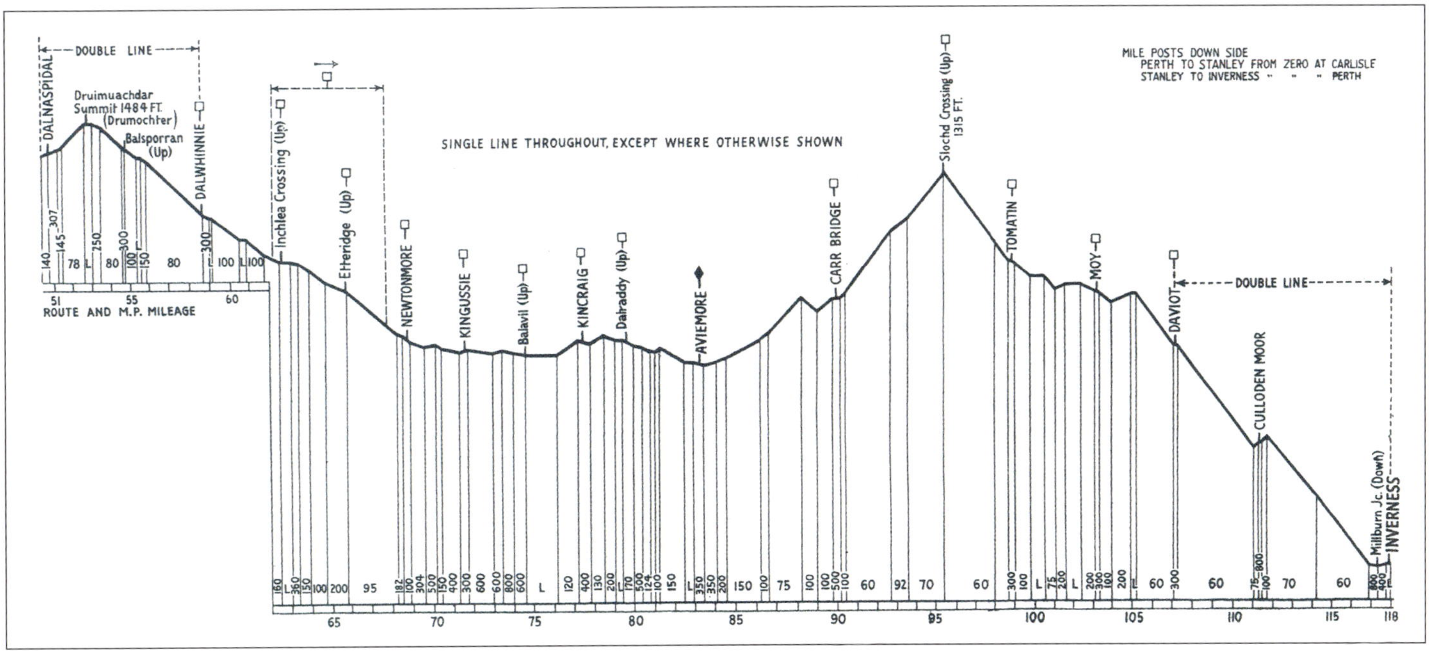

Britain's highest and third-highest standard-gauge summits dominate the Highland Main Line, but only on the 'Direct Line' from Aviemore to Inverness (opened in 1898) do gradients exceed 1 in 70.

The elements were not entirely overcome by the railway however. Snow has always been a hazard between Perth and Inverness, particularly at the exposed summits, and – despite small ploughs fitted to every passenger locomotive at the approach of winter, and the availability of large 'independent' ploughs pushed by as many as five locos – on 17 December 1880 two passenger trains were completely trapped by snow at Dava, one to a depth of no less than 69 ft.

The Highland Main Line became heavily congested during the First World War when it acted as a key supply route to the Grand Fleet in Orkney, to American mine-assembly bases in Inverness and Invergordon, and to the Grand Fleet's ammunition distribution centre in Inverness. After the exhausting war effort, the Highland Railway became part of the largest of the four companies created by the 'Grouping' of 1923 – the London, Midland & Scottish Railway (LMS) – but few major changes came to the Highland Main Line over the next thirty-five years.

Day and night trains continued to run from Inverness to Edinburgh and Glasgow along with night trains to London – even during the Second World War – and it was not until well after nationalization in 1948 that the operation of the railway was transformed. Diesel locomotives began to haul trains between Perth and Inverness in 1958 and, by 1962, steam traction had disappeared – bringing faster journey times and operating economies but some loss of railway romance too.

The 1963 'Beeching Report' brought the closure of the original main line route between Aviemore and Forres via Grantown-on-Spey in 1965 and, before the decade was out, all the double-track sections of the Highland Main Line – other than the first six miles at 1 in 60 or 70 from Inverness – had been singled. Ironically, there followed a surge of freight traffic growth, stimulated by North Sea oil and the new Invergordon aluminium smelter, and by 1980 the track had been redoubled from Blair Atholl to Dalwhinnie and additional crossing loops created between Aviemore and Inverness.

Following electrification of the West Coast Main Line in 1974, Inverness secured its first regular day-time train to London – 'The Clansman' – and in 1984 a faster service via the East Coast Main Line on 'The Highland Chieftain', which continues to provide a daily link between London and the Highland capital. Mechanical signal boxes still largely control the line, but in the interests of speed and economy physical tokens are no longer exchanged between signaller and driver, having been replaced by 'tokenless block' in the late 1960s. In the 1990s, the Highland Main Line pioneered the movement by rail of supermarket supplies with its 'Safeway Flier' from Central Scotland to Inverness and now Tesco customers in Inverness and beyond can do their shopping in the knowledge that much of what they are buying has 'let the train take the strain' – no doubt to the relief of car drivers on the A9 trunk road!

The Highland Main Line now carries more passenger trains than at any time in its history, including tourist charters and 'The Royal Scotsman' luxury land cruise train. It is the rail gateway to the Highlands, providing a fast connection from Glasgow and the Scottish capital to the Highland capital, through the grandest of largely unspoilt scenery. And with Britain's highest standard-gauge railway summit mid-way between Perth and Inverness, this railway offers a truly unique travel experience.

A Glasgow-Inverness service skirts Craig Vinean in the Tay valley between Dunkeld and Pitlochry in March 2012.

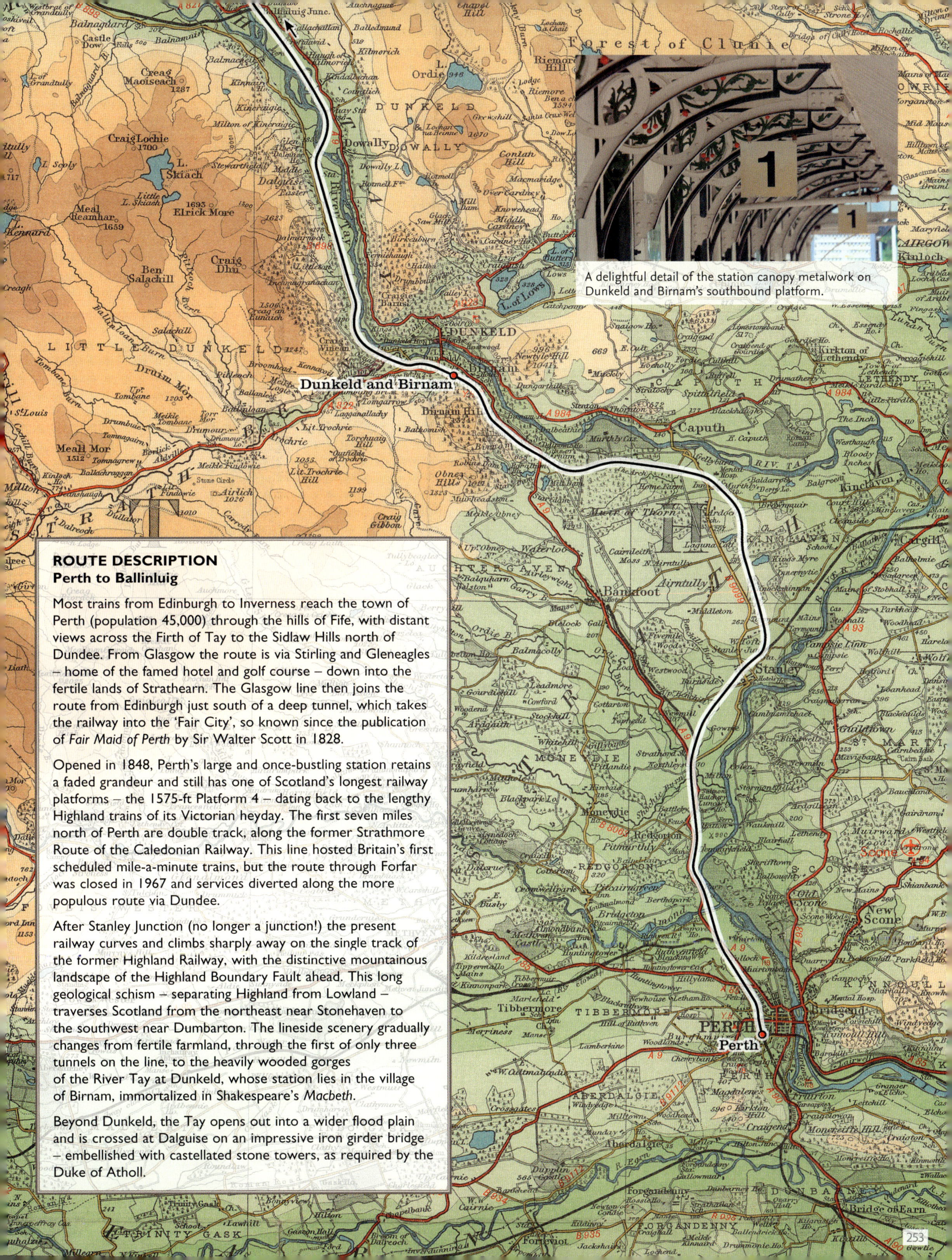

A delightful detail of the station canopy metalwork on Dunkeld and Birnam's southbound platform.

ROUTE DESCRIPTION
Perth to Ballinluig

Most trains from Edinburgh to Inverness reach the town of Perth (population 45,000) through the hills of Fife, with distant views across the Firth of Tay to the Sidlaw Hills north of Dundee. From Glasgow the route is via Stirling and Gleneagles – home of the famed hotel and golf course – down into the fertile lands of Strathearn. The Glasgow line then joins the route from Edinburgh just south of a deep tunnel, which takes the railway into the 'Fair City', so known since the publication of *Fair Maid of Perth* by Sir Walter Scott in 1828.

Opened in 1848, Perth's large and once-bustling station retains a faded grandeur and still has one of Scotland's longest railway platforms – the 1575-ft Platform 4 – dating back to the lengthy Highland trains of its Victorian heyday. The first seven miles north of Perth are double track, along the former Strathmore Route of the Caledonian Railway. This line hosted Britain's first scheduled mile-a-minute trains, but the route through Forfar was closed in 1967 and services diverted along the more populous route via Dundee.

After Stanley Junction (no longer a junction!) the present railway curves and climbs sharply away on the single track of the former Highland Railway, with the distinctive mountainous landscape of the Highland Boundary Fault ahead. This long geological schism – separating Highland from Lowland – traverses Scotland from the northeast near Stonehaven to the southwest near Dumbarton. The lineside scenery gradually changes from fertile farmland, through the first of only three tunnels on the line, to the heavily wooded gorges of the River Tay at Dunkeld, whose station lies in the village of Birnam, immortalized in Shakespeare's *Macbeth*.

Beyond Dunkeld, the Tay opens out into a wider flood plain and is crossed at Dalguise on an impressive iron girder bridge – embellished with castellated stone towers, as required by the Duke of Atholl.

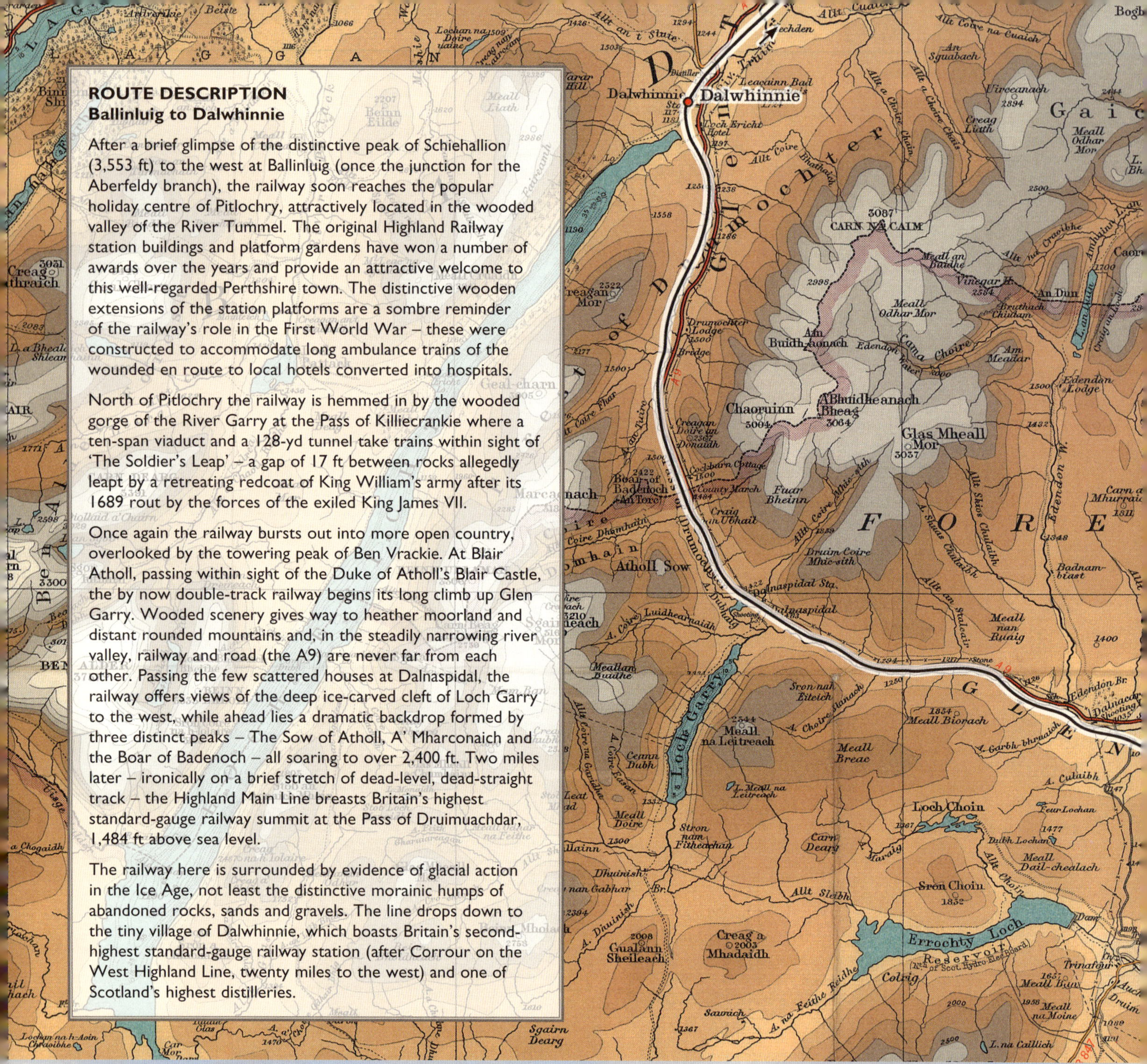

ROUTE DESCRIPTION

Ballinluig to Dalwhinnie

After a brief glimpse of the distinctive peak of Schiehallion (3,553 ft) to the west at Ballinluig (once the junction for the Aberfeldy branch), the railway soon reaches the popular holiday centre of Pitlochry, attractively located in the wooded valley of the River Tummel. The original Highland Railway station buildings and platform gardens have won a number of awards over the years and provide an attractive welcome to this well-regarded Perthshire town. The distinctive wooden extensions of the station platforms are a sombre reminder of the railway's role in the First World War – these were constructed to accommodate long ambulance trains of the wounded en route to local hotels converted into hospitals.

North of Pitlochry the railway is hemmed in by the wooded gorge of the River Garry at the Pass of Killiecrankie where a ten-span viaduct and a 128-yd tunnel take trains within sight of 'The Soldier's Leap' – a gap of 17 ft between rocks allegedly leapt by a retreating redcoat of King William's army after its 1689 rout by the forces of the exiled King James VII.

Once again the railway bursts out into more open country, overlooked by the towering peak of Ben Vrackie. At Blair Atholl, passing within sight of the Duke of Atholl's Blair Castle, the by now double-track railway begins its long climb up Glen Garry. Wooded scenery gives way to heather moorland and distant rounded mountains and, in the steadily narrowing river valley, railway and road (the A9) are never far from each other. Passing the few scattered houses at Dalnaspidal, the railway offers views of the deep ice-carved cleft of Loch Garry to the west, while ahead lies a dramatic backdrop formed by three distinct peaks – The Sow of Atholl, A' Mharconaich and the Boar of Badenoch – all soaring to over 2,400 ft. Two miles later – ironically on a brief stretch of dead-level, dead-straight track – the Highland Main Line breasts Britain's highest standard-gauge railway summit at the Pass of Druimuachdar, 1,484 ft above sea level.

The railway here is surrounded by evidence of glacial action in the Ice Age, not least the distinctive morainic humps of abandoned rocks, sands and gravels. The line drops down to the tiny village of Dalwhinnie, which boasts Britain's second-highest standard-gauge railway station (after Corrour on the West Highland Line, twenty miles to the west) and one of Scotland's highest distilleries.

47 500 hauls the 'The Royal Scotsman' luxury land cruise train south from Pitlochry in May 2012, with Ben Vrackie in the distance.

Locomotive 66 108 hauls a Mossend-Inverness intermodal service between Druimuachdar Summit and Dalwhinnie in the summer of 2010.

Moody weather as an Inverness-bound Class 170 heads towards the camera in November 2005, on the twelve-and-a-half-mile single-track section from Dalwhinnie to Kingussie.

ROUTE DESCRIPTION
Dalwhinnie to Aviemore

Double track now reverts to single over a lengthy descent by the River Truim and increasingly wooded country towards the Spey Valley, with the Monadhliath Mountains forming a northern backcloth.

Along the broad and flood-prone valley of the Spey – with the Cairngorms looming to the southeast – the line serves stations at Newtonmore and Kingussie, and train drivers open up the throttle for the only stretch of 100-mph railway in the Highlands. Aviemore – formerly no more than a small railway village at the junction of the original and later direct lines to Inverness – is now a popular all-year holiday centre close to the Cairngorm ski slopes and, with 125,000 passengers annually, is by far the busiest intermediate station on the Highland Main Line. Despite the closure of the Dava route in 1965, Aviemore is still a railway junction, with interchange to the Strathspey Railway and its nine-and-a-quarter-mile route to Boat of Garten and Broomhill (see page 259).

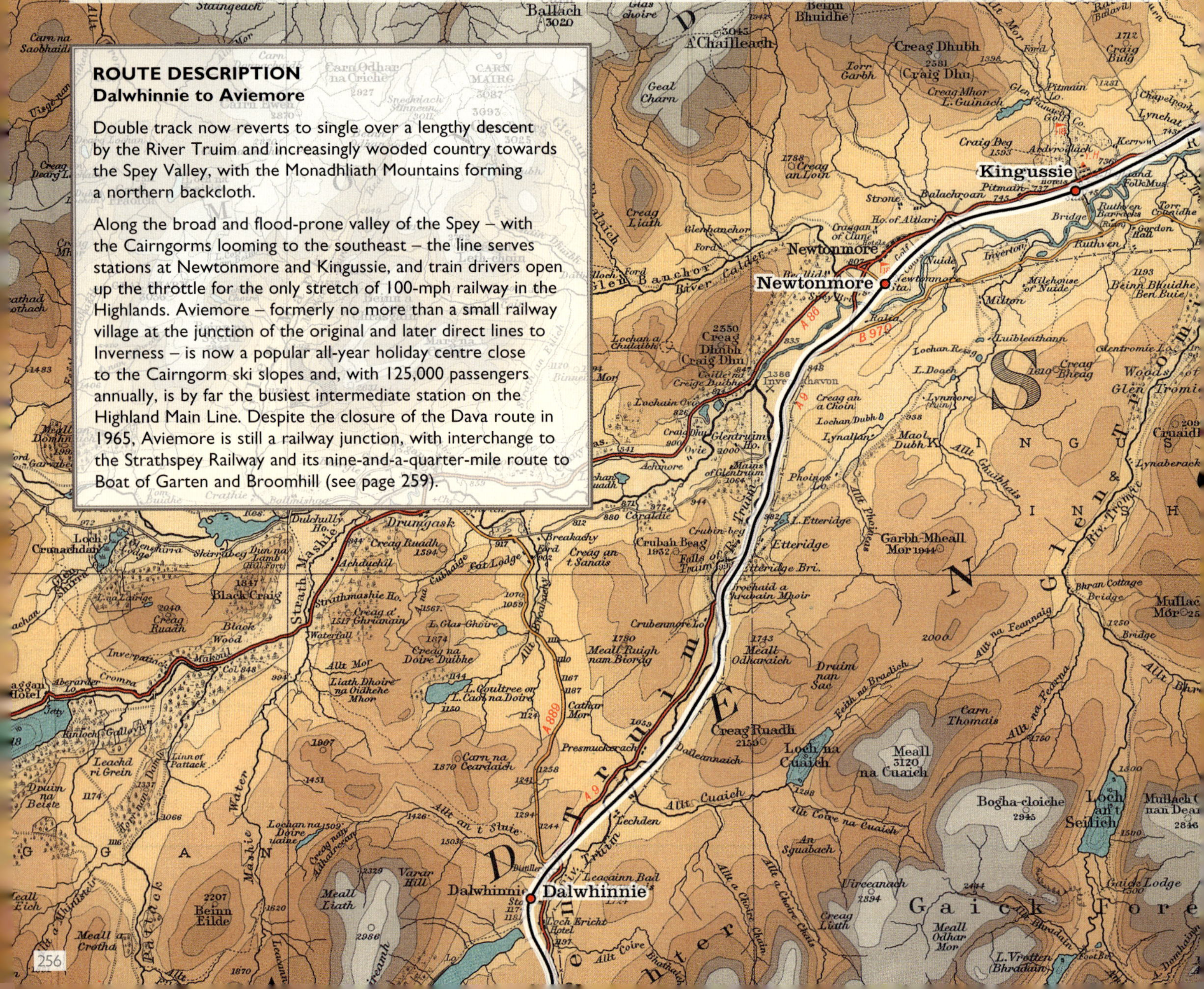

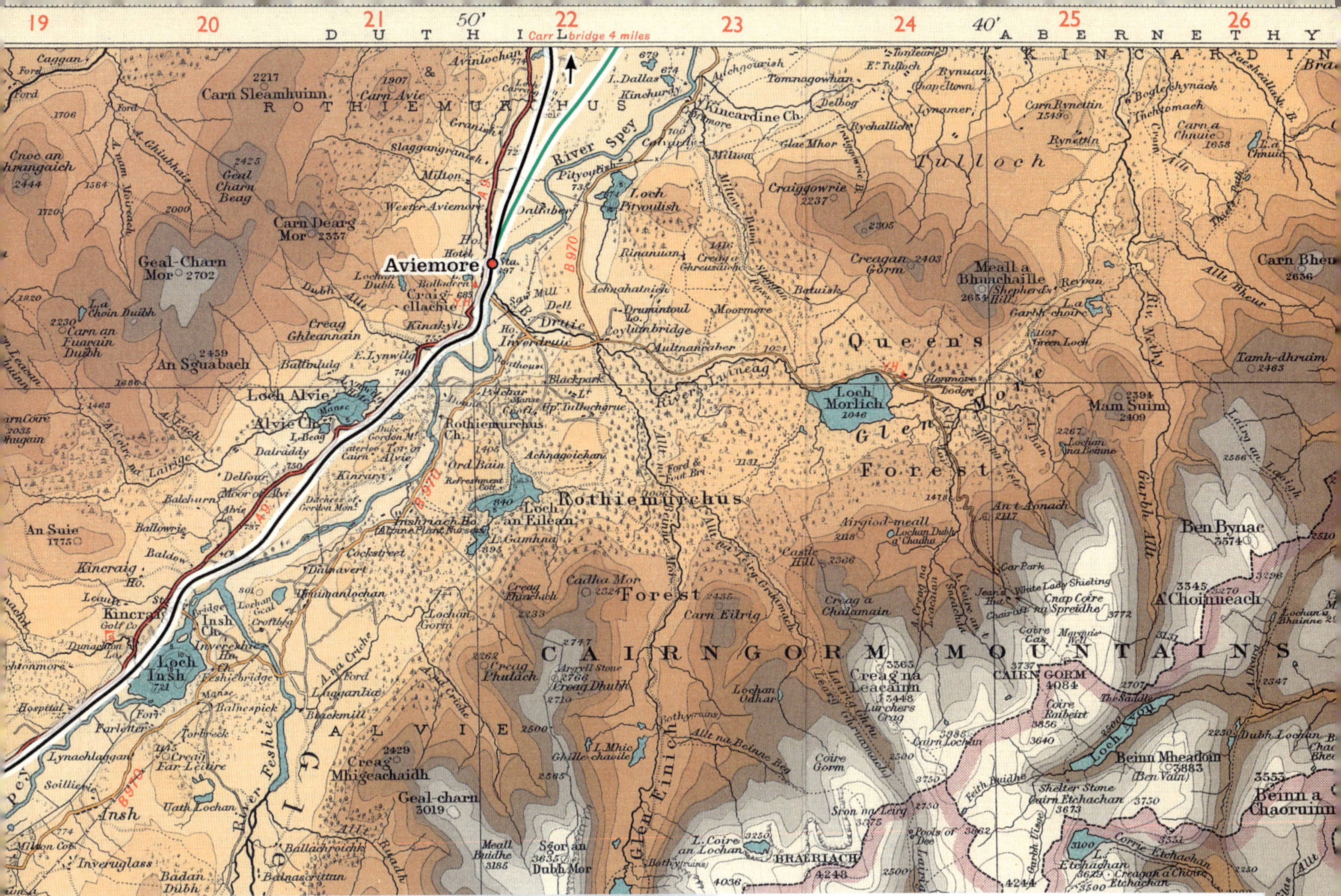

A4 No. 60009 'Union of South Africa', built in 1937, heads an Inverness-bound rail tour north past Dalwhinnie distillery in July 2009.

ROUTE DESCRIPTION

Aviemore to Inverness

North of Aviemore the railway leaves behind a view dominated by Cairn Gorm (4,084 ft) and Braeriach (4,248 ft), separated by the deep cleft of the Lairig Ghru pass which takes determined hill walkers and climbers through the mountains to Deeside in Aberdeenshire. Immediately north of Carrbridge station (reprieved in the 'Beeching era', together with Dalwhinnie) the eastward view – on a clear day – is to the distinctive peaked shape of 2,759-ft-high Ben Rinnes, more than twenty miles distant in the heart of Speyside whisky territory.

Now northbound trains face their stiffest climb of the 118-mile route – five miles mostly at 1-in-60 and 1-in-70 gradients to the 1,315-ft Slochd Mhuic summit. The late nineteenth-century construction of the 'Direct Line' involved exceptionally heavy engineering works, including an eight-arch 100-ft-high masonry viaduct a mile south of the summit. North of Slochd the line drops down to cross the valley of the River Findhorn at Tomatin on an imposing viaduct of nine steel spans supported on masonry columns. Just beyond the village of Moy, trains pass over the last timber bridge in Scotland to carry a main-line railway, dating from 1897; in 2003 a new reinforcing structure was sympathetically constructed within the framework of the bridge itself, and the completed project subsequently received no less than five heritage, design and engineering awards.

At Daviot the line swings east on a wide sweep to avoid gradients beyond the capability of any main-line railway – the only part of the route from Perth where the railway strays more than a couple of miles from the A9 road. To the north, beyond the steep cleft of the valley of the River Nairn – and Drummossie Moor, where Prince Charles Edward Stuart (Bonnie Prince Charlie) prepared his army on 15 April 1746, the eve of the Battle of Culloden – is the unmistakeable mass of Ben Wyvis (3,433 ft) more than twenty miles distant.

Down another 1-in-60 gradient, easing briefly to climb across the Nairn valley on a twenty-eight-arch viaduct constructed entirely of local red sandstone, the railway again becomes double track at the site of Culloden Moor station before falling once more at 1 in 70 and 1 in 60 over the six remaining miles to Inverness. The immediate approaches to Inverness are blighted by sprawling retail and warehousing developments, but close to the town's 1855 station the original stone-built Lochgorm Works of the Highland Railway continues in use as a maintenance depot for diesel units.

Over recent decades Inverness (population 59,000) has become one of the fastest-growing settlements in Britain, but it retains much of its Victorian and Edwardian charm by the River Ness and in the idyllic 'Ness Islands'. For the rail traveller, it is the crossroads of the Highlands.

'Black Fives' Nos. 45231 and 45407 haul the 'Great Britain' excursion towards Slochd Summit in April 2009, with Slochd Viaduct and the Cairngorms as backcloth.

STRATHSPEY RAILWAY
Aviemore to Broomhill

Britain's most northerly steam railway follows the route of the original 1863 Highland Main Line, nine and a quarter miles from Aviemore to Broomhill, along the valley of the River Spey, set against the dramatic backcloth of the Cairngorm Mountains. The Strathspey Railway is particularly redolent of the 1950s, in the closing years of steam, with original buildings beautifully restored and new structures faithfully following the original Victorian styles.

In 1967 the Edinburgh and Falkirk-based Scottish Railway Preservation Society (SRPS) was contacted by a young enthusiast (one of the authors) whose father had just taken up a post with the new Highlands & Islands Development Board (HIDB). David Spaven suggested that the recently closed Aviemore-Grantown section of the original Highland Main Line would make an ideal 'preserved railway', being at the heart of a tourist area with scenic and railway heritage interest – and potentially eligible for grant aid from the HIDB.

SRPS took up the idea enthusiastically and began negotiations with British Rail and HIDB, but distance from Central Scotland was always a concern and in 1971 SRPS pulled out. A new company – the Strathspey Railway Co. Ltd – was formed, with volunteer support from the Strathspey Railway Association. With financial backing from HIDB, the company purchased the solum and intact track from Aviemore to Boat of Garten in 1972, and subsequently the entire solum of the dismantled line through to Grantown.

After years of restoration work, the line opened as far as Boat of Garten in 1978, operated and maintained almost entirely by volunteers, using former London, Midland & Scottish Railway and industrial locomotives. In the early years the company was not permitted to run its trains directly into the spare Platform 3 at Aviemore main-line station, and had to construct its own 'Aviemore Speyside' terminus using the redundant Highland Railway buildings from Dalnaspidal station. In 1998, however – with assistance from the Railway Heritage Trust and Historic Scotland – the whole main-line station was refurbished as part of a wider regeneration project, and the Strathspey was allowed to operate its trains directly into Platform 3.

Since the 1980s the Strathspey Steam Railway (as it is often now known) has been a regular feature of the 'The Royal Scotsman' itinerary, with the luxury train stabling overnight in the period surroundings of Boat of Garten station. In 2002, the line to Broomhill was re-opened with a new station, in traditional style, built at the terminus – doubling as Glenbogle in the BBC TV drama series *Monarch of the Glen*. Plans are now in hand to extend the railway the final three miles through to Grantown-on-Spey, the original destination envisaged back in 1967.

Former Caledonian Railway No. 828 – built in Glasgow in 1899 – returns south from Broomhill loop to Broomhill station in October 2010.

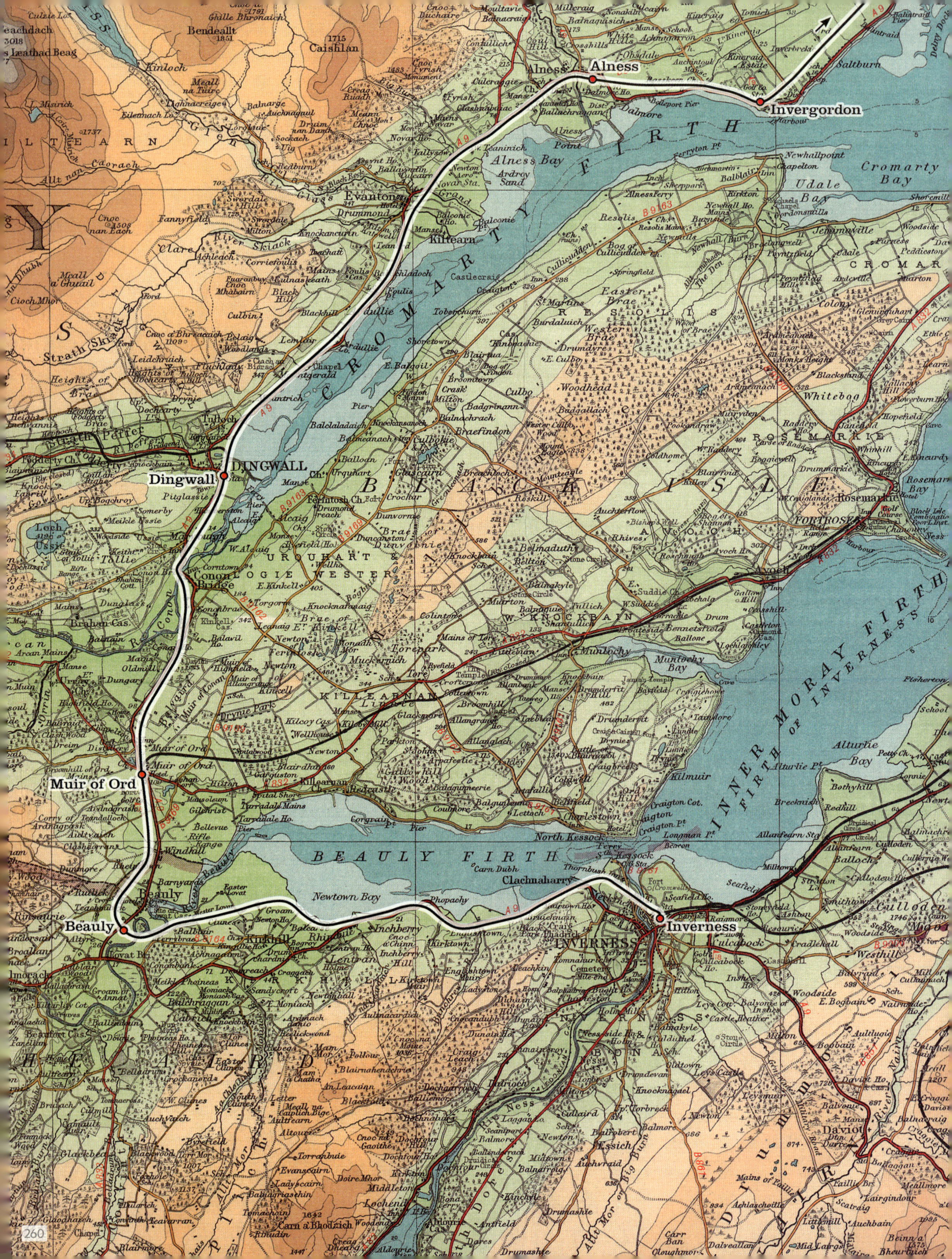

Alness
Invergordon
Evanton
Dingwall
DINGWALL
Muir of Ord
Beauly
Inverness
INVERNESS
CROMARTY FIRTH
Cromarty Bay
BEAULY FIRTH
INNER MORAY FIRTH
FIRTH OF INVERNESS
BLACK ISLE
ROSEMARKIE
Rosemarkie
FORTROSE
Avoch
Munlochy
Munlochy Bay
Alness Bay
Udale Bay
Newtown Bay
Clachnaharry
North Kessock
Kiltearn
Kilcoy Cas.
Redcastle
Conon Bridge
Brahan Cas.
Strath Peffer
Culloden
Alturlie Bay
Saltburn
Kirkhill
Phopachy

The River Ness lies at the heart of Inverness – the rail crossroads of the Highlands – with snow-capped Ben Wyvis to the north.

INVERNESS TO THURSO AND WICK

FAR NORTH LINE

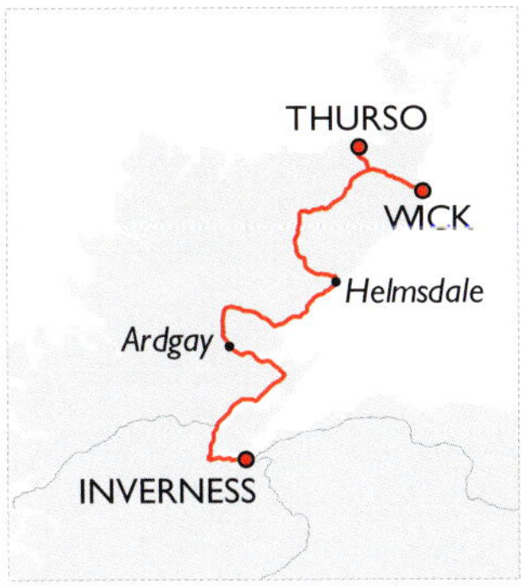

The 168 route miles, from the capital of the Highlands to the Caithness towns of Thurso and Wick, is by far the longest rural railway in Britain. The journey's diverse coastal and inland scenery takes the discerning traveller to some of the loneliest locations on the entire British network.

The first railway in the Highlands opened between Inverness and Nairn in 1855, but it would not be until 1874 that the seventh and final phase of construction that began on the outskirts of Inverness in 1860 brought a line to the 'Far North' termini in the only towns of any size in Caithness, Scotland's most northerly county. Wick was a flourishing fishing port and Thurso was (and is) the nearest mainland town to the Orkney Islands.

The Duke of Sutherland, that county's biggest landowner, played a key role in pushing the railway northwards. When the line had opened as far as Helmsdale in 1871, a new company was formed to take the railway on from Helmsdale to Caithness, but the difficult coastal geography, coupled with a wish to open up more country, took the new line inland up the Strath of Kildonan and across the wild moorlands of the Sutherland / Caithness county march. What could have been a railway journey of thirty-five miles to Wick (albeit one involving very heavy engineering works and a slow, curved alignment in places) became one of sixty miles, a poor competitor with the coastal road. The railway finally reached Wick, with a short branch line to Thurso, in 1874 – covering no less than 161 miles from Inverness (plus six and a half miles on the Thurso branch), compared to significantly less by road and just 80 miles as the crow flies. Given the sparsity of population and potential traffic, the railway was single track throughout (with relatively frequent passing places) although a six-mile section between Inverness and Dingwall was doubled in 1914.

The line was worked by the Highland Railway Company, which had been created in 1865, although the original Sutherland Railway, Sutherland & Caithness and Duke of Sutherland's Railway all retained a separate existence until 1884 – and the Duke had his own train and a private station at Dunrobin Castle (which remains open to the public to this day).

Dornoch, the county town of Sutherland, had been completely bypassed by the Far North Line, and pressure for a local connection to the national network brought the opening in 1902

ROUTE DESCRIPTION
Inverness to Invergordon

With a population of 59,000, Inverness is one of the fastest-growing settlements in the UK, and its seven-platform station is busier than at any time in its history. Just a few hundred yards beyond the station, the Far North Line crosses the River Ness on a new bridge built in 1990 to replace the original structure swept away by floods in 1989. Trains soon slow to cross the Clachnaharry swing bridge across Thomas Telford's Caledonian Canal, completed in 1822 to link the North Sea with the Atlantic by way of the Great Glen and Lochs Ness, Oich and Lochy. The bridge is controlled by the last mechanical signal box to survive on the railway network north of Inverness.

The railway skirts the southern shore of the Beauly Firth – renowned for its prolific breeding grounds for wild fowl – with views across to the peninsula known as the Black Isle and westwards to the mountains of Ross-shire. As the single-track railway turns north at the head of the firth it passes the priory town of Beauly (or *beau lieu*, the French for 'beautiful place'), one of five stations to be re-opened on the Far North Line since 1960. At Muir of Ord – another re-opened station – the 3,433-ft bulk of Ben Wyvis looms up to the north, but around here the railway is running close to sea level over generally moderate gradients, with the real operational challenges lying further ahead. Dingwall – the one-time Viking capital of the North – has the only staffed intermediate station on the Far North Line and, as Michael Pearson records in *Iron Roads to the Far North and Kyle* (2003), a plaque at the station "records that 134,864 servicemen were given a cup of tea here during the First World War"!

The railway divides at Dingwall, bearing westwards for Kyle and northwards for Wick and Thurso. Hugging the northern shore of the Cromarty Firth, the Far North Line soon reaches Invergordon – a deep-water location popular with visiting cruise liners – where the station is adorned with murals commemorating the story of local men's role in the Second World War.

ROUTE DESCRIPTION
Invergordon to Ardgay

Passing through the middle of the distillery complex which supplies spirit for several of Scotland's famous blended whiskies, the railway runs dead straight across some of the most fertile land in the Highlands, eventually turning north west to reach the historic town of Tain, then – with views of Dornoch and the hills of Sutherland to the north – passing close by Glenmorangie Distillery.

Thanks to financial support from the Duke of Sutherland in the nineteenth century the line then – rather than heading directly north across the firth to Dornoch – heads inland towards Ardgay. The circuitous route from Tain to Golspie via Lairg provides a delightful scenic journey for the traveller, but the addition of twenty miles to the through rail journey hinders the ability of the railway to compete for daily traffic between Caithness and Inverness in the era of fast road transport.

In April 2001 two 'Black Fives' – Nos. 44871 and 45407 – ease a southbound charter train across Invershin Viaduct, high above the Kyle of Sutherland.

of a 'light railway' – built to less costly engineering standards – from a junction at The Mound near Golspie to a terminus in Dornoch, eight miles to the south. This was not the only settlement to be reached by a very circuitous route from the south – with assistance from the Government, a light railway was opened from Wick to the coastal fishing village of Lybster in 1903. Both lines were fated to have short lives, closing completely in 1960 and 1944 respectively.

The Far North Line played a critical role in the First World War, with Britain's Grand Fleet – 96 ships and 70,000 men – based at Scapa Flow in Orkney, and a large repair port established at Invergordon. As PJG Ransom records in *Iron Road* (2007), "For the Highland Railway, the effect was as though a new city had materialised at its northernmost terminus." Special trains for naval personnel ran regularly between London Euston and Thurso, scheduled to run every weekday between February 1917 and April 1919, at a timing of twenty-one and a half hours for the 717 miles. These 'Jellicoe specials' – named after the Admiral in Chief of the Grand Fleet – usually consisted of fourteen vehicles, including sleeping accommodation for officers and prison cells for naval personnel who had out-stayed their leave or were guilty of other offences.

The regular winter threat of snow added to the difficulties of handling double-track traffic volumes on a single-track railway. In January 1918 a northbound troop special became stuck in a deep snowdrift near Scotscalder. It had to be abandoned by its 300 passengers, who tramped across the moor to Thurso, and a week would elapse before the line could be returned to normal service. The snow hazard had long been appreciated – after a particularly severe winter in 1864-65, the Highland Railway erected artificial barriers, including 'snow fences', at the most vulnerable points in its network.

The Far North Line again carried a vast amount of additional traffic during the Second World War, by this time as part of the larger London, Midland & Scottish Railway. Recovery from the war effort was slow, but – as road competition began to bite – the Far North Line saw major changes three years ahead of the drastic national upheaval brought by Dr Beeching's infamous report. In June 1960 no less than twenty of the forty intermediate stations between Inverness and Wick/Thurso were closed. By this time the Far North Line – far from sources of suitable coal – had become one of the first regions in Britain to see diesel traction replace steam, and the combination of fewer stops and more powerful haulage brought reduced journey times. But a major shock was in store – the 1963 'Reshaping Report' proposed the withdrawal of all passenger services north of Inverness.

Highlanders were quick to react. Just two days after Beeching had published his report, the Inverness Courier thundered that: "Today this country is facing the greatest crisis it has ever had to face in peace", and two weeks later a conference held in Inverness unanimously opposed the withdrawal of passenger services. This immediate reaction to Beeching soon led to the formation of the 'MacPuff' campaign which was instrumental in securing a Ministerial reprieve of all 231 miles of railway north of Inverness (the Kyle and Far North lines) in 1964.

Despite the growth of lorry competition, rail freight at the southern end of the Far North Line saw a major boost in the 1970s, serving a new aluminium smelter at Invergordon, an oil platform construction yard, a pipe coating plant and the large grain whisky distillery complex sited adjacent to the railway in Invergordon in 1961. Today, the smelter is closed, and the railway is overwhelmingly focused on passenger trains, with a new 'Invernet' commuter service stimulating a doubling of rail passengers between 2005 and 2011.

To reduce costs, the traditional semaphore signalling was replaced in 1988 by the innovative Radio Electronic Token Block system, controlled from a single signalling centre in Inverness. Amongst the biggest challenges of operating the Far North Line are the long stretches of track between crossing loops, including the longest single-track section in Britain, twenty-four miles from Helmsdale to Forsinard.

The line has never been busier with passenger trains, opening up more opportunities to explore a corridor of endless scenic variety – from the rolling moorlands of Caithness and the delightful Sutherland coast to the rugged straths around Lairg and the fertile farmland of Easter Ross. The railway offers the best of 'slow travel', allowing the traveller to savour the wonderful scenery of a route corridor less well known than most of its counterparts elsewhere in the Highlands, but worthy of careful exploration.

A northbound service traverses the distinctive crofting landscape of Strath Fleet, between Lairg and Rogart, in August 2007.

8F No. 48151 heads north away from Dunrobin Castle towards Helmsdale on the April 2007 'Great Britain' rail tour.

ROUTE DESCRIPTION

Ardgay to Helmsdale

Skirting the southern shores of the Dornoch Firth, the railway continues northwestwards towards Culrain (for Carbisdale Castle youth hostel) and Invershin – two stations less than half a mile apart, joined by the mighty Shin or Oykel Viaduct. Beyond Invershin begins one of the stiffest climbs on the entire route, over the five miles to Lairg, much of it at 1 in 72. Lairg – the railhead for much of northwest Sutherland – also handles one of the farthest-flung flows of freight by rail in Britain, in the shape of oil traffic from the Grangemouth refinery.

In one of its many geographical peculiarities, the railway then turns southeastwards down Strath Fleet, through the scattered village of Rogart – whose station was closed in 1960, re-opened a year later, and is now a railway-themed holiday spot – onwards past the old junction for Dornoch at The Mound, to reach the North Sea coast once again at the town of Golspie and its listed station building. This is the territory of the landowning Dukes of Sutherland, the most infamous of whom – the First Duke – has gone down in history for his part in the 'Highland Clearances', shifting sheep on to crofting lands and precipitating mass emigration to the New World. The Duke is 'commemorated' by a prominent statue at the summit of Beinn a' Bhragaidh overlooking Golspie.

North of Golspie comes the first stretch of the steepest gradient on the line – 1 in 60 – taking the railway through the half-timbered formerly private station (now incorporating a small museum) at Dunrobin Castle, where trains halt by request to the train guard or by signalling to the driver. This is one of six request stops on the railway, all located north of Culrain. Beyond Dunrobin Castle is the once industrial village of Brora, where coal was extracted as recently as 1974, and the railway then runs along raised beaches close by the sandy shores of the North Sea to the village of Helmsdale, formerly the main centre of railway operations between Inverness and Wick.

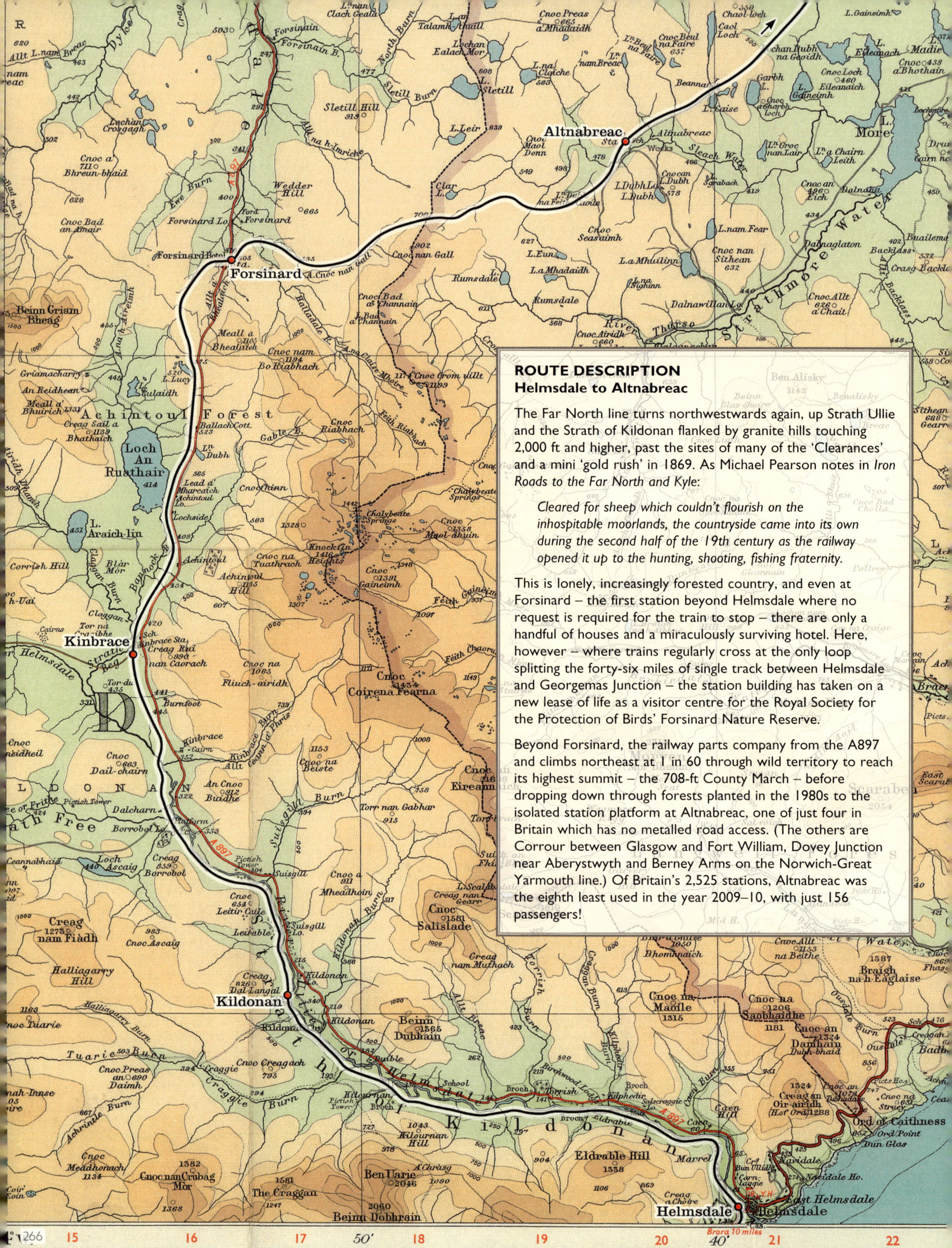

ROUTE DESCRIPTION
Helmsdale to Altnabreac

The Far North line turns northwestwards again, up Strath Ullie and the Strath of Kildonan flanked by granite hills touching 2,000 ft and higher, past the sites of many of the 'Clearances' and a mini 'gold rush' in 1869. As Michael Pearson notes in *Iron Roads to the Far North and Kyle*:

> *Cleared for sheep which couldn't flourish on the inhospitable moorlands, the countryside came into its own during the second half of the 19th century as the railway opened it up to the hunting, shooting, fishing fraternity.*

This is lonely, increasingly forested country, and even at Forsinard – the first station beyond Helmsdale where no request is required for the train to stop – there are only a handful of houses and a miraculously surviving hotel. Here, however – where trains regularly cross at the only loop splitting the forty-six miles of single track between Helmsdale and Georgemas Junction – the station building has taken on a new lease of life as a visitor centre for the Royal Society for the Protection of Birds' Forsinard Nature Reserve.

Beyond Forsinard, the railway parts company from the A897 and climbs northeast at 1 in 60 through wild territory to reach its highest summit – the 708-ft County March – before dropping down through forests planted in the 1980s to the isolated station platform at Altnabreac, one of just four in Britain which has no metalled road access. (The others are Corrour between Glasgow and Fort William, Dovey Junction near Aberystwyth and Berney Arms on the Norwich-Great Yarmouth line.) Of Britain's 2,525 stations, Altnabreac was the eighth least used in the year 2009–10, with just 156 passengers!

Looking north to the simple but attractive and well-kept Forsinard station building – now an RSPB visitor centre – in April 2011.

8F No. 48151 heads north from Helmsdale on the 'Great Britain' tour in April 2007 – only the second time the line had seen steam since the early 1960s.

Thurso – the most northerly town on the British mainland – has a Norse name meaning 'Thor's River', and looks out over Thurso Bay and the Pentland Firth.

ROUTE DESCRIPTION

Altnabreac to Thurso and Wick

Caithness is predominantly a county of flattish rolling moorland and farmland, but by far its highest and most distinctive peak – the 2,316-ft Morven – can be seen from the train at Altnabreac, before the line returns to 'civilization' at Scotscalder and reaches Britain's most northerly railway junction at Georgemas. Here pipes from northeast England arrive in lengthy trainloads for the offshore oil industry, and southbound trains will convey materials associated with the dismantling over the next decade of the failed nuclear complex at Dounreay on the nearby coast.

Trains from Inverness formerly divided at Georgemas, but nowadays, as an economy measure, the complete train firstly heads north up the six-and-a-half-mile branch line to Thurso, stops briefly at the sandstone-built train shed, then returns south, pausing at Georgemas again before completing its 175-mile transit at the original single-platform Caithness flagstone train shed at Wick (population 7,000). This may seem an odd routeing, but it makes sense for the train to head first for Thurso (population 9,000) as the interchange for road transport to Scrabster harbour and the ferries to Orkney. With its wide streets, coastal walks and many bus services, Thurso makes an excellent touring base.

Few rural railways in Britain can match the diversity of the Far North Line for scenic quality. With four trains a day to Wick and Thurso, and additional services at the southern end of the line, the leisure traveller has all kinds of rail journey options to make the most of this far-flung outpost of the British railway network.

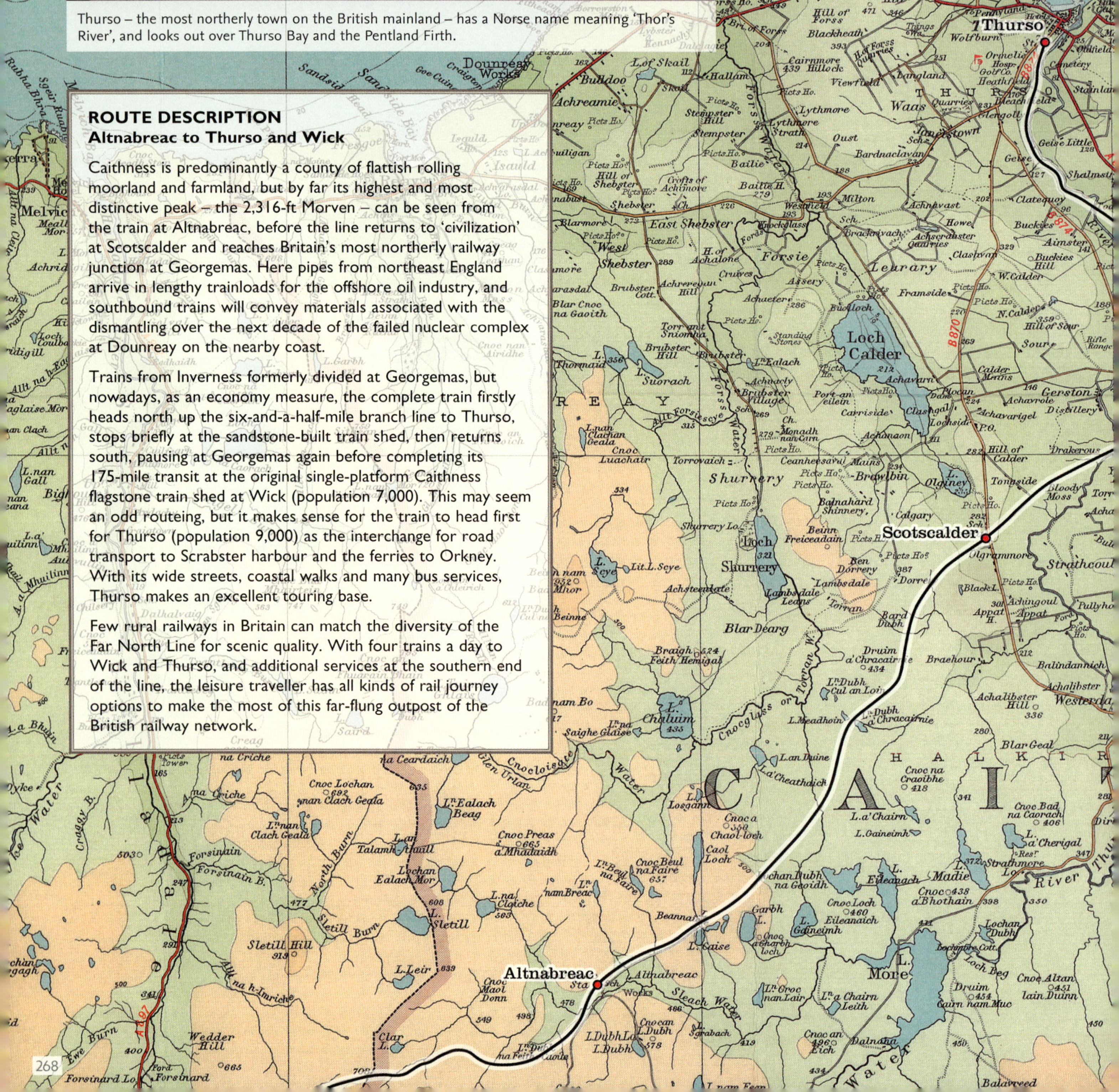

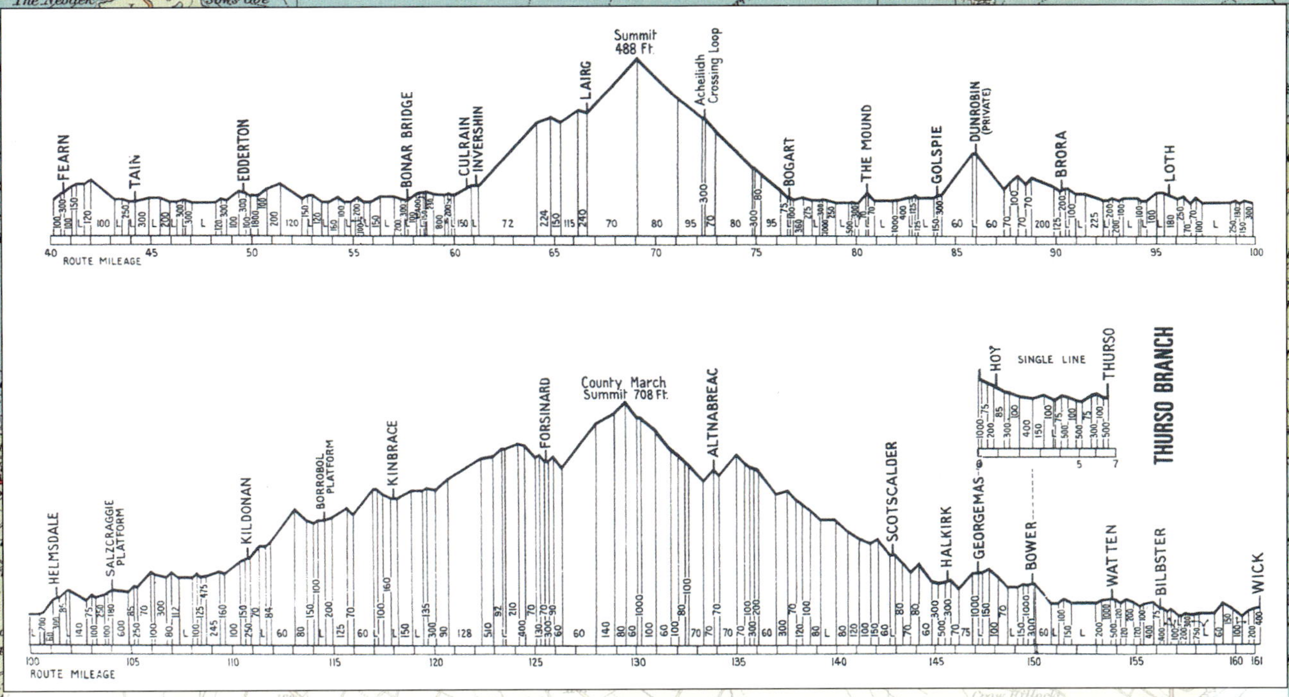

The Far North Line's gradient profile illustrates the many steep climbs and frequent changes of gradient north of Invergordon.

DINGWALL TO KYLE OF LOCHALSH

DINGWALL & SKYE LINE

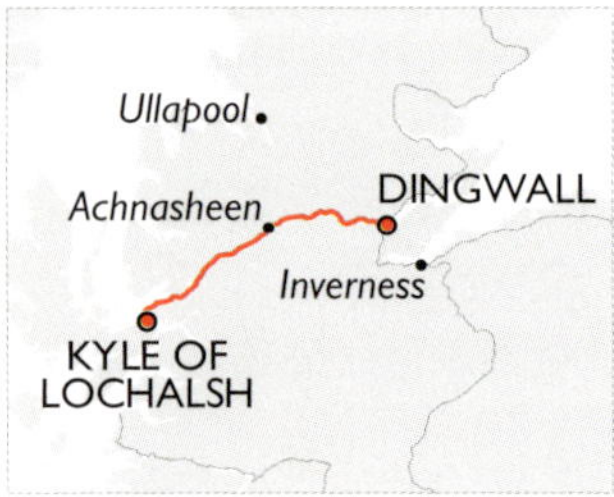

The rail journey from east coast to west coast is one of *the* scenic rail experiences in Britain, taking in delightful pastoral country, bleak mountain country and a gorgeous coastal climax.

Once the Inverness & Ross-shire Railway had reached Dingwall in 1862, attention turned to the economic opportunities which could be opened up by a rail route through the mountains to the west coast – allowing livestock to be sent south without the loss of meat quality from 'droving' across such rough terrain and opening up the western fishing grounds to the London market for the first time.

In 1865 powers were granted for the sixty-four-mile Dingwall & Skye Railway, from Dingwall via the spa town of Strathpeffer and lonely Straths Bran and Carron to the west coast village of Kyle of Lochalsh, immediately opposite the island of Skye on the sheltered waters of the Inner Sound. However, opposition from landowners in the east and the enormous cost of blasting through hard Torridonian sandstone along the final ten miles to Kyle, meant that the single-track railway, which eventually opened in 1870, took an expensive and steeply graded detour away from Strathpeffer and terminated at Stromeferry. A branch line was opened from the outskirts of Dingwall to Strathpeffer in 1885 but its use declined with the advent of the motor era and closed completely as early as 1951.

The Dingwall & Skye merged with the Highland Railway in 1880 and the combination of difficult currents for shipping at Stromeferry, together with the threat posed to Skye traffic by the authorization of the West Highland Railway in 1889, led to the Highland securing powers in 1893 for the final extension. Heavy engineering works eventually took the railway to Kyle, where the station site was blasted out of solid rock and a new pier provided for ships to Skye and the Outer Hebrides.

A Kyle-Inverness service skirts Loch Carron on the approaches to tiny Attadale station, and the adjacent grounds of Attadale House, in October 2009.

During the First World War, the Admiralty commandeered the whole railway west of Dingwall and the Highland Railway was permitted to run just one train daily in each direction for passengers and mail. The Kyle line's key role was in transferring components shipped from America to Kyle to the Invergordon assembly base for the huge minefield – known as the Northern Barrage – stretching from Orkney to Norway.

Throughout its history, the Kyle line has been a popular route for tourist traffic, culminating in the summer operation of observation cars at the tail end of trains intermittently for some three decades from 1961 onwards. By the latter year, steam locomotives – many of them more than twenty-five years old – had been almost completely replaced by modern diesels.

Perhaps the most dramatic period of the Kyle line's history followed the 1963 publication of the 'Beeching Report' which proposed withdrawal of all passenger services north and west of Inverness. If enacted, this would have led to complete closure of the Kyle line, as the daily freight train could not have justified the cost of upkeep of sixty-four miles of railway. However, the spirited 'MacPuff' campaign mobilized powerful political forces

– at a time when the railway was still the main carrier of passengers, mail, parcels and freight – and both the Kyle and Far North lines were reprieved.

This was not to be the last threat to the Dingwall & Skye line, as the combination of parallel road improvements and the switch of the Stornoway ferry from Kyle to rail-less Ullapool, prompted the Government in 1972 to propose withdrawing the social grant aid required to maintain the passenger train service. Once more, a strong campaign was mounted and, with strategic arguments for rail strengthened by a planned oil platform construction yard on Loch Kishorn near Stromeferry, a second reprieve came in 1974.

To secure the future of the line, economies of operation in the 1980s saw the replacement of locomotive-hauled trains by diesel units and traditional semaphore signalling by the Radio Electronic Token Block system, since when services have grown, with weekday frequency increased from three to four trains daily and the introduction of all-year Sunday trains for the first time. The Kyle line is one of the top Scottish attractions for charter train operators and is now a regular part of the itinerary of the luxury land cruise train, 'The Royal Scotsman'.

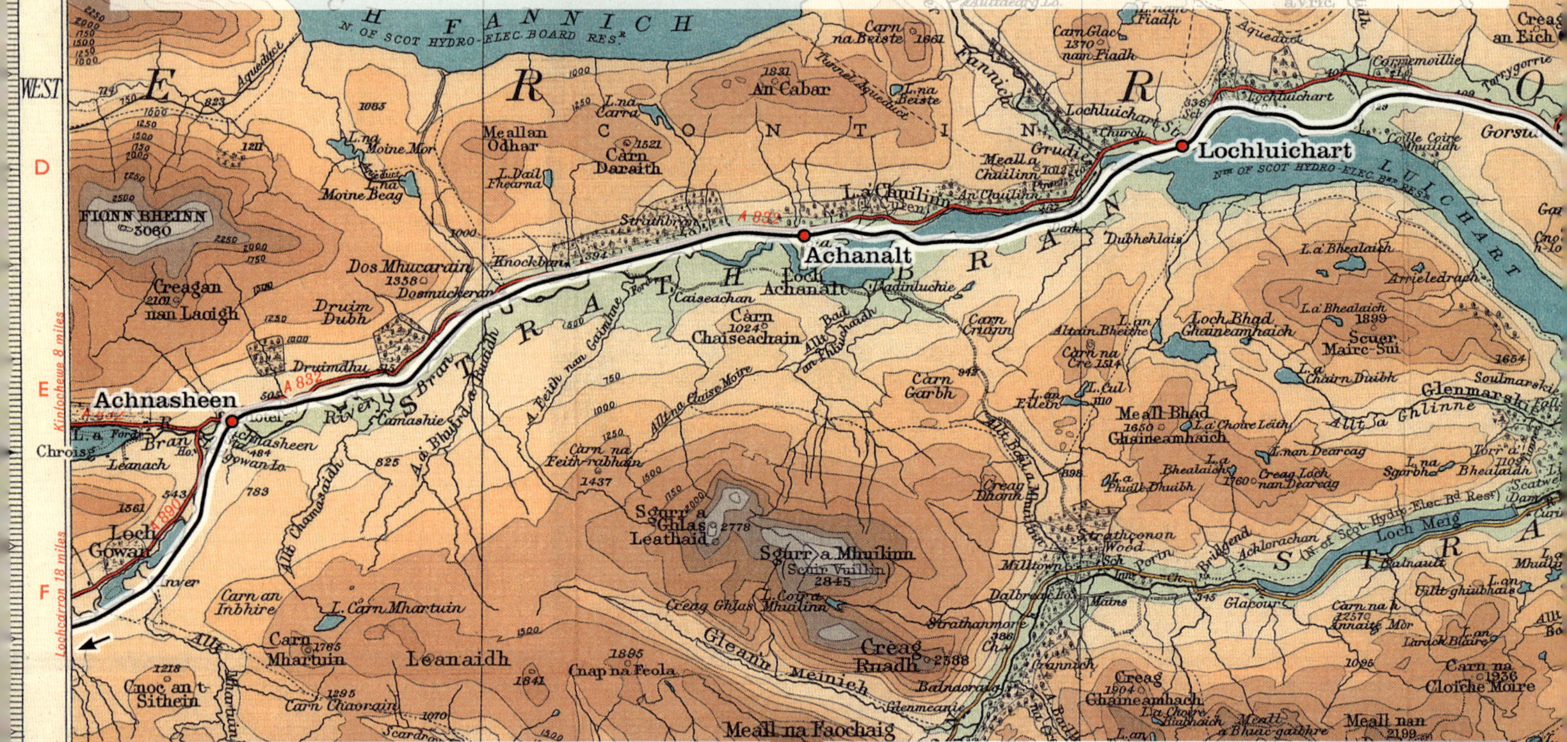

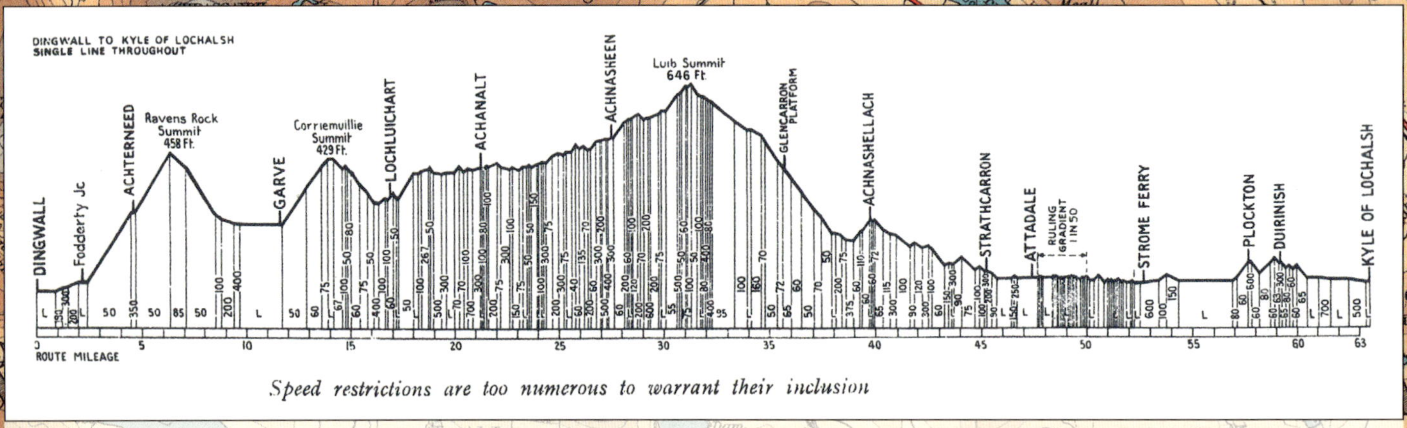

The switchback character of the Kyle line is clearly illustrated in the route's gradient profile.

K4 No. 61994 'The Great Marquess' tackles the 1-in-50 gradient to Raven's Rock Summit, between Dingwall and Garve, on the 'Cathedrals Explorer' charter train in May 2012.

'Black Five' No. 45305 negotiates the switchback route approaching Loch a' Chuilinn, west of Lochluichart at the head of the 'Great Britain' rail tour in April 2012.

Crossing boggy moorland, a Sunday service from Inverness to Kyle heads west (towards the camera) from Achanalt in March 2012.

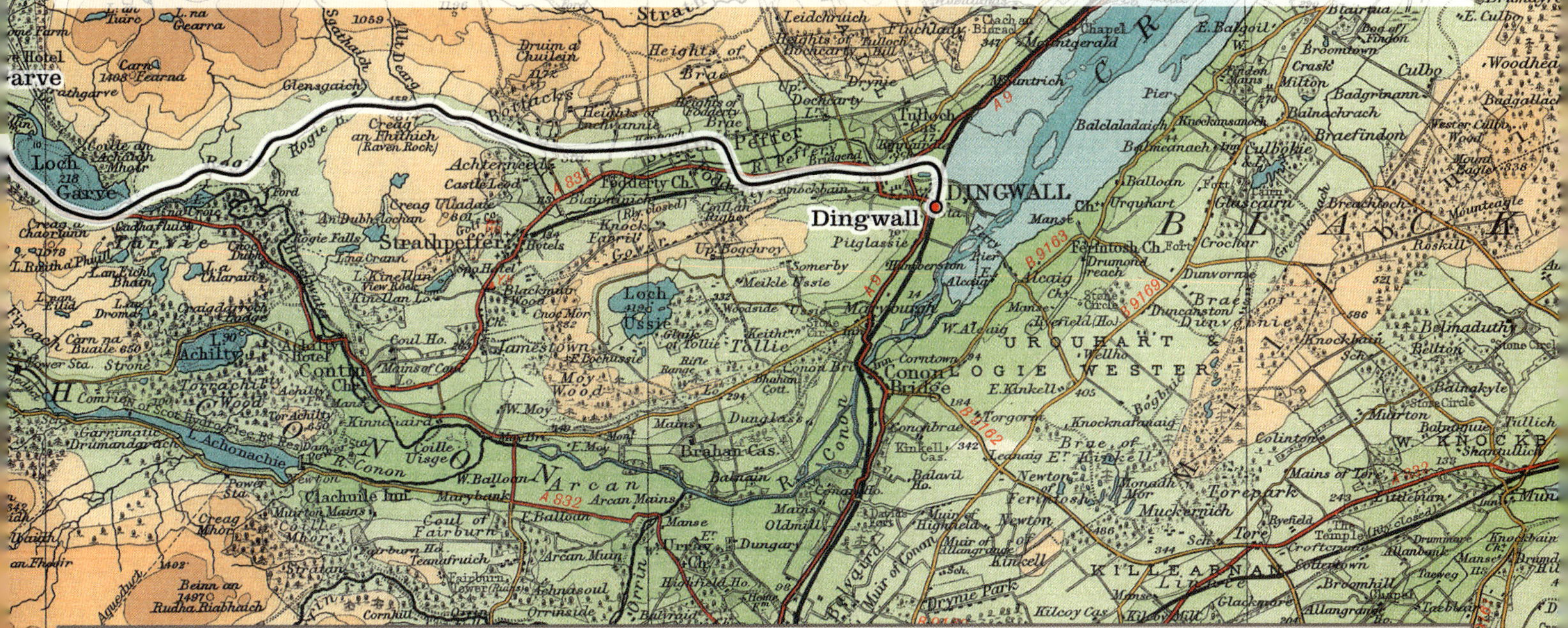

ROUTE DESCRIPTION
Dingwall to Achnasheen

The Kyle line veers west away from the Far North Line immediately north of Dingwall station and just two miles later swings across to the north side of Strath Peffer, beginning four miles of almost unbroken 1-in-50 gradient to Raven's Rock summit. To the south can be seen the distinctive shape of the 'Cat's Back' ridge and the route which the through line to Kyle should have taken – but the climb to the line's second-highest summit, at 451 ft, is a dramatic railway experience, not least for charter train passengers hauled by a powerful throaty diesel. A precipitous cliff towers above the summit, where it is said that ravens hold their annual ball. A surviving description of a nineteenth-century 'footplate' (locomotive cab) journey memorably records that "The re-echoing of the heavy blasts from the chimneys while in the cutting are like continuous thunder peals."

The railway drops down through heavily forested terrain to the shores of Loch Garve and the line's first station at the small village of Garve, where passengers and mail formerly detrained for the connecting bus to Ullapool. Immediately after this stop the railway begins another fierce climb at 1 in 50 towards the 429-ft Corriemuillie Summit beyond which, over its remaining forty-nine miles, the Kyle line is an almost continuous switchback of changing gradients and sharp or sinuous curves. As the gradient profile on page 272 succinctly notes: "Speed restrictions are too numerous to warrant their inclusion"! So a journey to Kyle may not be swift but it is certainly one to be savoured.

At Lochluichart a major hydroelectric scheme led to the re-alignment of the railway and construction of a new station in 1955. Beyond here the line climbs into Strath Bran ('valley of the drizzle'), at whose western end is the isolated staging post of Achnasheen. Traditionally the main trains of the day crossed at the loop here, giving regular passengers time to duck into the adjacent Achnasheen Hotel (with a direct entrance from the station platform to the bar) for a swift dram, confident in the knowledge that the driver or guard – often fellow bar customers – would give them the nod when the trains were ready to leave! Regrettably, the hotel burned down some years ago and nowadays, other than the junction where the roads to Gairloch and Kyle part company, the station effectively *is* Achnasheen.

ROUTE DESCRIPTION
Achnasheen to Kyle of Lochalsh

The line reaches its highest point at Luib Summit – 646 ft above sea level – a few miles west of Achnasheen, before dropping almost continuously downhill through the forests of Glen Carron (where the local landowner had his own private station, complete with hand-operated signals) and Achnashellach to the village of Strathcarron, near the head of Loch Carron. Connecting buses operate from here to the west coast villages of Lochcarron and Shieldaig, while the eighteen remaining railway miles to Kyle provide a delightful coastal journey over continuous curves, across causeways and beside sea defence works – with increasingly dramatic landscapes culminating in fine views across the Inner Sound towards the mighty Cuillins.

The first station after Strathcarron is Attadale (one of six request stops on the Kyle line), just a few yards from the gardens of Attadale House, where the mild west coast climate allows a wide variety of conifers, rhododendrons and exotic plants to flourish in twenty acres of sheltered grounds. Beyond Attadale the road was extended in the late 1960s to avoid the A890's ferry crossing at Stromeferry, forcing realignment of the railway along a sinuous route over the narrow coastal strip, and – after intermittent rock falls caused by the road works – necessitating the construction of the UK's only road and rail avalanche shelter. At Stromeferry there is now little sign of the station's former status as the terminus of the line – nor of the railway sidings serving the Kishorn oil platform construction yard, which helped to save the route from closure in 1974.

The ten and a half miles from here to the terminus at Kyle challenged the Victorian railway engineers but have bequeathed the modern traveller a twisting, turning journey of changing vistas and scenic surprises. The first section takes the railway along 'raised beaches', demonstrating how sea levels have changed over geological time. Then, amongst heavily forested coastal slopes, comes the tiny private halt serving Duncraig Castle – closed in 1964 (one of just three stations on the line to close) but re-opened to the public twelve years later. Next comes Plockton, where the original station building serving the picturesque village and sailing resort has been converted into a popular restaurant. The railway then passes through the crofting settlement of Duirinish, with its tiny station, before carving through solid rock to make its dramatic arrival at Kyle's railway pier opposite the island of Skye – now of course connected to the mainland by a bridge, conveying bus services to take the rail traveller on to further explorations of the magical west coast.

Kyle station houses a small railway museum, offering a glimpse of Highland life from a century ago, and the 'Friends of the Kyle Line' group is raising funds for the planned restoration of the original Kyle signal box, still complete with traditional levers and point rodding.

The mountains of Skye form a dramatic backcloth for the terminus of the 'Dingwall & Skye' line at Kyle of Lochalsh station.

Hauled by K4 No. 61994 'The Great Marquess', the 'Great Britain' rail tour climbs westwards to Luib Summit in April 2009.

GLASGOW TO FORT WILLIAM

WEST HIGHLAND LINE

Over the ninety-nine and a half miles from Craigendoran Junction (twenty-three miles west of Glasgow) to Fort William, the single-track West Highland Line crosses some of the most isolated and beautiful territory in the British Isles – a magnet for charter trains as well as the regular ScotRail service.

The West Highland was originally intended as a competitive main line to Inverness via Fort William and the Great Glen, challenging the Highland Railway's route north from Perth. After several decades of proposals and counter-proposals, the West Highland Railway – with support from one of Scotland's 'big two' companies, the North British Railway (NBR) – finally secured its Act of Parliament in 1889.

The labour force employed to build this lengthy railway eventually swelled to no less than 5,000 men. As John Thomas reflects in *The West Highland Railway* (1965), the employees of the Lucas & Aird contracting company must have felt at home on this job, as the firm had specialized in colonial railways, which shared many characteristics with the West Highland:

> *In view of the sparse traffic likely to be offered the directors wanted it built as cheaply as possible; it was intended to carry light loads at moderate speeds. The territory which the contractors set out to master was as tough as any they had encountered overseas. There were passes over 1,000 [ft] above sea level through which the railway had to be led. There were innumerable gorges to be spanned, rivers to be bridged and miles of boggy moor to be crossed.*

Men and machinery were both crucial to this difficult project, which took five years of gruelling construction work, using pack horses to carry construction materials to locations without road access. It could be dangerous work, as had been demonstrated in January 1889 when a party of ill-clad railway surveyors came close to disaster in the remote and inhospitable territory north of Bridge of Orchy, only narrowly surviving wild weather conditions while exploring the planned route of the railway.

Said to have been the greatest mileage of railway ever opened on a single day in Britain, the line's at that time 100 miles opened for traffic on 7 August 1894, with the North British Railway as operator. Through its years of operation by the North British Railway, the West Highland Line became a considerable

After an easy run along the shores of the Clyde, the West Highland Line rises steeply from Craigendoran Junction to Helensburgh Upper and beyond.

ROUTE DESCRIPTION

Glasgow to Garelochhead

Most rail journeys to Fort William begin at Queen Street station in the heart of Glasgow (population 592,000), once the 'Second City' and industrial powerhouse of the British Empire. A ferocious climb up Scotland's steepest railway gradient – the 1 in 45 of the Cowlairs Incline – is shared with the route of 100-mph inter-city trains to Edinburgh before West Highland trains branch away westwards through Glasgow's suburbs, running by the shores of the River Clyde at Bowling.

On the eastern fringes of the attractive dormitory town of Helensburgh, the character of the railway changes dramatically. From electrified double track on a virtually flat alignment, at Craigendoran Junction the single-track West Highland Line swings away to the north on a fierce 1-in-58 gradient and the first of a serpentine succession of curves that characterizes much of the next ninety-nine and a half miles of this magical railway.

Beyond Helensburgh the railway hugs the contours of the map on wooded slopes high above Gare Loch and Faslane. Built as a military port in the Second World War in case Glasgow's docks were disabled by bombing, Faslane is now a heavily fortified base for the UK's nuclear submarines.

Peaks of up to 2,300 ft – in a militarized 'danger area' – overlook the West Highland Line at Garelochhead.

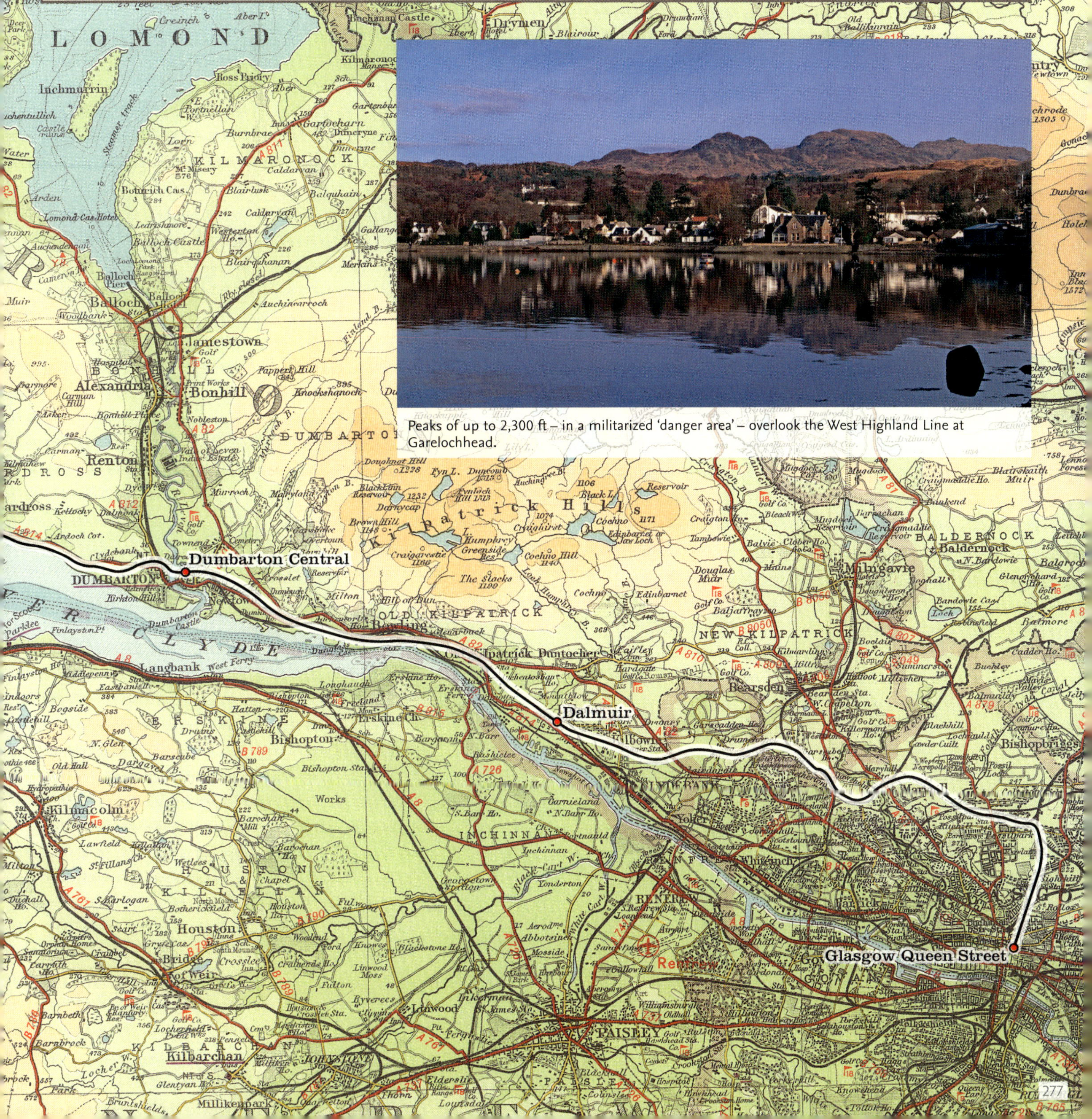

tourist attraction, boosted by the publication of the travel guide *Mountain, Moor and Loch* in 1894. The NBR became part of the London & North Eastern Railway (LNER) in the 'Grouping' of 1923. In 1933 it first featured as a regular part of the itinerary of *The Northern Belle*, the luxury touring train which ran several times a year on week-long excursions from King's Cross – complete with sleeping and restaurant cars – until the outbreak of war in 1939.

The tourist trade resumed after the war and the railway's role as a freight carrier – notably of fish, as well as aluminium from the Fort William smelter – was then boosted in the early 1960s by the announcement that a new pulp and paper mill was to be built at Corpach on the shores of Loch Linnhe near Fort William. This industrial development was to save the West Highland Line from any threat of closure under Dr Beeching's infamous report of 1963.

With its steep gradients and often sharply curved alignment, the West Highland has always been a challenging railway to operate. Traversing sparsely populated countryside and serving a terminus town with a population of just 10,000, the line was single track from its inception – now with just twelve intermediate stations and eleven 'crossing loops', most of them at the same locations.

To reduce the cost of operating the railway the traditional semaphore signalling, with staffed signal boxes at each crossing loop, has been replaced by the innovative Radio Electronic Token Block system, controlled from a single signalling centre at Banavie near Fort William. With all the intermediate stations now unstaffed, many of the railway buildings have been put to good alternative use – including bunkhouses at Bridge of Orchy, Corrour and Tulloch.

The railway carries not only passenger services but also freight trains, which continue to thread their way through the glens. On Rannoch Moor eagle-eyed observers rub their eyes in bemusement on spotting convertible road-rail Landrovers plying the line on track maintenance duties.

A team of railway maintenance staff waits to head north from Rannoch station in a RoadRail Land Rover in May 2003.

Class 4 2-6-0 No. 76079 steams across Rannoch Moor at the head of a charter train in October 2008.

A glorious view of Ben Lomond opened up by clearance of lineside vegetation, as a southbound ScotRail service approaches Arrochar and Tarbet in January 2007.

ROUTE DESCRIPTION
Garelochhead to Crianlarich

After Garelochhead the railway turns briefly inland to avoid the steeply graded mountainside by Loch Long, then swings northwest towards the distinctive shape of Ben Arthur ('The Cobbler') high above the western shore of the loch.

From the 1960s onwards, lineside vegetation, which had previously been carefully managed, became neglected and the immediate wooded surroundings of parts of the West Highland Line gradually became a barrier to viewing the landscape from trains in summer months. Recently – in a partnership between Friends of the West Highland Lines, Network Rail and the Loch Lomond and The Trossachs National Park – tree clearance funded by the Park has opened up views on a number of stretches of line, notably around Arrochar, in Glen Falloch and north of Crianlarich, providing stunning views of the National Park's lochs and mountains for the tens of thousands of visitors travelling the line every summer.

North of Ardlui, the railway climbs through Glen Falloch to Crianlarich. This small village – nestling beside the River Dochart, which drains eastwards to Loch Tay, the River Tay and eventually the North Sea – has long been an important base for railway operations and maintenance, as well as the junction for Oban since the former Caledonian Railway route from Central Scotland via Dunblane and Callander closed in 1965.

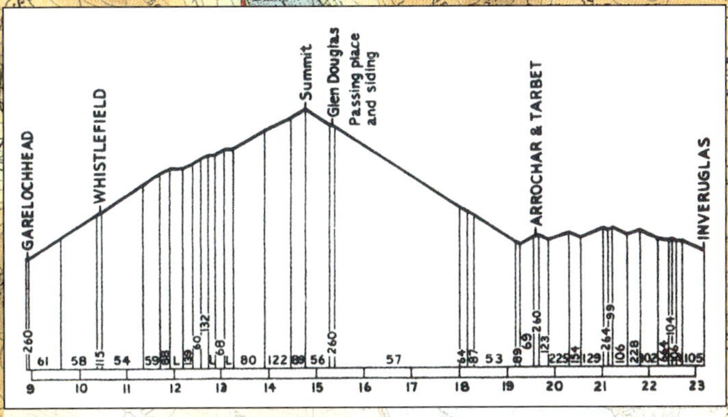

The inland detour via Glen Douglas to avoid the steep slopes of Loch Long still involves a stiff climb and frequent changes of gradient.

No other railway in Britain traverses as much dramatic scenery as the West Highland Line as it heads northwards from the shores of the Clyde to Britain's second-highest main-line summit at Corrour, before turning west and then south to reach Fort William. The line has long been a lifeline for both passenger and freight traffic and owes much of its character to the operational requirements of running a railway through lonely territory and in sometimes extreme weather conditions. Most of the line's distinctive Swiss chalet-style station buildings have survived as an attractive addition to the stunning natural backcloth of mountain, moor and loch. As British Rail's promotional material put it in the 1970s, it truly is 'a Line for All Seasons'.

B1 No. 61264 heads a southbound charter train away from the Horseshoe Curve on the climb to County March summit in October 2003.

The northbound Caledonian Sleeper crosses Auchtertyre Viaduct between Crianlarich and Tyndrum in March 2012, passing a stretch of lineside recently cleared of trees to open up superlative views of Strathfillan and beyond.

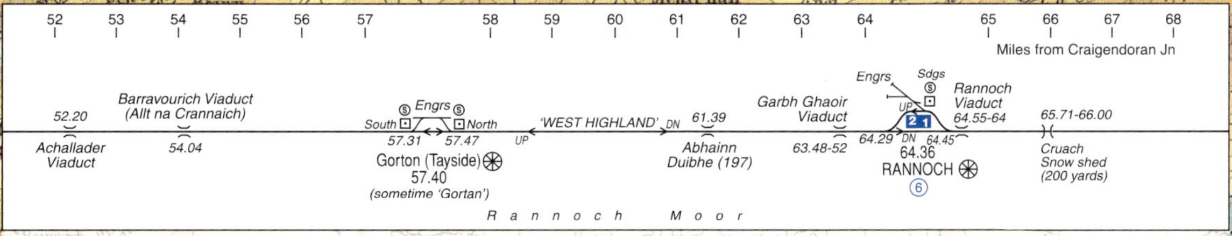

This track diagram illustrates the simplicity of rail infrastructure across Rannoch Moor – but also the importance of engineers' sidings in road-less terrain.

ROUTE DESCRIPTION

Crianlarich to Rannoch

Immediately beyond Crianlarich, the line begins the almost continuous six-mile ascent – at gradients of up to 1 in 60 – through Upper Tyndrum station to the 1,024-ft-high County March Summit. The outstanding feature between Tyndrum and Bridge of Orchy is the famous Horseshoe Curve, which takes the railway on a wide sweep across the Allt Kinglass Burn with the distinctive bulk of Beinn Dorain towering 3,523 ft high above.

North of Bridge of Orchy the railway parts company from the A82 road on its route towards Glencoe, turning northeast to traverse thirty desolate miles over Rannoch Moor to Tulloch in Glen Spean. After the long climb up the valley of the Water of Tulla, the train speeds through the lonely railway outpost of Gorton. Here there was once a signal box and a house for the signalman's family – probably the most isolated place of railway employment in the British Isles. The family who staffed the signal box retired in the mid 1960s and when the job was advertised throughout the railway network, there were no takers so the points were disconnected and signal arms removed in 1968. Then nineteen years later – to aid track maintenance work over Rannoch Moor – a simple loop and short sidings were re-laid for railway engineers' track vehicles.

Beyond Gorton lies the wide expanse of the saturated high-lying bogland of Rannoch Moor. As John Thomas notes in his introduction to the 1972 reprint of *Mountain Moor and Loch*, during the five years of construction of the West Highland Line:

> *The Moor of Rannoch was particularly troublesome. There was difficulty in persuading traditional navvies to go to the remote camps on the moor, and crofters were brought in from the islands and trained in railway work. There were weeks on end when work was impossible because of weather conditions. At the worst places the line had to be floated across the bogs on a raft of brushwood and heather.*

Today this is still dramatically bleak territory, with not a road or car in sight.

The unusual 1949 signal box at Spean Bridge still survives – despite being closed in 1988 – and the station building now houses a restaurant.

'Black Five' Nos. 45407 and 44871 head south through Monessie Gorge – between Roy Bridge and Tulloch – in October 2011.

The old signal box at Rannoch station survives as a classic example of the West Highland Line's many 'chalet' style buildings.

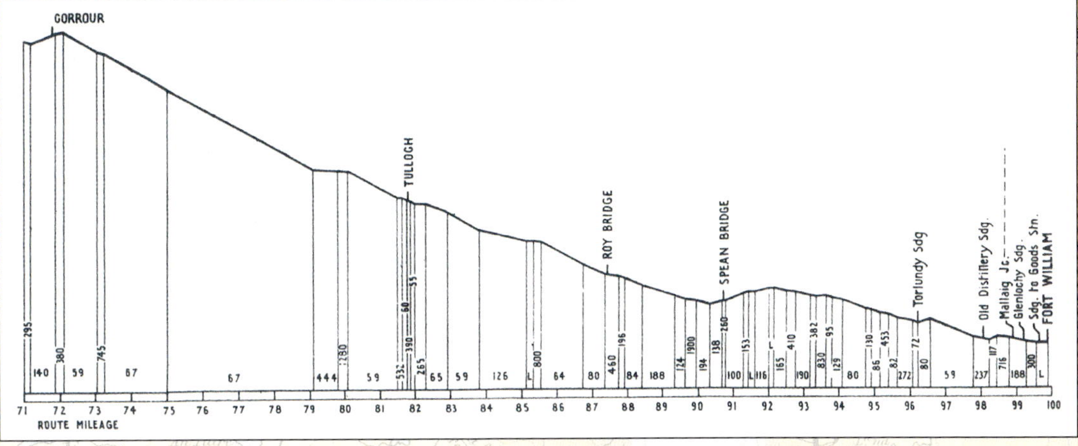

The severity of the long climb from Spean Bridge to Britain's second-highest standard-gauge summit (and highest standard-gauge station) at Corrour is all too apparent.

ROUTE DESCRIPTION

Rannoch to Fort William

Rannoch station, on the edge of the moor, has just a few houses, a charming small hotel and pub and a café in the old station building. A mile after Rannoch the train plunges into Cruach cutting and Britain's only 'snow shed', built after drifts had blocked the line during the severe first winter of the line's operation in 1895.

Corrour, seven miles from Rannoch, boasts one of just four railway stations in Britain that have no metalled road access. This is the highest standard-gauge station in Britain and unquestionably the loneliest spot where you can board a 'main line' train, although Loch Ossian youth hostel is just a mile away. This is breathtaking walking country, surrounded by the granite peaks of the western Grampians. To the northeast is Beinn na Lap at 3,074 ft high – one of the most inviting of the Scottish 'Munros', as walkers can step off the train just a few yards before Corrour Summit, the highest point of the line at 1,350 ft!

After Corrour Summit the railway leaves behind the treeless wastes of Rannoch Moor, dropping down by the side of Loch Treig towards wooded Glen Spean and the next station at Tulloch. The loch was incorporated in a mammoth hydroelectric power scheme for the new Fort William aluminium smelter established in the 1920s and a narrow-gauge railway was built to facilitate construction of the dam and pipeline. The smelter – Britain's last surviving manufacturer of aluminium – is powered by virtually free energy and still generates freight traffic for the West Highland Line.

Beyond Spean Bridge – with its splendidly refurbished station building now housing a restaurant – there are magnificent mountain views on the approach to Fort William, skirting the lower slopes of Ben Nevis, Britain's highest summit at 4,406 ft. Fort William (named after William of Orange) was established in 1690 as a military outpost of the British Army, one of a chain of fortifications guarding the Great Glen. The 100 miles of the West Highland Line formerly terminated at the original Victorian station, by the shores of Loch Linnhe, but the station and the final half a mile of the railway were obliterated in 1975 by a new dual-carriageway road. The current terminus is a disappointingly utilitarian end to a wonderful journey, but nevertheless a good interchange for local buses and above all the unrivalled scenic splendours of the Mallaig line (see pages 284–289).

FORT WILLIAM TO MALLAIG

WEST HIGHLAND LINE

Throughout its history the forty-one-mile line from Fort William to Mallaig has attracted tourists drawn to its dramatic mountain, loch and coastal scenery and the many excursion explorations available on diesel and steam-hauled trains and connecting buses and ferries.

By the time the West Highland Railway had opened from Glasgow to Fort William in 1894, attention was already turning to the line's next step westwards – to the Atlantic coast and its rich fishing grounds. The bill for the 'Mallaig Extension' was uncontroversial in the context of the planned route but its financial provisions were unusual for the time, with the Treasury guaranteeing a minimum return on capital and allocating £30,000 (two-thirds of the cost) for a pier and breakwater at a new harbour in Mallaig. Political concerns about these provisions delayed the passage of the bill until 1896 and it was anticipated that the thirty-nine miles of arduous construction through predominantly rocky mountainous territory would then require at least five years to complete. In practice the contractor, Robert 'Concrete Bob' McAlpine – who would become famous for his pioneering use of mass concrete on the Mallaig Extension bridges – completed the job in just four years.

The new railway involved no less than eight major bridges or viaducts and eleven tunnels including the bridge over the Borrodale Burn, which at the time was the world's longest concrete span at 128 ft, and the now world-famous Glenfinnan Viaduct, 1,247 ft long and built on a curve with no less than twenty-one arches at a maximum height of 98 ft.

As Thomas and Turnock record in *A Regional History of the Railways of Great Britain Volume 15* (1993):

> *On the opening day, 1 April 1901, steamers and trains converged on the tiny village of Mallaig (population of Glenelg parish 1,843 in 1861, 1,670 in 1921 and 1,236 in 1981). Steamer links with Skye and Stornoway were provided by the simple process of extending the existing sailings from the terminus of Kyle of Lochalsh down the Sound of Sleat to Mallaig…This must have been very satisfactory to MacBrayne's [the shipping company], which for a very small additional expenditure was now able to tap two railheads.*

The railway transformed local and long-distance travel – with Glasgow to Mallaig quickly becoming recognized as one of *the* epic rail journeys for the discerning tourist. Trains brought in

coal to replace local peat as the main power source and fish found new markets – but this tended to be a seasonal traffic, like tourism. The Mallaig Extension would never be a busy railway, but it played a crucial social and economic role while the parallel A830 road remained largely single track with passing places (as the railway has always been).

'Black Five' No. 45231 'The Sherwood Forester' hauls a westbound 'Jacobite' service across Loch nan Uamh viaduct in July 2011.

The inadequacy of the alternative road ensured that the railway did not suffer from the 'Beeching Axe' – indeed the infamous report did not propose the line's closure and even its proposal to cut local services was not implemented. At various periods between the late 1950s and the late 1980s, British Rail (BR) operated tourist observation cars on the rear of trains between Fort William and Mallaig during the summer season and in 1983 – more than twenty years after the disappearance of steam on this line – BR revived the traditional method of haulage for summer specials, forerunners of today's highly successful 'Jacobite' train.

Heading eastwards away from the camera, a Network Rail Track Assessment & Recording Unit descends the 1-in-48 Beasdale Bank in October 2011.

ROUTE DESCRIPTION

This memorable journey begins modestly at Fort William's utilitarian station, which opened in 1975 – but urban surroundings are soon left behind as the Mallaig line veers away from the 'main line' at Fort William Junction, controlled by the last traditional electro-mechanical signal box in the West Highlands. The hills and mountains of the Lochaber district stand out on all sides as the railway approaches Banavie, the line's first station and location of the radio signalling centre that controls the entire West Highland rail system other than the immediate environs of Fort William. Immediately beyond the station is the swing bridge across the Caledonian Canal and the succession of locks known as 'Neptune's Staircase' that raise the canal towards Loch Lochy, Loch Oich, Loch Ness and the sister swing bridge at Clachnaharry, sixty miles distant on the Far North line.

After the stop at Corpach, the line passes the site of the major industrial development that saved the West Highland line from any closure threat in the early 1960s – the pulp and paper mill has now closed, but its replacement by a large saw mill promises to bring new freight traffic to the railway. Beyond, the line is sandwiched between the A830 road and the shores of Loch Eil, a sea loch with a towering backdrop of mountains fringing the remote Moidart district. A simple station of wooden construction was opened at Loch Eil Outward Bound in 1985, serving the nearby outdoor activity centre.

Just three miles further on is the lonely request stop at Locheilside station; then beyond the head of Loch Eil the railway begins the sharply curved switchback section of the line to Arisaig. The Mallaig line may not have the big summits of its sister line south from Fort William, but its most severe gradients are tougher – as steep as 1 in 40 – and combined with the route's curvature this has always demanded the highest driving skills. Its eleven tunnels in forty-one miles – compared to just two in the ninety-nine and a half miles of the West Highland line proper – tell their own story about the tough terrain.

The rail approach to the mid-way point of the line at Glenfinnan could not be more dramatic. The train curves round Glenfinnan Viaduct, now even better known through the *Harry Potter* films, with dramatic mountain-fringed views down Loch Sheil beyond the Glenfinnan Monument commemorating the 1745 Jacobite rising. Glenfinnan station – with its museum in the Victorian building and nearby restored dining and sleeping cars – is the first crossing loop after Fort William and a calling point not just for the all-year ScotRail service but also the steam-hauled 'Jacobite' train which operates twice daily in the summer season.

Westwards from Glenfinnan, the railway passes the idyllic Loch Eilt (a popular location for photographing trains in the landscape) and traverses a series of bridges, tunnels and cuttings, which carry the line through this most difficult rocky countryside. Lochailort (another request stop) is followed by a tantalizing first glimpse of the Atlantic from the imposing Loch nan Uamh viaduct, before the train plunges into tunnels that bore their way inland towards the formerly private station of Beasdale, the line's other request stop. At Arisaig – Britain's most westerly station – there is a crossing loop and the unstaffed station building is now used as a local office for HITRANS, the regional transport partnership for the Highlands.

Beyond Arisaig, glorious views open out towards the Sound of Sleat. At Morar the 'silver sands' to the west (as featured in the film *Local Hero*) and Britain's deepest loch – Loch Morar – to the east, provide a 360-degree panorama from the railway. The final stretch of the line has classic views over the sea to the islands of Eigg and Rum, before terminating at the tiny port of Mallaig, a settlement created by the arrival of the railway in 1901 and which continues to benefit from the trade it brings. The ideal way to complete this rail adventure through the West Highlands is to continue by ferry to Armadale on Skye, then onwards by bus via the Skye Bridge to Kyle of Lochalsh before resuming scenic rail travel by the rather different but equally stunning 'Dingwall & Skye' line to Inverness (see pages 270–275).

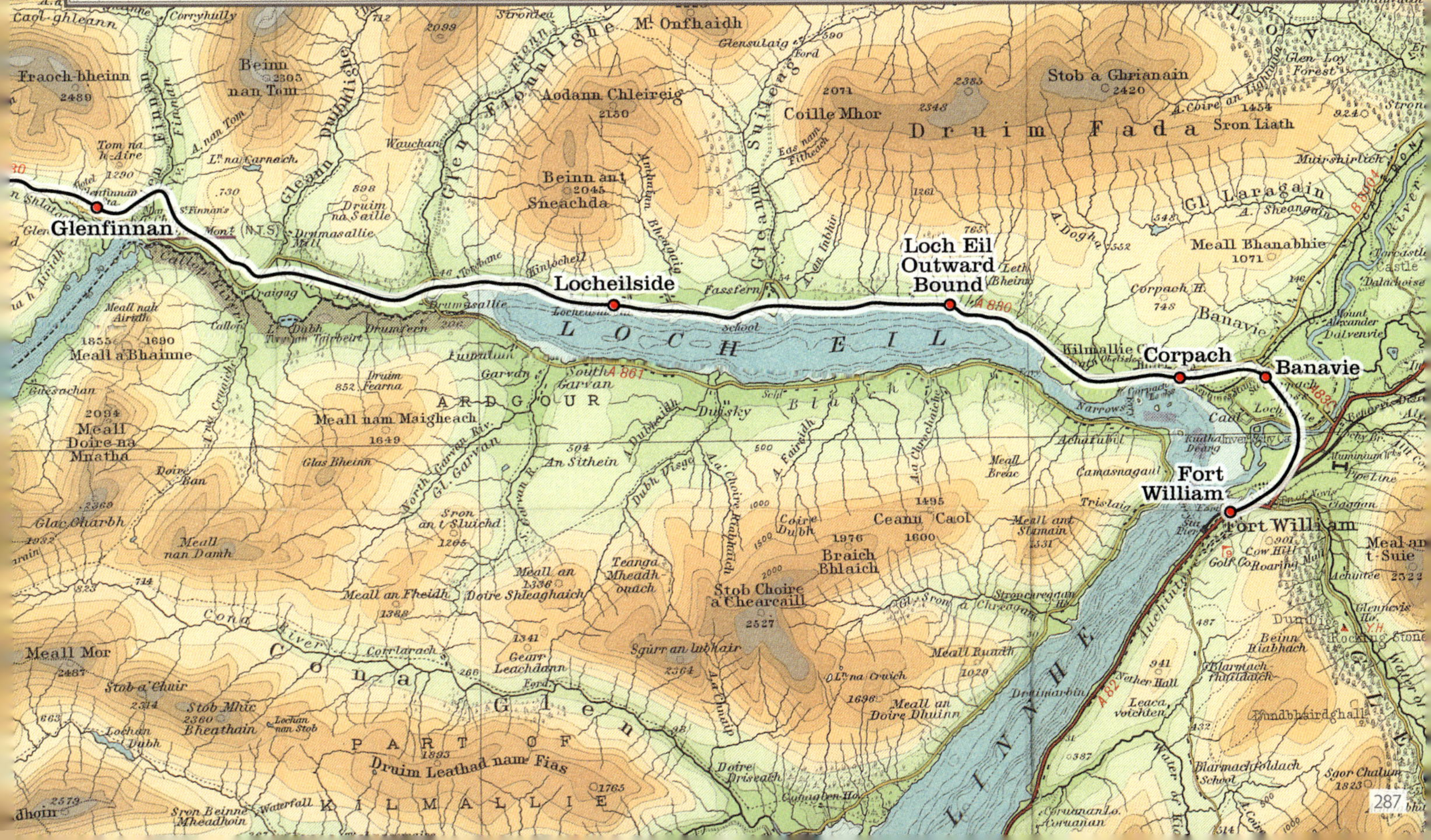

A sublime scene as Class K1 No. 62005 hauls a Jacobite service across 'Concrete Bob's' Glenfinnan Viaduct in October 2006.

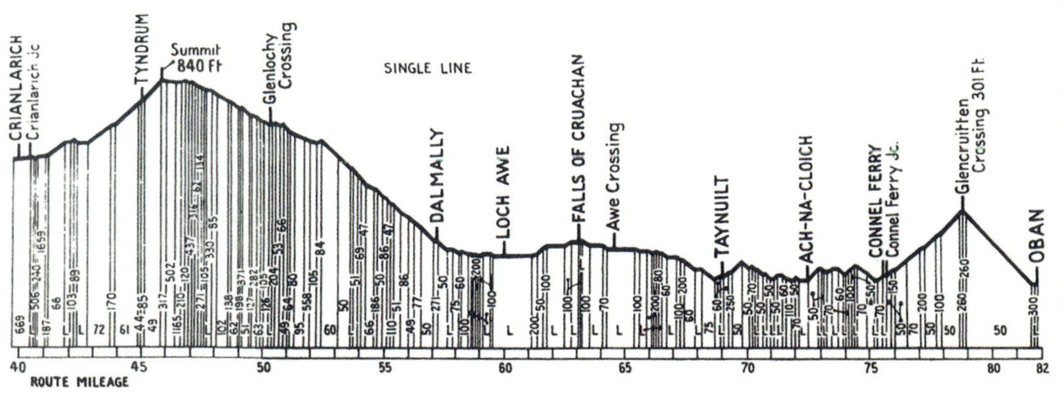

The steep climbs and very frequent changes of gradient on the surviving section of the Callander & Oban Railway are graphically illustrated here.

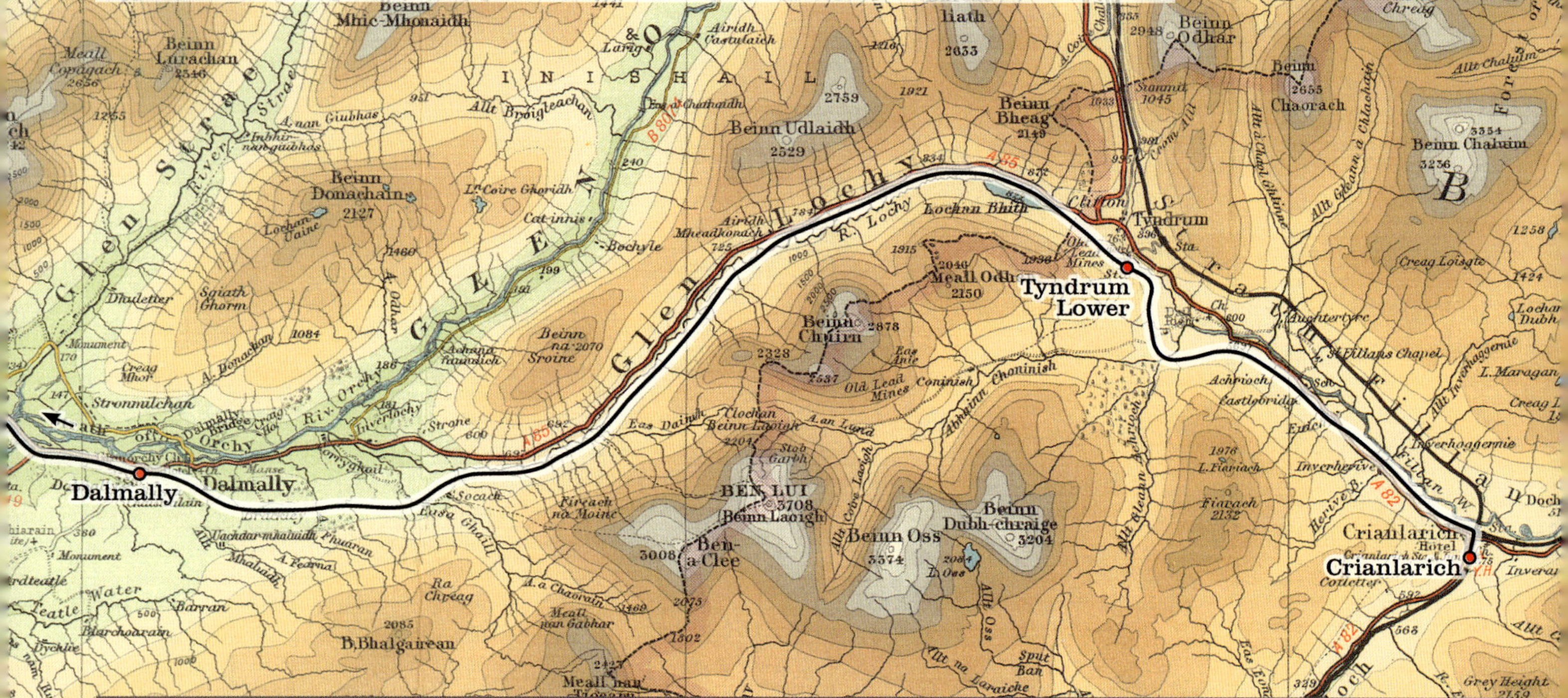

ROUTE DESCRIPTION

Crianlarich to Dalmally

At Crianlarich station (formerly known as Crianlarich Upper), trains for Fort William and Oban are split and the Oban branch diverges from the West Highland proper, dropping down along the 1897 chord to connect with the rusting remains (from a former timber terminal) of the final stretch of the abandoned line from Dunblane and Callander. Until nationalization in 1948 this chord was used only occasionally for freight traffic and excursions, but from 1965 it became the only rail route to Oban following the complete closure of the line between Dunblane and Crianlarich (Lower).

Heading west from Crianlarich along the valley of Strath Fillan, Oban train passengers can often view the slow uphill progress of the Fort William train as it climbs the steep gradient of the parallel West Highland line, just a mile to the north. The Oban line's 840-ft summit comes shortly after the first station at Tyndrum Lower, Upper Tyndrum being high above on the West Highland. A quirky feature of the locality is that from 1967 to 1971 this tiny village, with its population of around fifty, had just one less railway station than the city of Edinburgh – population circa 400,000!

The Oban branch follows a lower-lying course than the West Highland's climb over Rannoch Moor, but a distinctive feature of the line's geography is the very frequent change of gradient, providing a challenge for train drivers and much variety for the passenger. The longest stretches of unchanging gradient are just one and a half miles of level track by Loch Awe, and the two and a half miles of southbound 1-in-50 climb from near sea level in Oban to the 310-ft summit at Glencruitten. Despite this, trains often reach higher speeds than south of Crianlarich on the West Highland – the sweep of this remaining section of the Callander & Oban seems to reflect the typically grander approach of the Caledonian Railway to route construction than its North British Railway rival.

The steepest sections of the line – gradients of 1 in 49 – are encountered in lonely Glen Lochy, between Tyndrum and Dalmally, where extensive forestry plantations are reaching maturity. At Dalmally, the original and attractive station buildings still survive, as does a unique ornamental fountain (carved from local granite) featuring a sculpture of a heron.

CRIANLARICH TO OBAN

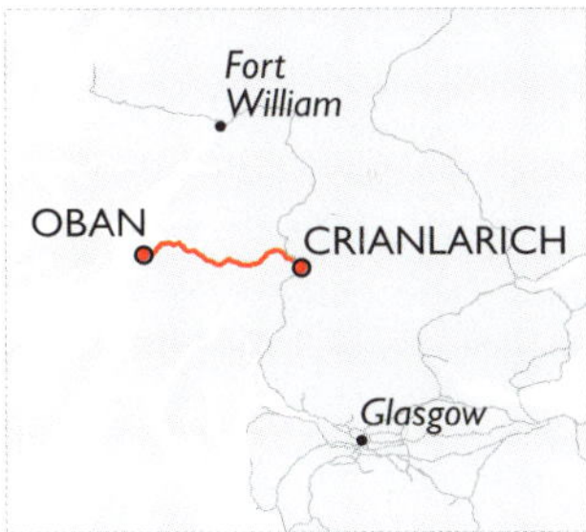

The forty-two-mile single-track branch line from Crianlarich on the West Highland Line takes the rail traveller through classic mountain, loch and coastal landscapes to the port of Oban, 'the Gateway to the Isles'.

The railway reached Oban in 1880 and was (by fourteen years) the first complete line to open in the West Highlands, but many obstacles – not just physical – had to be overcome before the original 1845 idea could be turned into reality. Argyll in general was attractive to railway speculators for its presumed mineral wealth and Oban specifically as a port for the islands and even Ireland.

Early schemes during the 'Railway Mania' – for a route from Callander through Glen Ogle to Crianlarich and onward to Oban as well as a rival plan for a line north from Glasgow by Loch Lomond and Glen Falloch to Crianlarich – foundered due to financial difficulties. But the rail network steadily pushed north from the Central Belt, reaching Callander from the Scottish Central main line at Dunblane in 1858. Parliamentary approval for the Callander & Oban Railway (C&O) was secured in 1865 but the absorption of the Scottish Central by the Caledonian Railway that same year delayed progress. The C&O was opened in various stages to Tyndrum by 1873, but – spurred on by John Anderson, the indefatigable secretary and manager of the C&O – it took a financial injection from the Caledonian's English ally, the London & North Western Railway, to ensure that the line finally reached the long-sought objective of Oban in 1880.

The thirty-seven miles from Tyndrum to Oban involved some of the biggest engineering challenges of the entire seventy-seven-mile length of the C&O, notably by Loch Awe and on the steep slope of the Pass of Brander. The railway runs on a ledge along the loch side and then climbs through the pass, where boulders falling from the slopes of Ben Cruachan were a regular hazard. A rock fall partially derailed a train in 1881 and thereafter, on the suggestion of John Anderson to the Caledonian Railway, an elaborate system of tripwires and connecting semaphore signals was constructed to act as an early-warning system for train drivers. As John Thomas records in *The Callander & Oban Railway* (1966): "Old-timers on the C&O referred to the boulder screen as 'Anderson's Piano' because the wind whistling through the taut wires produced a musical hum." The system survives to this day, but failed to prevent the derailment of a passenger train in 2010 in which fortunately nobody was seriously injured.

Oban was already a thriving town with a population of some 3,000 and the centre of an established network of steamer services by the time the railway arrived. However, trains brought many more visitors to what became known as the 'Charing Cross of the Highlands' and, as elsewhere in the region, cattle and fish traffic prospered. In the face of perceived competition from planned railways between Glasgow and Fort William, the C&O was soon proposing a branch line to Ballachulish, continuing across Loch Leven to Fort William. This railway – one of the last

The 'little Forth Bridge', linking Connel Ferry and North Connel high above Loch Etive, carried trains until 1966 but only road vehicles thereafter.

ROUTE DESCRIPTION

Dalmally to Oban

The next section of the route – by Loch Awe and through the Pass of Brander – is perhaps the most spectacular and features two stations re-opened in the 1980s. A Victorian stone staircase still leads down from the imposing Loch Awe Hotel to the station platform – until the latter's 1965 closure, guests could leave the hotel on foot and a few minutes later board the sleeper train direct to London. The station re-opened in more modest circumstances in 1985 but now boasts a privately owned static railway 'camping coach' which provides overnight accommodation with glorious views to the south, east and west.

Ben Cruachan, at 3,689 ft above sea level, dominates the landscape for scores of miles and more. The 'pump storage' hydroelectric scheme tunnelled inside the mountain remains a wonder of modern engineering and in 1988 a simple summer-only request stop was built at Falls of Cruachan to service the scheme's popular visitor centre. Beyond here the railway continues through the Pass of Brander, fringed to the north by 'Anderson's Piano', the early warning system for rock falls from the steep slopes of the Ben. The gradient then falls into gentler wooded country beside Taynuilt and Loch Etive, the long sea loch that penetrates far northwards towards Glen Coe.

Staying largely within sight of the loch, the railway passes through Connel Ferry – a once-grand junction of four platforms, now reduced to one – and there are glimpses of Connel Ferry's unmistakeable cantilever bridge, which on completion in 1903 was hailed as the biggest British rival to the Forth Bridge. The bridge, 1,044 ft in length, originally carried only the single-track railway, but there was enough room to squeeze in a narrow road and subsequently both trains and vehicles used the bridge – although not at the same time – until the railway's closure. Now it is purely a road bridge – straddling the Falls of Lora – but it remains an enduring reminder of the heyday of the Callander & Oban Railway.

Today's railway strikes inland and makes a steep and sweeping descent – with panoramic views across Oban Bay – before making its way through rock cuttings on its scenic approach to the town. As at Fort William, the modern rail terminus station is a disappointment; however, most leisure rail travellers will press on eagerly to the adjacent ferry terminal to catch a boat to any of a wonderful range of sea destinations – Colonsay, Mull, Lismore, Coll, Tiree, Barra and South Uist.

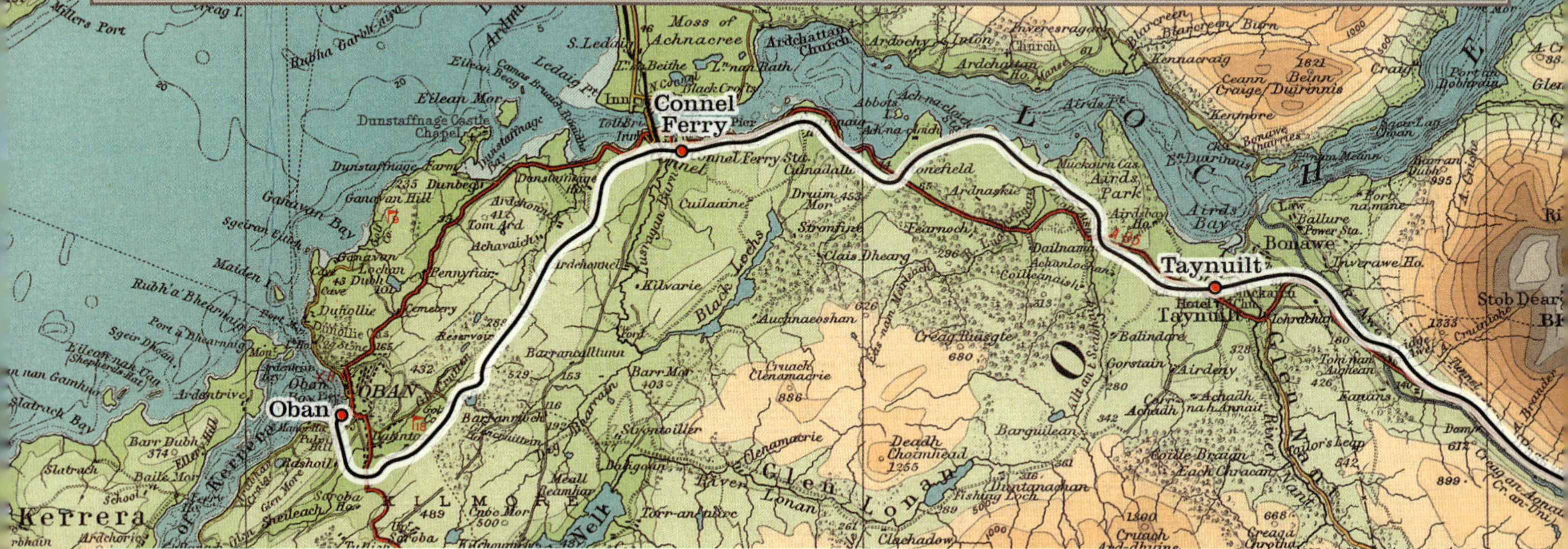

The Loch Awe Hotel – linked by stone staircase to Loch Awe station below – towers over the resident 'camping coach'.

The railway terminates in the heart of Oban, with views across Oban Bay to the islands of Kerrera and Lismore and mainland Morvern beyond.

conventional railways to be built in Scotland until the early twenty-first century revival – was opened from Connel Ferry in 1903, but only as far as Ballachulish. In the early 1960s a scheme was mooted to revive the proposed extension to Fort William – allowing sixty-three miles of expensively operated and maintained railway from Crianlarich to Fort William to be abandoned – but this intriguing idea got nowhere and the Ballachulish branch itself closed in 1966.

In its halcyon years the Oban line and tourism prospered together, often in conjunction with other modes of transport. John Thomas records in *The Callander & Oban Railway* that:

> *Through the years the Callander & Oban offered excursions in infinite variety, and at low cost, from brief evening outings to elaborate tours spread over several weeks. Their itineraries ranged from simple return journeys between two points to elaborate programmes involving numerous trains, steamers and coaches…A handbook of Caledonian Railway tours published in 1911 listed forty-one tours throughout the system, and thirty-two of them involved the C&O.*

After the Second World War, the railway to Oban faced its most severe challenge – many steamers had been laid up and were replaced by buses following broadly parallel routes. Growing affluence brought more cars and less rail travel and road haulage benefitted from improving technology, better roads and cheap oil. The railway did fight back, with summer trains conveying a tail-end observation car during the late 1950s and the 1960s. Also popular was the 'Six Lochs Land Cruise', a summer excursion formed of a six-car diesel unit from Glasgow to Crianlarich via Callander and Killin to Crianlarich, returning to Glasgow via the West Highland line.

But it was all to no avail – the Dunblane-Crianlarich section succumbed to the 'Beeching axe' in 1965 and trains to Oban were diverted to run via the West Highland line to Crianlarich. Train journeys from Glasgow were now quicker and cheaper, but those from Edinburgh were longer and more expensive and there was no longer an overnight sleeper from London.

Diesel had replaced steam in 1962 and the next big economy measure came in 1986 when, sadly, the imposing original station at Oban was demolished and replaced by a utilitarian modern structure. In 1988 the traditional signal boxes controlling crossing loops at Dalmally, Taynuilt and Oban were replaced by the Radio Electric Token Block system based in a new signalling centre at Banavie near Fort William, and the following year diesel units took over from locomotive-hauled trains.

Reinstating a link from the past, in 2011 ScotRail successfully introduced a new through service on summer Sundays from Edinburgh to Oban, avoiding a change of trains in Glasgow, and the Oban line continues to attract charter trains from far afield during the tourist season. While less renowned than its West Highland sister, the railway from Crianlarich to Oban offers a delightful train journey and an unsurpassed range of onward ferry connections to tempt the traveller even further west.

AYR TO STRANRAER

Scotland's rural railways are world renowned, but in the rush to the Highlands it is easy to overlook an intriguing and unsung outpost of the national network – the fifty-nine-mile single-track route over the hills from Ayr to Stranraer, close to Scotland's southwestern tip.

The Ayr-Stranraer railway was built in stages by a variety of different companies, the bulk eventually becoming part of the Glasgow & South Western Railway's empire. As John Thomas and Alan JS Paterson observe in *Scotland: The Lowlands and the Borders* (Volume 6 of *A Regional History of the Railways of Great Britain, 1984*):

> *The country south of Ayr and west of Dumfries had less to offer the railway promoters than Ayrshire itself. There were minerals and rich agricultural areas it is true, but there were also high, bare moorlands, and miles of empty coastline, and the few small towns were widely scattered. But beyond the western coast lay Ireland, and it was Ireland that lured the railway speculators.*

The railway from Ayr reached Maybole in 1856 and Girvan in 1860 and in 1877, after five years of construction work, the Girvan & Portpatrick Railway was opened to Challoch Junction on the Dumfries-Stranraer-Portpatrick line, which had been completed in 1862. While the rail network in this part of Scotland reached its zenith that year, by 1890 a number of schemes had been mooted for a dramatic extension – to create what would have been the world's longest rail tunnel, under the North Channel of the Irish Sea from Galloway to Northern Ireland!

With its steep gradients through wild inland country, the Ayr-Stranraer railway could be difficult to operate and on three occasions, in 1895, 1908 and 1947, complete trains have been trapped in snow south of Barrhill. During the Second World War, the drivers and firemen on this route played a big part in the war effort, as David Smith records in *Tales of the Glasgow and South Western Railway* (1962):

> *Passenger trains, troop trains, trains of evacuees, goods trains, coal trains, petrol trains, and finally, trains of gas-shells for dumping at sea via Cairn Ryan – they had their fill of them.*

The line's military role would have been even greater if bombing had disabled Liverpool's docks. A new military port was established at Cairnryan, seven miles closer to the open sea than Stranraer, and the Cairnryan Military Railway was opened for traffic from a junction east of Stranraer in 1943. Military use ceased in 1959 and, despite the presence of a ship-breaking operation, the railway itself closed completely in 1962.

Traditional features abound at lonely Barrhill station, including the distinctively tall and narrow signal box formerly located at Portpatrick in Galloway.

The Ayr-Stranraer line survived its threatened closure in the 1963 'Beeching Report', but in the face of increased competition from other modes of transport, the direct London trains from Stranraer and all freight services on the line were lost by 1994. On 20 November 2011 – after 134 years of rail-sea connection – the last Northern Ireland ferry left Stranraer Harbour for Belfast.

The next morning, Stena's shipping service resumed from the now rail-less former military port at Cairnryan. Freed from the constraints of the ferry timetable – but also losing those passengers – the Ayr-Stranraer line is now finding a new role, making the most of its scenic attractions and the access it gives to a charming but under-rated corner of rural Scotland.

The line between Girvan and Stranraer is one of the last in Scotland to be operated primarily by the traditional 'electric token block' system, with semaphore signals and staffed signal boxes at all crossing loops. Three of the line's six crossing loops lost their passenger stations many years ago – Kilkerran, lonely Glenwhilly (where the signal box has no mains water) and Dunragit. Girvan has Scotland's only surviving Art Deco-style

station building, dating from 1951, after a fire destroyed the previous structure.

Seen for 134 years as essentially the route to Northern Ireland, the line over the hills from Ayr to Stranraer is now being rediscovered as a surprisingly scenic railway, featuring some of the most traditional of railway operations. With its ready access to a variety of hill-walking and cycling routes and coastal explorations, this is a railway that is set to become part of the classic itinerary for discovering Scotland by rail.

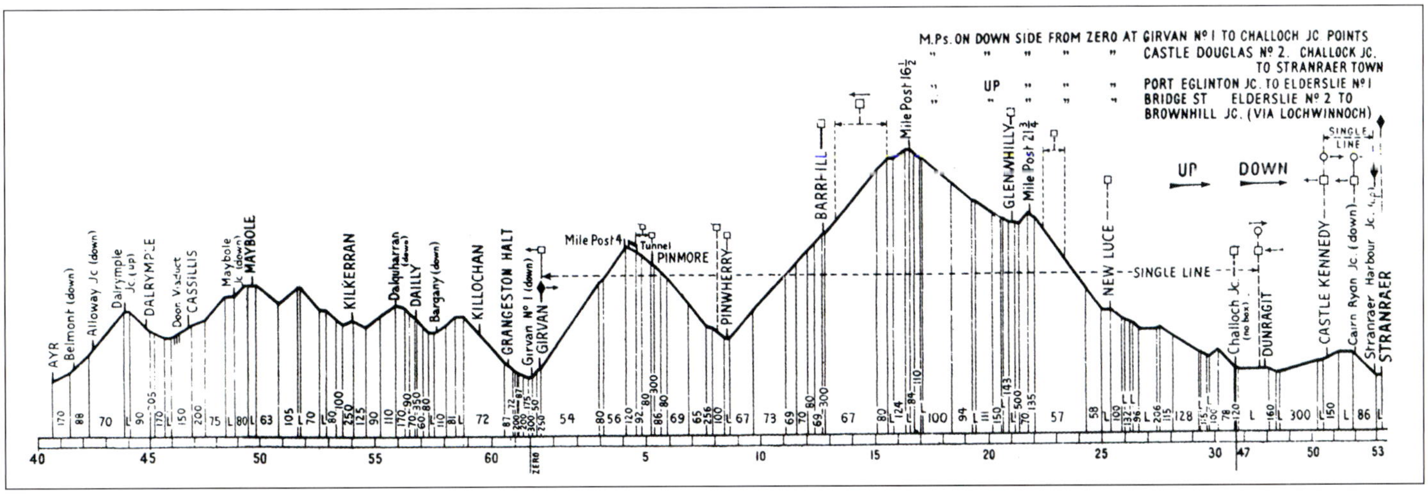

There are just a handful of level stretches of track on the steeply graded route over the hills from Ayr to Stranraer.

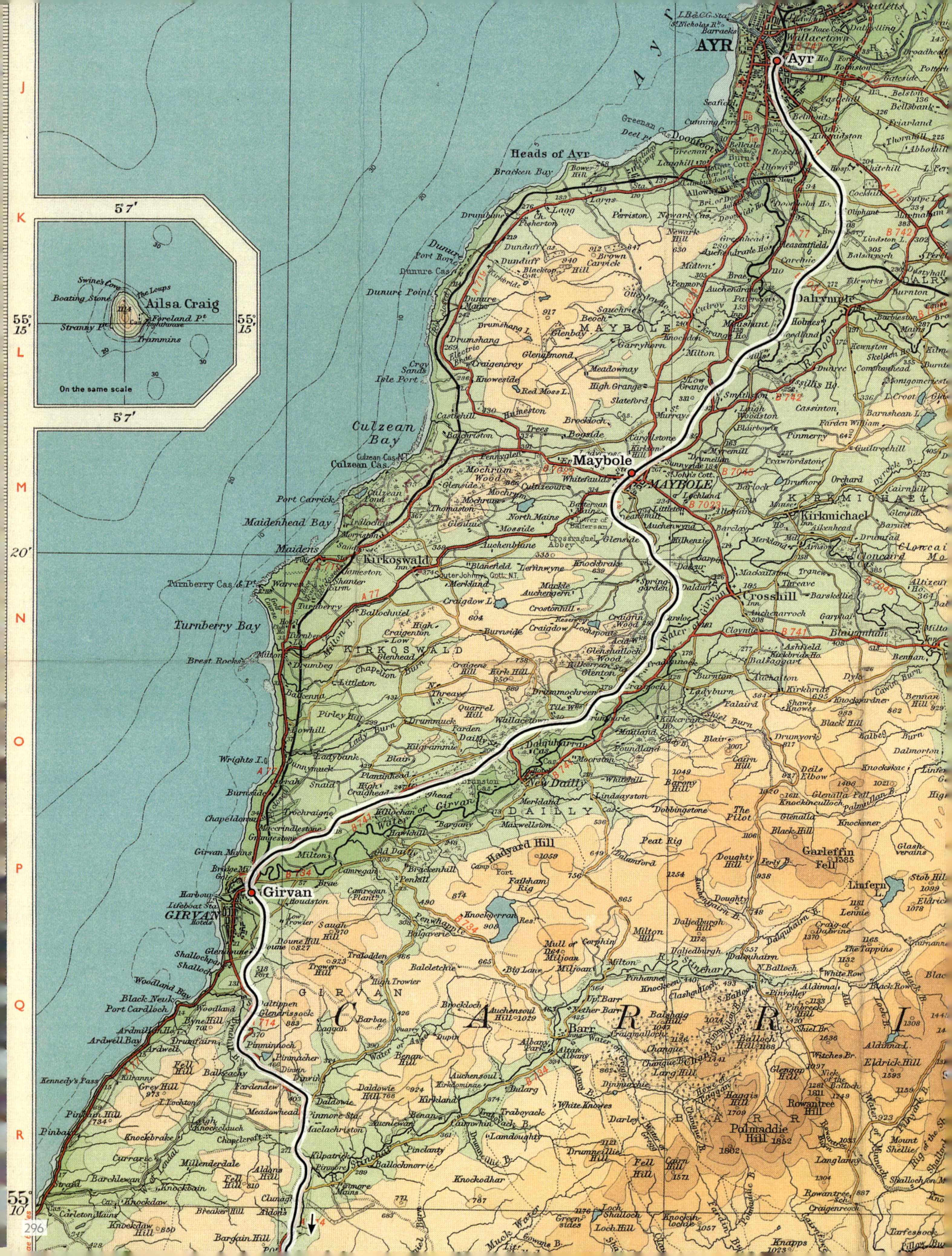
AYR
Ayr
Heads of Ayr
Bracken Bay
Dunure
Dunure Point
Isle Port
Culzean Bay
Culzean Cas.
Port Carrick
Maidenhead Bay
Maidens
Turnberry Cas & Pt
Turnberry Bay
Brest Rocks
Wrights I.
Girvan Mains
Girvan
GIRVAN
Shallochpark
Woodland Bay
Black Neuk
Port Cardloch
Ardwell Bay
Kennedy's Pass
Pinbain
Ailsa Craig
Swine's Cove
The Loups
Boating Stone
Foreland Pt
Stranny Pt
Trammins
On the same scale
5°7'
55° 15'
Maybole
MAYBOLE
Kirkoswald
KIRKOSWALD
Crosshill
Kirkmichael
KIRKMICHAEL
Dalrymple
Dailly
DAILLY
New Dailly
Barr
Doonfoot
Alloway
Newark Cas.
Mochrum Wood
Dunduff
Glenluie
North Mains
Mosside
Auchenblane
Crossraguel Abbey
Knockbrake
Hadyard Hill
Peat Rig
Garleffin Fell
Linfern L.
Doughty Hill
The Pilot
Black Hill
Milton Hill
Mull or Milroan
Big Lane
Balcletchie
Auchensoul Hill
Glengap Hill
Polmaddie Hill
Rowantree Hill
Haggis Hill
Eldrick Hill
Craig of Dabrine
The Tappins
Pinmore
Pinminnoch
Daltippen
Laggan
Fell Hill
Aldons Hill
Knockdaw Hill
Bargain Hill
Knockdolian
Fardenbreck
Meadowhead
Millenderdale
Barchlewan
Currarie
Knockbrake
Kilpatrick
Ballochmorrie
Lamdoughty
Pinclanty
Drumlamford
Knapps
Larg Hill
Knocknalling
Changue
Nether Barr
Carnochan
Dalquharran Cas
Killochan
Water of Girvan
Bargany
Penkill
Tradunnock
Balgarrie
A 77
B 741
B 734
B 7045
B 7023
B 742
CARRICK
J
K
L
M
N
O
P
Q
R
20'
55° 10'

ROUTE DESCRIPTION

Ayr to Pinmore

Many visitors to the Stranraer line will start their journey at Glasgow's grandly restored Central station whence direct diesel trains run several times daily to Galloway's only railway station. A fast journey to Ayr (population 46,000) takes the traveller through Paisley (of textile pattern fame), past Castle Semple Loch and Kilbirnie Loch to the shores of the Clyde by Troon, where on a good day you can see not just the Island of Arran but also the distinctive peaks of the Paps of Jura in the distant Inner Hebrides.

Beyond Ayr, long associated with Robert Burns, the Stranraer train leaves behind the overhead electrified wires of the Strathclyde suburban rail network and is soon climbing steeply, at gradients of up to 1 in 70, to the open countryside of the River Doon valley by Maybole. This small town is the first stop on the by-now single-track railway, a convenient railhead for the National Trust for Scotland's Culzean Castle on the Clyde coast. With the hills of Carrick Forest rising to the south, at wooded Kilkerran the line reaches the first of its crossing loops. The railway swings along the northern side of the valley of the Water of Girvan – through former coal-mining territory now thick with pleasant mixed woodland – and soon reaches the town of the same name, attractively located on the Firth of Clyde and home of William Grant's whisky distillery, one of Scotland's largest.

Throughout the line's history, train drivers have had to negotiate steep gradients, including two and a half miles of continuous 1 in 54 from the platform end at Girvan – with dramatic coastal views towards the island of Ailsa Craig – before plunging into the 543-yd tunnel at Pinmore and crossing four viaducts in just five miles.

Two 'Black Five' locomotives do battle with Glendoune Bank south of Girvan on the 'Great Britain' rail tour in April 2010.

A Class 156 heads south over Pinmore Viaduct in April 2010. The Saltire livery will be applied to all ScotRail trains on behalf of the Scottish Government.

Lendalfoot
Bennane Head
Colmonell
COLMONELL
Ballantrae
Ballantrae Bay
Barrhill
BALLANTRAE
Benerard
Beneraird
Smirton Hill
Auchencrosh Hill
Altimeg Hill
Milleur Point
LOCH RYAN
Kirkcolm
Cairn Ryan
Stranraer
STRANRAER
Glenluce
New Luce
NEW LUCE
Pinwherry
Glen Tig
Liberland
Craigairie Fell
Pinwhirrie Res.
Soulseat L.
Black L.
White L.
THE MOORS
INCH
LUCE
W
T
G

ROUTE DESCRIPTION
Pinmore to Stranraer

The railway no longer serves the village of Pinwherry, but its station featured in the 1931 detective novel *Five Red Herrings* by Dorothy L Sayers, a renowned murder story from the Lord Peter Wimsey series.

South of Pinwherry the line resumes its steep climb before reaching the isolated station and crossing loop at Barrhill, sitting high above the village of that name. Barrhill is the only remaining station in Scotland where the signaller exchanges a physical 'token' with northbound and southbound train drivers from the platform, giving them the authority to proceed safely on to the single-line sections.

Beyond here is another steep climb – the 1-in-67 gradient to the 690-ft Chirmorie Summit which, on a good day, offers panoramic views eastwards to Merrick, the highest mountain in the Southern Uplands at 2,765 ft. Then in the words of Michael Pearson in *Iron Roads to Burns Country* (2006), "you are engulfed by forestry and you cease to feel yourself so much in Scotland as in Central Europe; the Harz Mountains perhaps, or even further east in the Carpathians". The line then traverses some bleak country, sometimes described as being reminiscent of the interior of the Falkland Islands, before reaching another lonely railway outpost at Glenwhilly crossing loop. Immediately south of here the line is characterized by significant curvature, including the section known as the 'Swan's Neck', where the railway forms an 'S' curve sufficiently tight that it can be seen from the train window before reaching it.

Descending into the lusher valley of the Water of Luce, the railway passes close to the ruins of Glenluce Abbey – a Cistercian house dating from the twelfth century – and with brief views towards Luce Bay and the Mull of Galloway (and even the Isle of Man on a good day), turns westward over farmland towards Stranraer. After reaching the outskirts of Stranraer and the surviving signs of the original railway to Portpatrick, the train drops into a steep cutting with no hint of the dramatic final arrival at the coast, bursting out under the A77 road bridge to emerge on to a long causeway towards the old harbour station, with commanding views across the water of Loch Ryan towards Cairnryan and the South Ayrshire hills.

Stranraer (population 11,000) is an ideal stepping-off point for the many garden attractions featuring exotic plants in this distinctive corner of Scotland, renowned for its mild Gulf Stream-influenced climate. Twenty miles to the south is the Mull of Galloway lighthouse and visitor centre at Scotland's dramatic southwestern tip.

Beyond the long causeway to Stranraer station, the waters of Loch Ryan stretch towards the replacement ferry port of Cairnryan.

INDEX

ACKNOWLEDGEMENTS

t = top; b = bottom; r = right; l = left; m = middle

Photo credits:
Colin Alexander: 180bl, 180br
Jim Allan/Alamy: 274
Ange/Alamy: 277
ATGImages/Shutterstock.com: 198
Hugh Ballantyne: 97
Hugh Ballantyne/Colour-Rail: 110, 139b
Don Bishop/Colour-Rail: front cover, 6/7, 8/9, 99, 288/289
Christina Bollen/Alamy: 169b
Jack Boskett: 72/73, 88, 91t, 91b, 92, 93, 96, 98l, 144/145, 235, 258
Jon Bowers/Colour-Rail: 24, 45, 47, 84/85b
John Chalcraft/Colour-Rail: front endpaper, 15b, 27, 29, 43, 63, 75t, 194, 196, 209t, 249
Paul Chancellor/Colour-Rail: 64, 95t, 95b, 100/101b, 106b, 107
Ben Collier: 263, 265, 267b, back endpaper
Dave Collier: 257, 259, 272bl, 272br, 280t, 282br
Colour-Rail: 203tr
Ewan Crawford: 244, 256, 270/271
Ian Dinmore: 186bl, 253
Gordon Edgar/Colour-Rail: 4/5, 38, 48/49, 66, 87, 108/109, 116, 120/121t, 126, 147, 150/151, 172/173b, 193, 202, 206, 207, 216/217, 219, 278b
Mike Esau: 61, 62, 67
Stephen Finn/Shutterstock.com: 261
Brian Forbes: 233, 252
John Furneval: 185, 186br, 189br, 220, 226ml, 264, 278mr
John Goss: 50, 52t, 129b, 142
John Gray: 242/243, 267t, 273, 282bl, 284/285, 286, 297b
Julian Holland: 255, 282tl, 292, 299
Brendan Howard/Shutterstock.com: 229
Image Rail: 10/11, 13, 15m, 20, 22, 30/31, 35, 37b, 40, 41
imagesef/Shutterstock.com: 240
International Photobank/Alamy: 184ml
Gail Johnson/Shutterstock.com: 139t
Julietphotography/Shutterstock.com: 293
Keith & Dufftown Railway: 247
Mike Kipling Photography/Alamy: 162
Andy Kirkham: 294/295
Anna Kucherova/Shutterstock.com: 230/231
Ian Lothian: 189t, 297t
Norman McNab: 279, 280b
Stephen Meese/Shutterstock.com: 291
Milepost 92½: 19, 52b, 59, 71, 102, 128, 129t, 133tr, 136, 140/141, 157tr, 158, 161, 163, 165, 167, 173br, 178, 190/191t, 195, 197, 200, 205, 210, 213, 236
Gavin Morrison: 54, 56/57, 68, 70, 104, 106t, 125, 131, 133b, 156/157b, 168, 169t, 171, 182, 192, 203tl, 209b, 221, 224/225, 237
Will Newitt/Alamy: 268
Colin Palmer Photography/Alamy: 153br
Peresanz/Shutterstock.com: 143
r.nagy/Shutterstock.com: 69
Brian Sharpe: 74/75b, 78, 80, 81, 82, 83, 85br, 103, 112, 114, 115, 117, 119, 123br, 134/135, 148, 152/153b, 154, 170, 175, 176/177b, 177t, 184b, 212, 222, 250/251, 254
Brian Sharpe/Colour-Rail: 32, 101t
Bob Sweet/Colour-Rail: 227, 275
Brian Taylor: 232
Phil Wills/Alamy: 77
John Yellowlees: 238

Maps and additional material courtesy of:
Ian Allan Publishing Ltd (gradient profiles): 9, 17, 28, 47, 53, 57, 94, 97, 123, 159, 198, 214, 221, 228, 245, 247, 251, 269, 272, 276, 279, 283, 290, 295
© Collins Bartholomew (UK map and location maps): 4, 9, 13, 20, 22, 24, 33, 38, 40, 50, 54, 56, 62, 67, 68, 70–71, 74, 78, 83, 84, 89, 96, 100, 104, 110, 120, 128, 131, 134, 142, 144, 152, 156, 163, 168, 170, 172, 176, 180, 186, 191, 201, 208, 219, 224, 231, 242, 250, 261, 270, 276, 284, 291, 294
Julian Holland collection: 17t, 17ml, 17mr, 37t, 58, 98r, 123bl, 218, 226tl, 241
National Library of Scotland (Bartholomew's Half-Inch Series): 9, 12–14, 16, 18–19, 21, 23–29, 32–34, 36, 39–44, 46–47, 51–53, 55, 58–61, 64–66, 70–71, 76–82, 86–90, 92–99, 102–103, 105, 107, 111, 113–118, 120–130, 132, 136–139, 143, 146–149, 154–155, 158–160, 162, 164, 166, 169–171, 174–175, 178–179, 181, 183–204, 206–208, 210–211, 213–215, 218–220, 222–223, 226–234, 236–239, 241, 244–249, 253–262, 264–269, 272–277, 279, 281–283, 286–287, 290, 292–293, 296–299, back cover
Trackmaps, reproduced by kind permission (track diagrams): 9, 23, 55, 63, 80 (two), 83, 155, 174, 182, 214, 238, 281

With thanks for research assistance/advice to:
Chris Fleet, Rae Montgomery.

Cover image:
Ex-GWR 'Hall' Class 4-6-0 No. 4965 'Rood Ashton Hall' and ex-GWR 'Castle' Class 4-6-0 No. 5043 'Earl of Mount Edgcumbe' cross Cornwood Viaduct in South Devon with a return Plymouth to Bristol rail tour on 15 May 2010.

Front endpaper:
A ScotRail service over the Highland Main Line south from Inverness crosses Slochd Viaduct on a crystal-clear afternoon in December 2009.

Back endpaper:
Snow-dusted Carn Chaiseachain towers above Loch a' Chuilinn, east of Achanalt, as a rail tour heads towards Kyle of Lochalsh in April 2008, hauled by 'Black Five' No. 45407.